The Japanese trajectory: modernization and beyond

The Japanese trajectory: modernization and beyond

edited by

GAVAN McCORMACK

and

YOSHIO SUGIMOTO

The right of the
University of Cambridge
to print and sell
all manner of books
was granted by
Henry VIII in 1534.
The University has printed
and published continuously
since 1584.

CAMBRIDGE UNIVERSITY PRESS

Cambridge

New York   New Rochelle   Melbourne   Sydney

CAMBRIDGE UNIVERSITY PRESS
Cambridge, New York, Melbourne, Madrid, Cape Town, Singapore, São Paulo, Delhi

Cambridge University Press
The Edinburgh Building, Cambridge CB2 8RU, UK

Published in the United States of America by Cambridge University Press, New York

www.cambridge.org
Information on this title: www.cambridge.org/9780521345156

First published 1988
This digitally printed version 2008

A catalogue record for this publication is available from the British Library

Library of Congress Cataloguing in Publication data

The Japanese trajectory: modernization and beyond /
edited by Gavan McCormack and Yoshio Sugimoto.
    p.   cm.
Bibliography.
Includes index.
1. Japan – History – 1868– . 2. Japan – Social conditions – 1868– .
3. Japan – Popular culture.
I. McCormack, Gavan.   II. Sugimoto, Yoshio, 1939– .
DS881.95.M63   1988
952.03 – dc19   88–1722

ISBN 978-0-521-34515-6 hardback
ISBN 978-0-521-10075-5 paperback

Contents

Contributors

Aoyama, Tomoko. Lecturer, Department of Japanese and Chinese Studies, University of Queensland, Brisbane, Australia.

Arnason, Johann. Senior Lecturer, Department of Sociology, La Trobe University, Melbourne, Australia.

Bolitho, Harold. Professor of Japanese History, Harvard University, Cambridge, Massachusetts, U.S.A.

Buckley, Sandra. Assistant Professor, Department of East Asian Languages and Literature, McGill University, Montreal, Canada.

Inkster, Ian. Senior Lecturer in Economic History, University of New South Wales, Sydney, Australia.

Irokawa, Daikichi. Professor of Japanese History, Tokyo University of Economics, Japan.

Kawamura, Nozomu. Professor of Sociology, Department of Sociology and Social Anthropology, Tokyo Metropolitan University, Japan.

Kogawa, Tetsuo. Social critic. Lecturer, Wako University, Tokyo, Japan.

Large, Stephen. Lecturer in Modern Japanese Studies, Faculty of Oriental Studies, Cambridge University, Cambridge, U.K.

Mackie, Vera. Senior Lecturer in Japanese, Swinburne Institute of Technology, Melbourne, Australia.

Matsuzawa, Tessei. Professor of Japanese History, Tokyo Women's University.

Tada, Michitarō. Professor, Meiji Gakuin University, Tokyo, Japan.

Tanaka, Yuki. Lecturer in Japanese, Centre for East Asian Studies, University of Adelaide, Adelaide, Australia.

Ueno, Chizuko. Associate Professor, Heian Women's College, Kyoto, Japan.

JAPANESE NAMES

The Japanese convention that family name precedes personal name is generally followed in this book. However, those Japanese who either live in the West or have chosen to adopt the opposite (Western) convention of personal name followed by family name in their writings in English are identified accordingly. In the table of contents, for example, Aoyama, Sugimoto and Tanaka are listed in Western order, but all other Japanese names are in Japanese order, surname first. Uncertainty may be resolved by reference to the index where all persons named in the text are listed in alphabetical order according to their family names.

Introduction: modernization and beyond

JAPAN AND MODERNIZATION THEORY

Historians, sociologists, and political scientists continue to debate the fine points of what it means for a society to 'modernize', even as they agree that the so-called advanced countries have now completed the process and begun to move into the even more unknown territory 'beyond modernization'. Western social science theory for long understood 'modernization' as a basically unilinear process of transformation of the world which stretched from the cultural and intellectual world of seventeenth-century Europe to the post-1945 United States; the Japanese experience was never easily incorporated within such a model, and the emergence of Japan as an economic superpower calls the whole theory in question. Likewise the theoretical extension of the 'modern' trajectory into the 'post-modern' will remain unsatisfactory so long as such theorization remains exclusively Euro-American oriented.

As the first society outside the Western cultural tradition to achieve a high level of industrialization Japan was, and still is, the crucial case for the debate over whether industrialized societies have a tendency to become increasingly alike in structure and values – the socalled convergence debate.[1] In the context of Japanese studies, there are two further interrelated problems. Is modernity to be defined by Western, Japanese or universal (and somehow culturally neutral) standards? And however the concept is defined, does Japan qualify as 'modern'?

As Japan develops into an economic superpower, whose trade performance appears to threaten other industrialized societies, it has become common to suggest that it is the country which others must emulate. In other words, Japan has come to be seen as the most modern society on earth, and as a beacon lighting the path which others are destined to follow. In a remarkable paradigm shift the scholarly consensus has moved from seeing Japan as backward, semi-feudal and undemocratic to seeing it

1

as post-industrial, super-efficient and 'more Western than the West'. The present volume is an attempt to evaluate these claims by examining Japanese cultural and political life – historical as well as contemporary – both at national and local levels.

This brief introduction is designed to provide a sketch of the modernization debate in the context of Japanese studies and consider chapters of the volume in the light of that debate. The essays collected here are reflections on different aspects of our theme. They are designed to introduce and comment on complex problems, rather than assay any formal proposal on how such problems should be resolved.

JAPANESE STUDIES AND MODERNIZATION THEORY

The western model

The extent to which Japan is perceived to be modern has been conditioned primarily by the political context of the U.S.-Japan relationship. Immediately after World War Two, Japan was depicted as a society where almost everything was unmodern, the government autocratic, social relations hierarchical and unequal, and the national mentality (in the words of General Douglas MacArthur) that of a 'twelve-years old'. The Western approach was essentially anthropological: Japan was analyzed as a compact, homogeneous and exotic unit, qualitatively different from and fundamentally inferior to Euro-American industrial societies. The Japanese scholarly view at the time was dominated by the indictment of 'feudalistic elements' of Japanese society, abhorrence of the 'dwarfishness' of the Japanese pre-war and wartime character, and the necessity of 'democratization', which most assumed would be patterned after the American model.

In this context, the notion of 'citizen' was inseparably related to that of modernization: ideally, modern society was defined in terms of self-sufficient, autonomous and independent individuals committed to universalistic, rational, libertarian and egalitarian values. In the democratization debate which took place immediately after World War Two among Japanese intellectuals, the majority took the view that despite rapid industrialization Japanese society had so far failed to produce modern citizens,[2] although this view was rejected by a wide range of intellectuals who contended that the concept of 'modern citizen' was the product of Western experience, and that Japanese reality should not be interpreted by using such a culturally biased concept; whether or not particular Western attributes were discernible in Japan was of no particular significance, they argued.[3]

The extent to which modernizing Japan produced democratic citizens was debated again as historical studies accumulated to show the activities of individuals who appeared to conform to such a pattern. Most notably, research on Meiji Japan unearthed popular drafts for a national constitution drawn up in the latter half of the nineteenth century, articulating ideas of popular sovereignty, human rights and egalitarian values; the ideal type of the 'citizen' would therefore seem to have been present in early Meiji Japan.[4]

Furthermore, since the 1960s, Japan has witnessed the rise of citizens' movements which do not rely on the established political parties and labour union movements. The citizens' movements range from anti-war, anti-nuclear, peace movements through pollution-conscious environmental movements to community-based resident movements; they tend to cut across company and party lines. This suggests that modern citizens may have emerged as significant political actors on a large scale in post-war Japan. It remains a moot point, however, whether the tradition and the recent upsurge of citizens' movements represent mainstream or anti-mainstream currents of life in Japan.

The explicit application of modernization theories to Japan followed the rise of structural-functional theories in sociology, especially the Parsonian theory of social evolution. As publications on Japan based on this model flourished in the late 1950s and the 1960s, it became evident that the modernization framework was being applied to the Japanese case with the conscious aim of providing an alternative paradigm to that developed by Marxists in their historical and sociological analysis of Japan. To a considerable extent, this trend reflected changes in U.S.-Japan relations as the United States establishment increasingly found it profitable to show Japan to the world as the successful case of industrialization without discernible social disruption, an example of modernization based on smooth evolution rather than abrupt revolution.[5] Since such a claim could not be made of any other country Japan was assigned a central role in modernization theory.

The dominant (American academic) paradigm for understanding Japan at this time was avowedly 'productionist', that is to say it saw modernization exclusively in terms of indices of production. In a striking passage, one scholar especially active in representing Japan as a model of non-Communist success wrote:

the important thing is *that* people read, not *what* they read, *that* they participate in the generalized functions of a mass society, not whether they do so as free individuals, *that* machines operate, and not for whose benefit, and *that* things are produced, not *what* is produced. It is quite as modern to make guns as automobiles, and to organize concentration camps as to organize schools which teach freedom.[6]

This is perhaps the classic statement of 'modernization equals rationalization'; the attribution of modernization value to *means* without consideration of social or human ends amounts to a representation of modernization/rationalization as an end-in-itself. Fascism, imperialism and war were reduced to neutral referant points on the graph of steadily rising levels of modernization.

The Japanese pattern

As Japan astonished the world with its phenomenal economic growth in the 1960s and 1970s, theories of Japanese modernization were given another twist: where the Japanese tradition had till then been seen as an obstacle in the path of rationalization/modernization, to be gradually transcended as Westernization advanced, now the Japanese tradition was redeemed and proclaimed a uniquely suited vessel for modernization by a new school that developed around the 'unique Japan' hypothesis. In recent years, at least five variants are identifiable within this school.

The first of these is the so-called 'post-modern society' thesis, in which Japan is portrayed as leading the process of surpassing the modern, and is seen as outperforming Western societies. This argument tends to be intertwined with the long-standing theme of *kindai no chōkoku* (transcendence of modernity), which can be traced back to slogans generated in the 1940s to justify Japan's war activities against the Allied Powers. In the 1980s, Japan is depicted as becoming post-modern as it gets rid of 'modern' elements originating in the West, and places renewed emphasis on the traditional, indigenous and 'pre-modern' ingredients of Japanese culture. The literature which takes this line of argument has been produced by Japanese writers who tend to rely on Western theorists such as Michael Foucault, Gilles Deleuze, Jacques Derrida and Felix Guattari who are critical of aspects of Western civilization. Self-criticism by Western scholars reflecting on the crisis of Euro-American culture is twisted by Japanese critics into justification for seeing Japanese society as *beyond* criticism.[7]

The second variant is the *ie* society theory formulated by Murakami, Kumon and Sato.[8] They characterize Japanese society as being governed by an emphasis on *aidagara* (the interpersonal) rather than the Western notion of an autonomous self, and they stress the trans-temporal principle of group formation based upon kinship lineage. Against the background of Japan's phenomenal economic expansion these writers argue that patterns of *ie* society are compatible with modernization and industrialization; in fact they argue that such patterns are most conspicuous in company organizations. While organization of Japanese corporations along family lines was viewed as a tool of control and manipulation immediately after World War Two, the *ie* theorists have now redefined those principles as conducive to Japanese-style corporate democracy. Significantly, they also believe that parliamentary democracy is essentially incompatible with the consensual Japanese character in so far as it is based on competition between opposed political parties. To the extent that this argument disfavours political pluralism as being incongruent with the *ie* society tradition, the theory is open to justifiable concerns that it may be consistent with a soft form of neo-fascism.

The third variant is to be found in the renewed upsurge of *bunmei-ron*,[9] or civilizations theory. This arises in part out of a desire to correct the culturological biases built into the *bunka-ron* theories. Civilization theorists, including Umesao and Ito, attribute primary importance to technology and institutions, which they see as 'hardware'. Their models specify not only the geographical and ecological distribution of civilizations but also their developmental rank-ordering. While this approach uses a geographical zone larger than a nation state (Islamic civilization, Christian civilization), as the unit of analysis, 'Japanese civilization' is the single and exceptional case where a national boundary is equated with a boundary of civilization. That civilization is then portrayed as both unique and superior on the basis of a model in which material culture is crucial.

The fourth variant is a Confucianism model. With the economic rise of (South) Korea, Taiwan, Hong Kong and Singapore, the concept of a Confucian zone has been injected into the modernization debate.[10] Together with these countries, it has been argued, Japan shares the Confucian ethic emphasizing the virtues of austerity, hard work and submission to authority, all of which contribute to rapid economic growth. Ironically Confucianism, which was perceived in the 1950s and sixties as an obstacle to modernization, is in the 1980s given the status of its facilitator. The extent to which the Confucian zone theory may be used as an ideological tool to justify an authoritarian political system remains to be seen.

The fifth variant, a 'cultural physiology' model, has racist implications. It enjoys rather extensive support in Japan, such that it was little surprise when in 1986 Mr Yasuhiro Nakasone, reputed to be the most internationally minded Prime Minister in post-war Japan, made public remarks which implied the racial superiority of the Japanese. According to one widely publicized theory, the Japanese brain is qualitatively different from the Western brain and more susceptible to subtleties in nature and in emotional feelings.[11] Others claim that the Japanese who are traditionally herbivorous therefore think and behave differently from westerners who are mostly meat-eaters.[12] Such writings mesh nicely with the xenophobic elements in folk belief, still prevalent in Japan, to reinforce the popular notion that the Japanese are not only unique but superior to other nationalities. The Nakasone remarks were the tiny tip of a substantial chauvinist iceberg.

Presented in the context of the emergence of Japan as an economic superpower, these arguments are not really free from the undertone of various types of ethnocentrism, provincialism and even bigotry. Their ideological and political functions must be closely scrutinized. On the other hand, it must be noted that major proponents of the 'unique Japan' hypothesis, including Nakane and Doi,[13] contributed to the modernization debate by articulating some key concepts endemic to Japanese culture and language. They brought to the analysis Japanese words such as *tate* and *amae* (terms referring to the supposed 'vertical' structure of Japanese society and the psychology of 'dependence' which is said to characterize it) without which, they argued, Japanese society and psychology could not be understood. In proposing conceptual tools particular to Japanese culture, they took a position of cultural relativism and, by implication, brought into relief the ethnocentric bias of modernization theories based almost exclusively on a Western conceptual vocabulary.

Using the dichotomy between *emic* (culture-specific) and *etic* (transcultural) concepts,[14] cultural relativists such as Nakane and Doi maintain that the *etic* concepts used in Western modernization theories are, in fact, derived from Western *emic* experiences. Clearly they are correct. The most *etic* categories, therefore, are Western *emic* concepts which happened to prevail as trans-cultural, universal notions simply because Western societies were politically stronger than others.

The literature which set the stage for the 'learn from Japan' campaign in the late 1970s and the early 1980s went so far as to claim, directly or indirectly, that what earlier theorists had regarded as uniquely Japanese social characteristics were in fact patterns of modernity which other societies, especially Western industrial societies, would in due course reproduce. Symbolically, Vogel put Japan at the top rung of the modernization ladder by entitling his book *Japan as Number One*.[15] A similar point was put by Dore, in more technical language, in his 'reverse convergence hypothesis'.[16] Modernization was in the process of being freed of the bias implicit in the assumption of convergence on a Western pattern; instead the West, in particular the U.S., would have to 'Japanise'. The message has been taken up in many ways and at different levels of sophistication since then as officials, politicians, businessmen and academics in one after another Western country urge adoption of the

Japanese way of management, labour organization, education, or bureaucracy.

Political considerations

As the 1980s began to wane, some of the problems that had earlier been set aside were reopened, in particular the relationship between economic and political aspects of modernization. Japan was undoubtedly highly industrialized, but if democracy was also a mark of modernization, where did Japan stand in the ranking order of democracy? The distinction between modernization as industrialization and modernization as democratization came to be highlighted; there was no doubt that Japan had been modernized in terms of industrialization, but democratization was a different matter altogether. The same consensus system of Japanese society which the 'Japan as Number One' theorists praised might be seen instead as a sophisticated mechanism of control, through which the masses of the people were regulated in concrete, specific and situational contexts.[17] From this perspective, group-oriented behaviour, which an overwhelming majority of Japan specialists regarded as indicating 'consensus from below', might be seen as showing 'control from above'. Specifically, according to the control theorists, the *ringi* system, the total quality control circles, the company excursion practices and so forth – which the groupism theorists interpreted as part of a voluntary democratic participatory process – could be seen as part of an elaborate and intricate structure of manipulative control.[18]

In a separate book which the editors of the present volume compiled,[19] scholars in Australia, Japan and Canada explored these issues in some detail with particular attention to the state, society, education, labour, citizens' movements, women, human rights and science. The present volume reflects the same problem consciousness though its focus is both broader and deeper. In emphasizing the democratization dimension of modernization we are in effect calling for a reopening of the democratization debate of the late 1940s and the 1950s. The sweeping bird's eye view of indices of economic productivity, technological innovation and bureaucratic efficiency reveals very little about the state of human rights, freedom of expression, equality among people, and the general quality of Japanese life. To examine the life conditions and lifestyles of ordinary men and women with respect to these areas, it is necessary to develop a 'worm's-eye' perspective. Japan has to be observed not only 'from above' but 'from below' as well. It is in this spirit that the Japanese Studies Association of Australia chose 'Japan from Down Under' as the theme of its fourth national conference. This volume is based on the revised versions of papers delivered at this conference which was held at La Trobe University, Melbourne, from 22nd to 25th May 1985 and attended by about one hundred scholars, including six leading intellectuals from Japan.

ISSUES IN JAPAN'S MODERNIZATION

The contributions to this volume are built around explorations of four main sets of questions. They are (1) the extent to which political manipulation may be present under the guise of popular culture in Japan; (2) the extent to which Japanese *emic* or culture-specific categories are useful in the analysis of the Japanese way of life; (3) the extent to which the process of modernization in Japan has been accompanied by conflict and social disorganization; and (4) the extent to which traditional Japanese social structure and values may contribute to the country's post-modernization.

1 Politics in the name of culture. A crucial issue in interpreting the level of democratic modernization (as distinguished from industrial modernization) is the degree to which apparently voluntarily shared culture actually manifests patterned political control. Defining Japan as a 'manipulated society', Kogawa is explicit about the politically managed nature of Japanese popular culture. He stresses the power of mass media, especially electronic media, in shaping mass consciousness and creating and disseminating a mass culture that is as homogenous and pervasive as the traditional *tennō* (emperor) system. Kogawa's assessment is pessimistic: the prevailing 'simulated reality', where weekly magazines, television gossip programmes and comic strips mould popular perceptions of the world, has stripped the media-soaked masses of the sense of human rights, spontaneous communication and the capacity to understand and respond to fundamental problems. Post-industrial Japan, dominated by the large-scale information industry, (according to Kogawa) may show a 'friendly' and non-repressive face but nevertheless controls the mass psyche in a manipulative way by inducing mass political apathy.

On the other hand, Kogawa sees some hope in the development of 'free communion' among youth who have found ways to make use of electronic devices for the formation of counter-culture. He interprets the 'Mini FM' networks spreading across Japan in the mid-1980s as a community-based and spontaneous grass-roots cultural phenomenon, a symbol (however apparently insignificant) of a post-modern reality that is not homogenized, packaged and controlled.

Bolitho looks at the evolution of Sumō wrestling. The popularity of

Sumō, a quintessential component of Japanese popular culture for over 300 years, was purchased only at a price. First, it could not have been obtained without explicit submission to state control. This involved externally imposed regulation, the imposition of an organizational hierarchy, and the express recognition of approved codes of morality and forms of religiosity. Secondly, it could not have been achieved without the economic intervention of professional organizers, and the patronage of the feudal lords. In origins and character, the sport of Sumō contrasts sharply with Western pugilism, and its evolution as described by Bolitho constitutes a nice study of gentrification, professionalization and aesthetic refinement – in short, of modernization – to which Western contact seems peripheral. From Bolitho's account, it would seem that Sumō might lay claim to be the world's first fully professional spectator sport.

2 *Utility of emic concepts.* Since the process of concept formation is crucial in modernization theory, it is necessary to consider who it is in Japan that defines what is modern. In this volume, several scholars have emphasized regional variation in the classification scheme – progress, evolution and development. In contrast to Kogawa who evaluates the current situation mainly from the vantage point of Tokyo, Tada focusses on the religious, festive and dietary tradition of Osaka mass culture, which is qualitatively different from that of Tokyo, thus reminding us of the complexity of categories used in the assessment of modernity, especially its culture elements. As an 'emicist', Tada implies that local culture is often counter-culture *vis-à-vis* the culture of the centre which masquerades as national culture, and suggests that the classificatory categories of Tokyo culture may not be adequate for the analysis of Osaka culture. In Tada's analysis of Osaka vaudeville culture, both notions imported from overseas and vocabulary used in Tokyo are given meanings different from their original usage; theories based on Tokyo categories may therefore be seen as a form of Tokyo-based cultural imperialism.

Emic concepts in studying Japan are especially prevalent in Japanese literature. Aoyama's investigation into contemporary women writers unravels a Japanese classificatory vocabulary for the aesthetic analysis of male homosexuality. She traces a perspective on Japanese values and culture that, while intrinsically 'Japanese', is yet doubly removed from the dominant conventions of the culture – from the male and heterosexual.

It remains to be seen if and how these and other Japanese *emic* concepts can be used in the reconstruction of modernization theories. If the Osaka-type 'low-level' universalism based on the philosophy of one measure of modernity is characterized by the prevalence of the Pure Land Buddhist belief that 'in the long run people are all the same', what implications would this have for the ranking of various nations? Should the type of communication present in Osaka *manzai* be an index of modernity, then where would Western societies be located in terms of this indicator? Regional diversity of *emic* practices in a nation sensitizes us to difficulties in using national averages as the guidepost for the evaluation of the levels of modernization of different countries.

Following a similar tack, Arnason attempts to situate the Japanese modernization phenomenon within a theoretical framework which allows for distinctive versions of and autonomous roads to modernity. After an introductory overview of the Japanese trajectory and consideration of ways of comparing it with the West, he discusses some salient features of the process of state formation in Japan and their relationship to the peculiar structure of Japanese feudalism. He argues that a comparative analysis should focus on three main contrasts: a different structure of the feudal principle, different relations to non-feudal components of society, and different transformations of the feudal elite.

3 Level of social cohesion. Modernization theories suggest that the process of modernization tends to be accompanied by social dislocation, cultural lag and political disruption. In the field of Japanese studies, however, most scholars have argued that Japan is exceptional in that little large-scale disorganization took place despite the rapid tempo of modernization. Inkster in his chapter argues that the early phase of modernization in Meiji Japan was carried out without creating serious non-elite disturbance because it was legitimized in terms of traditional, late Tokugawa values. Japan first adapted pre-modern East Asian values and attitudes to the modernization process; in later years the 'Four Little Tigers' of East Asia followed suit. Protestant Europe had no monopoly on the work ethic.

Matsuzawa reconsiders the thesis of successful conflict management in Japan by scrutinizing street labour markets and the struggle of day labourers. According to him, Sanya in Tokyo, Kamagasaki in Osaka, and other *yoseba* markets where day labourers, cut off from their families, are concentrated and work in the construction, civil engineering, transport and stevedoring industries, represent the bottom line of modernization; outright exploitation and overt repression are rampant in the *yoseba*, and mass grievances run deep. Matsuzawa argues that these realities are either hidden or played down in discussion of the nature of Japanese society, although the lower strata of Japanese society to which these workers belong constitutes as much as 10 percent of the Japanese population.

Taking a similar tack, Tanaka also examines another 'dark shadow' of Japan's 'high-tech development': the so-called nuclear power plant gypsies who work under sub-contractors and wander from one nuclear plant to another, being exposed to high dosages of radiation. Mackie's research on the economics of sex tours reveals how such 'integration' entails the sexual division of labour at the international level. The work of prostitution is carried out by poor women in poor Southeast Asia for wealthy men from affluent Japan. Her study affords glimpses of the way in which Japan's modernization gave rise to a complex system of multi-dimensional international stratification based on sex, class and race.

The papers by Matsuzawa, Tanaka and Mackie serve as a reminder that the same Japanese industry which appears to be a model of the post-modern in terms of commitment to computer, robot, and nuclear technology, maintains a labour system which has the hallmarks of the most primitive capitalism.

4 *Indigenous elements in post-modernization.* In place of the conventional dichotomy between traditional and modern, Kawamura suggests instead the framework of a trichotomy, the 'post-modern' being the third phase. The central point of his argument is that higher forms of such *traditional* characteristics as Gemeinschaft, affectivity, and collectivity-orientation are part and parcel of post-modern societies. Should this be the case, Kawamura submits, the viable ingredients of Japan's tradition ought to be studied thoroughly, not in the spirit of the right-wing nationalism of wartime intellectuals, but as part of the process of Japan's post-modernization, leading to greater humanism and liberation.

Some writers in the present volume go so far as to argue, apparently paradoxically, that the pre-modern indigenous forces in Japanese society are in fact the principal strength to achieve post-modern liberation from the current framework of 'modern Japan'. They argue that in the industrialized, advanced and modern Japan where the conservative ideology of industrialism prevails, the 'modern' is the status quo, and that to reform that existing conservative structure one must draw inspiration from the traditional.

Irokawa's extensive overview of the characteristics of Japanese popular movements points out several aspects of their weakness and suggests that autonomous and independent movements tend to flourish when traditional (local, moral and particularistic) elements are fused with universalistic (international, ideological, and modern) elements. In focussing upon resident movements which are especially relevant in the current context, Irokawa makes a distinction between indigenous residents and external 'drifters' participating from outside. He maintains that, when the principle of the latter subjecting themselves to the former is observed, a creative perspective for radical social change is generated. Such a suggestion runs against the modernist assumption – long prevalent in Japanese social movements – that modern, well-educated outsiders can and should enlighten and lead traditional, ignorant local residents. Irokawa's argument compels attention in part because it is the fruit of his reflection on direct involvement in the movement in support of the victims of mercury pollution in the southern Japanese town of Minamata. His argument points to the hypothesis that indigenous values, perhaps even values negated or suppressed during modernization, may be revived as central components in the culture of post-modern Japan.

Irokawa's insistence on affirming the integrity of the core values of the Japanese peoples' movements and their relevance for the future is complemented by Large's analysis of Buddhist socialism in the post-war Japanese peace movement, which sheds light on the way in which indigenous traditional values and Western Marxism were intricately interwoven in pursuit of peace programmes designed to complement and transcend 'mere' industrial modernity.

Focussing on women's problems in Japan, Ueno takes the view that, so long as the framework of male-dominated values of modernity is not challenged in a fundamental way, women will inevitably remain second-class citizens. Instead of catching up with the world of men based on capitalist competition and work ethic, Ueno maintains, women can liberate themselves by affirming womanhood, sexuality and reproductive power. She insists on the feminization of men rather than the masculinization of women, and argues that the Japanese tradition may be a significant cultural resource to accomplish this goal. Because of the conventional separation of the worlds of women and men, women have been able to develop their own networks of 'sisterhood', and such networks can now be used for their own liberation. Because of the traditional family system in Japan where parents-in-law live under the same roof, men are increasingly required to take care of them at home under the inadequate welfare services. Traditional structural arrangements and value orientations, Ueno argues, can now serve as a force to counter 'modern' industrialism.

Her review of the political struggle over abortion law reform in Japan leads Buckley to be less sanguine. She maintains that such an appeal to traditional symbols paves the way for consolidation of the 'Japanese style welfare state'. Buckley seems to favour more modernizing reform of Japan's oppressive traditional political forces, whereas Ueno appears to see present-day Japan as already essentially modernized and needing a leap towards *post*-modernity. Traditional elements may serve as (at least) a stepping stone towards this happy future condition (according to Ueno).

Though their emphases differ, Irokawa, Large, Ueno, and Buckley all appear to agree on a prescription for a post-modern Japan which is broadly egalitarian, pacifist, feminist, and in which elements of Japan's traditional culture, reacting with external Western (or other) value systems, attain true universalism.

THE POST-MODERN PROJECT

Twice in the past half-century Japan has declared the transcendence of modernity to be an official state project – first under Premier Tōjō Hideki in 1942 and more recently under Premier Ōhira Masayoshi in 1979 (the central themes of the latter having then been adopted by Ōhira's successor, Nakasone Yasuhiro). Both projects were expressions of a deep-seated Japanese desire to define historical progress in ways that reduce its Western and European focus. The legitimacy of that project cannot be doubted, as there is no gainsaying the distortions of racism, the depradations of imperialism and the carnage of world war that accompanied European modernization, or the contemporary nuclear confrontation and ecological crisis it has generated. But the project in Japan has been defined on both occasions from above, by elements most committed to increased productivity and ceaseless rationalization, maintenance of central bureaucratic control, and of Japanese might in relation to the world economy. Ironically, such definitions tend to reproduce the core notions (rationalization as end-in-itself) and the key problems (bureaucratic absolutism, militarism, and alienation) most problematic in the model to be transcended; in short the project has been essentially contradictory.

Whether Japan will yet be able to find a path beyond modernization that is defined instead from below, and a formula that will allow it not just to redefine modernization by finding ways to outproduce the West and formulas to reinforce social control but really to transcend it, are matters of considerable significance yet to be resolved.

While it is often said that Japan in the 1980s is embarking on the path toward a post-modern society, there is sharp division over how such a transition should be evaluated. From Daniel Bell to Alvin Toffler, Herman Kahn to Hayashi Yujiro, optimists have presented a rosy picture of post-modern society in which most work is conducted by automated machines and electronic computers, leisure time is maximized, and everyone is able to enjoy a high level of creative activity. On the other hand, for pessimists from George Orwell to C. Wright Mills, H. Marcuse to Jurgen Habermas, a post-modern society is subject to sophisticated and technologically advanced forms of control and exploitation: psychic repression and manipulated regulation of everyday life constitute the defining characteristics. No discussion of the problem can afford to ignore the lessons to be learned from the Japanese experience, even if the Japanese case proves as difficult to assimilate within mainstream theories of post-modernization as it has in the case of modernization.

NOTES

1 Two recent volumes edited by Merlin B. Brinkerhoff and published by Greenwood Press (Westport, Connecticut) are highly relevant to this debate: *Work, Organizations, and Society: Comparative Convergences*, 1984; and *Family and Work: Comparative Convergences*, 1984.

2 The best known figures taking this position were Maruyama Masao, Ōtsuka Hisao and Kawashima Takeyoshi. See Maruyama's *Thought and Behavior in Japanese Politics*, Oxford, Oxford University Press, 1963.

3 Yoshimoto Takaaki, Tanigawa Gan, and other indigenously oriented intellectuals on the left took this position.

4 See Irokawa Daikichi, *et al.*, *Minshu kenpō no sōzō* (The Creation of a Popular Constitution), Hyōronsha, 1970.

5 This was the general theme of a series of six volumes published by Princeton University Press for the Conference on Modern Japan of the Association for Asian Studies, Inc., especially, R. P. Dore (ed.), *Aspects of Social Change in Modern Japan*, 1967.

6 Marius B. Jansen, 'On studying the modernization of Japan', *Asian Cultural Studies* (Tokyo: International Christian University), October 1962, quoted in John W. Dower, 'Introduction' to *Origins of the Modern Japanese State: Selected Writings of E. H. Norman*, New York, Pantheon, 1975, p. 54.

7 On these points, see Takeuchi Yoshirō, 'Posuto-modan ni okeru chi no kansei' (Intellectual pitfalls of post-modernism), *Sekai*, November 1986, pp. 92–115.

8 Murakami Yasusuke, Kumon Shunpei and Satō Sezaburō, *Bunmei to shite no ie shakai* (*Ie* society as civilization). Tokyo, Chūō kōronsha 1979.

9 See, for example, Umesao Tadeo (ed.), *Bunmei-gaku no kochiku no tame ni* (Constructing civilization studies), Tokyo, Chūō kōronsha; and Itō Shuntarō, *Hikaku bunmei* (Comparative civilizations), Tokyo, Tokyo daigaku shuppankai, 1985.

10 See, for example, Michio Morishima, *Why has Japan Succeeded?* Cambridge University Press, 1982; Dai Guo-hui, '"Jukyō bunka-ken" ron no ichi kōsatsu' (A thought on the 'Confucian cultural zone' theory), *Sekai*, December 1986, pp. 137–49; and Han Seung Soo, 'Economic success and confucianism', *Far Eastern Economic Review*, 20th December 1984.

11 Tsunoda Tadanobu, *Nihonjin no nō* (The brain structure of the Japanese). Tokyo, Taishūkan, 1978.

12 Sabata Toyoyuki, *Nikushoku no shisō* (Thought patterns of meat-eaters). Tokyo, Chūō Kōronsha, 1966.

13 Chie Nakane, *Japanese Society*, Ringwood, Penguin, 1970; and Takeo Doi, *Anatomy of Dependence*, Tokyo, Kodansha International, 1973.

14 See, Harumi Befu, 'Emic and etic approaches to the study of Japanese society', forthcoming in Ross Mouer and Yoshio Sugimoto (eds.), *Japanese Society: Alternative Ideas*, London, Kegan Paul International.

15 Ezra Vogel, *Japan as Number One*, Cambridge, Harvard University Press, 1979.

16 See his *British Factory–Japanese Factory*, Berkeley, University of California Press, 1973.

17 Ross Mouer and Yoshio Sugimoto, *Images of Japanese Society*, London, Kegan Paul International, 1986; and Yoshio Sugimoto, *Chō-kanri rettō Nippon* (The ultra-controlled Japanese Archipelago), Tokyo, Kōbusha, 1983.

18 For example, S. Prakash Sethi, Nobuaki Namiki, and Carl L. Swanson, *The False Promise of the Japanese Miracle*, Boston, Pitman, 1984; Jomo K. S., *The Sun Also Sets*, Institut Analisa Sosial, 1983; and Rokuro Hidaka, *The Price of Affluence*, Tokyo, Kodansha International, 1985.

19 Gavan McCormack and Yoshio Sugimoto (eds.), *Democracy in Contemporary Japan*, Sydney, Hale and Iremonger, 1986, and New York, M. E. Sharpe, 1986.

I Popular culture: tradition and 'modernization'

1 Sumō and popular culture: the Tokugawa period

HAROLD BOLITHO

Wrestling is one of the oldest of sports. Together with running, leaping, and the lifting or throwing of weighty objects, wrestling is made up of precisely those elements likely to appeal to the earliest forms of human society: it is satisfyingly competitive, it calls on physical skills of a kind valued in communities where strength and agility are at a premium, and those who engage in it need no equipment beyond that with which nature has endowed them. Wrestling emerged at an early stage in all civilizations, and there is no reason to doubt that it did so in Japan as well. Indeed the earliest of the Japanese chronicles, compiled in 720, says as much. In the year 23 B.C., according to the *Nihon shoki*, the Emperor Suinin came to learn of a certain strong man living nearby in the village of Taima, a man called Taima no Kehaya who boasted of being the strongest man in the world. On the face of it, this was perhaps not an unreasonable claim, since Kehaya was said to be capable of straightening iron hooks – and, into the bargain, breaking the horns of oxen – with his bare hands. The Emperor Suinin was obviously prepared to accept this claim, so the *Nihon shoki* records, until one of his courtiers chanced to mention yet another strong man – Nomi no Sukune, a 'valiant man', as he is described, from the distant land of Izumo, on the Japan Sea coast of Honshū. Immediately the Emperor arranged for the two men to meet, and the *Nihon shoki*'s account of that meeting, apparently like the meeting itself, is simple and direct:

The two men stood opposite to one another. Each raised his foot and kicked at the other, when Nomi no Sukune broke with a kick the ribs of Kehaya and also kicked and broke his loins and thus killed him. Therefore the land of Taima no Kehaya was seized, and was all given to Nomi no Sukune. This was the cause why there is in that village a place called Koshioreda, the field of broken loins.[1]

There are reasons for doubting the strict accuracy of this account. It was, after all, set down some seven and a half centuries after the alleged event, so it can hardly be judged an eye-witness account. Beyond that, too, is the context in which it appears. Perhaps there was such an Emperor as Suinin, but I think one is nevertheless obliged to question the *Nihon shoki*'s description of the events of his reign, since it claims that he ascended the throne in 29 B.C., at the age of forty, and was to occupy it for the next ninety-nine years, until his death at the age of 139. We know that there was a village of Taima – indeed the place name still survives – and that it was of some antiquity,[2] but there is no way of verifying the existence of either Kehaya or Nomi no Sukune. All that can be salvaged from the *Nihon shoki* account is the fact that wrestling matches of a kind were not unknown in Japan, certainly as early as 720, when this particular chronicle was compiled, and, allowing for an oral tradition from which the compilers gleaned their stories, presumably earlier as well. So much can be accepted quite readily.

What cannot be accepted is any putative link between wrestling of this kind and sumō wrestling, whether the sumō wrestling we know today, or any of its historically verifiable ancestors. As described in the *Nihon shoki*, the bout between Taima no Kehaya and Nomi no Sukune was brief, violent and deadly, and it is therefore difficult to accept this as in any way ancestral to sumō, which is not, nor has it ever been, any of these things. Instead, throughout its long history sumō has been curiously gentle, aimed not at hurting, much less killing, an opponent – aimed not even at immobilizing him, or extorting from him some gesture of submission, as is the case for example in Greco-Roman wrestling – but simply aimed either at bringing him to the ground or forcing him outside a circle. Of all body contact sports, sumō is surely the most considerate of its practitioners, being concerned to spare them as much pain and indignity as possible. The *Sumō densho* rules, laid down in the eighteenth century, were very precise on this point, excluding anything designed to hurt, or even disorient, one's opponent.[3]

Nevertheless, even though the world of primitive wrestling, such as that described in the *Nihon shoki*, bears no real relationship to sumō wrestling, the world of sumō has always been inclined to look to the bout described in this ancient account, despite its questionable authenticity, as a direct ancestor. Nomi no Sukune, the winner, was effectively accepted as the patron saint of sumō, his name commemorated at a shrine, the Sukune Jinja, located in the Ryōgoku area of Tokyo, where sumō wrestling has been centred for at least 300 years, and that shrine is maintained by contributions from the Japan Sumō Association.[4] Historians, too, those few who deal with the history of sumō, may discount the Nomi no Sukune story, but will nevertheless dignify it with inclusion in their accounts.[5] It is, after all, undeniably colourful. It is also – and this is very important –

an easy way of claiming antiquity, and hence, in a society where the richer the patina of the past the greater the degree of public respect – it is a way of achieving respectability, and this has long been a matter of crucial concern in the world of sumō wrestling.

It might be thought difficult, if not impossible, for a sport characterized essentially by fat, sweat and muscle to strive for respectability, but that is in fact what sumō has always done. Not content with claiming links with the remote Japanese past, tracing the sport's origins back to the earliest emperors, those involved in sumō have also been anxious to draw attention to another ancestry which, if not quite so venerable, was rather more genteel – that is, to assert an affiliation with the ceremonial wrestling, the *sechie-zumō*, of the successive imperial capitals of Heijō-kyō (710–84), Nagaoka-kyō (784–94) and finally Heian-kyō (794 onwards). Such wrestling matches, originally held (like the Chinese ceremonies on which they were modelled) on the day of the *tanabata* festival, the seventh day of the seventh month, were a regular part of court life until 1174.[6] The custom may or may not have begun precisely in 734, the year to which their inception is formally ascribed, but it would not be unreasonable to accept the introduction of *sechie-zumō* as somewhere around the early eighth century, when so much Chinese ritual was introduced. What would be unreasonable would be to see any real link between the wrestling of the imperial court and the sumō wrestling we know today. To begin with, it was strictly limited in duration, the whole ceremony being concluded in just one day. It was even more restricted in audience, being seen only by the emperor and his senior courtiers, all of whom viewed the wrestling from the confines of the Shishinden, a pavilion deep in the recesses of the imperial palace. It was in no way a public spectacle, therefore, but rather yet another of those rituals by which emperors carried out their primary function which was the guaranteeing of good harvests.[7] In keeping with its serious purpose, *sechie-zumō* was accompanied by a degree of ceremony which the sumō world has never been able to emulate, including a procession to the sound of drums, gongs, conch shells and a gagaku orchestra.[8] Instead of the hardbitten commercial promoters who were to organize sumō wrestling, *sechie-zumō* was in the hands of courtiers of a kind seldom if ever associated in the public mind with the activities of strong, fat men. One such courtier, indeed, the noted poet Ariwara no Yukihira (818–93), an organizer of *sechie-zumō* ceremonies under the Emperor Seiwa (r. 858–76), is far more readily associated with the delicate imagery of the poems he left to posterity in the *Kokinshū*.[9] In fact, poetry, dancing and feasting, all of which were carried on simultaneously, were as much a feature of the *sechie-zumō* ceremony as the wrestling itself.[10] Perhaps only in the nature of its competitors, strong men recruited each year by officials sent out to the provinces for precisely that purpose, did *sechie-zumō* resemble the sumō wrestling of later periods. The *Utsubo monogatari*, which refers to two such men by name – Naminori from Shimotsuke and Yukitsune from Iyo – makes it quite clear that some men wrestled regularly at *sechie-zumō*, year after year, but there is no indication that they were fully professional or otherwise wrestled regularly.[11]

The search for respectability has also extended into the medieval period, the years from 1185 to 1600, when wrestling was one of the battery of skills a samurai was expected to acquire, a useful adjunct to proficiency with bow, sword and javelin. Military leaders from Minamoto no Yoritomo to Oda Nobunaga had their *sumō bugyō*, men whose duty it was to train samurai in the art of unarmed combat. It is recorded that in 1570 Nobunaga called together all the wrestlers in the province of Ōmi, and had them wrestle before him at Jōraku-ji, the temple at which he was currently residing. From among them he selected the two best – Namazue Mataichirō and Aoji Youemon – declared them his vassals, and made them his *sumō bugyō*.[12] Clearly not all those Nobunaga assembled were professionals – in fact none of them may have been, although with names as unlikely as Namazue and Hakusaiji no shika some of them at least had professional pretensions, since such names were already an emblem of the nascent wrestling profession – but there were clearly professional sumō wrestlers abroad in Japan during this period. Just ten years earlier, in 1560, a daimyō in Kyūshū had imported from Kyoto a group of professionals, with names like Ikazuchi (Thunder), Ōarashi (Great Storm), and Inazuma (Summer Lightning) to have them perform at a fund-raising tournament in honour of a local shrine.[13] What interested military leaders like Nobunaga, however, was not so much wrestling as a sport as wrestling as a military skill, cognate to skill with sword, bow or spear. By employing the two best wrestlers of Ōmi province, and incorporating them into his formal military hierarchy, Nobunaga was in fact separating them from their previous profession, and attaching them to the far older *jūjitsu* tradition, already a well-established part of samurai training.

The office of *sumō bugyō* did not survive into the Tokugawa period (1600–1868). Instead, the barrier between samurai unarmed combat and professional wrestling, still crossable in 1570, was not to be crossed again, as the two different traditions – *jūjitsu* on the one hand, and sumō on the other – separated finally and irrevocably. Military leaders of the Tokugawa period turned their attention to *jūjitsu*, employing exponents of any of the more than a hundred different schools of unarmed combat to teach their skills to a samurai population less and less concerned with the prospect of military action.[14] They maintained their interest in wrestling, but as a pastime only, not as a response to military necessity. For a time they continued to keep professional wrestlers in their exclusive employ, much as they might also maintain Noh actors and musicians, if they were sufficiently wealthy, but this custom disappeared rapidly during the seventeenth century.[15]

More than anything else, the cause of such changes was to be found in shifting social circumstances. The sudden mushrooming of towns and cities which took place in the early seventeenth century, and which by 1700 was to leave Japan one of the most highly urbanized countries in the world, had also injected a new element into Japanese cultural life – large numbers of city dwellers with the leisure, the money, and the inclination to carve out their own entertainments. Their patronage was responsible for the emergence of a number of cultural forms which for the first time competed with elite culture on its own terms. The latter half of the seventeenth century saw the appearance of popular fiction, produced by printing presses in Osaka and elsewhere. Simultaneously, the wood block print was developed, in a process ultimately capable of bringing art to the masses for a few cents a sheet. In the theatre, where the first Ichikawa Danjurō (1660–1704) was striking heroic poses and for which Chika-matsu Monzaemon (1653–1724) was writing his puppet plays, common people were, for the first time, at the centre of a cultural life of their own, rather than, as previously, at the periphery of something controlled by the elite. Precisely the same developments affected sumō wrestling at precisely the same time, turning this most ancient of sports into something quite new. In the process, the promoters of the new sport were to retain – and in some cases deliberately to revive – aspects of earlier forms of wrestling, wishing to share some of the glow associated with the wrestling of the mythological period, or of the court aristocracy, or of the mediaeval warrior class. But this was simply camouflage, for the wrestling developed to amuse the common people of Tokugawa Japan was of a totally different kind.

Sumō wrestling, as it developed in the latter part of the seventeenth century, and as we know it today – that is, a sport carried on by professionals, men equipped by both physique and training to participate in it (and, incidentally, disqualified by both physique and training from doing much else), and carried on, moreover, for the entertainment of a paying public – had very little in common with those earlier forms of wrestling so often described in history books. These, if not haphazard (like the encounter between Taima no Kehaya and Nomi no Sukune), or restricted in audience (like the *sechie-zumō* performed before a small group of nobles in the imperial palace) were practised as a means to some other and more worthy end (as in the unarmed combat between samurai, aimed at increasing their professional military competence). The sumō wrestling of the Tokugawa period had none of these features, for it had developed out of a rather different, and far humbler tradition. Its origins lay in contests between village strong men who, on the few festival days each year, would pit their strength against each other in a variety of ways – tug of war, lifting stones, or grappling with each other. The common feature uniting them was their basically amateur and unskilled nature, for the village strong man, like everyone else in his community, was a farmer, not a professional athlete. Wrestling of this kind, known as *kusazumō*,[16] may have intersected with *sechie-zumō* and the warrior wrestling of the later middle ages – Naminori, the wrestler referred to in the tenth century *Utsubo monogatari*, like Namazue, the victor in the contest of Ōmi wrestlers held before Oda Nobunaga in 1570, would have come from just such a background – but it was nevertheless unambiguously plebeian, far more associated with the fairground than the courts of aristocrat or warrior.

It was from these antecedents that sumō wrestling emerged, and its immediate origins are to be found in the gradual development of troupes of itinerant wrestlers, in the latter part of the sixteenth century, who would tour country towns and, for a fee, take on all-comers, or *yo-sekata*.[17] They were also available, again for a fee, to put on fund-raising displays, like those who wrestled in Kyūshū in 1560.[18] In the early seventeenth century, as both Edo and Osaka grew in size, attracting villagers from the surrounding countryside, they provided ideal circumstances for professionals of all kinds, from the carpenters who constructed the buildings, to the prostitutes who peopled Shimmachi and Yoshiwara, and to wrestlers, some prepared to take on all comers in casual bouts on street corners (known as *tsuji-zumō*) and others with more ambitious projects in mind. It was the latter, together with their promoters, who developed sumō wrestling. In their hands, sumō appeared in the seventeenth century as a fully professional sport mounted for a paying public.[19] The major cities – Kyoto and Osaka, each with a population of half a million, and Edo, with a million inhabitants – were able to support troupes of wrestlers. Edo in particular could provide anything up to three thousand paying spectators for each day of a ten day tournament, and these, by 1843, could offer a livelihood to a total of 226 professional wrestlers.[20] Such relatively easy pickings effectively put an end to country tours by itinerant groups of athletes, and from the mid-eighteenth century sumō became almost exclusively an urban entertainment, with tournaments in Edo in the spring and winter, in Kyoto in the summer, and in Osaka in the autumn.[21]

At the same time that sumō wrestling developed as a popular sport, and not by coincidence, its reception by its erstwhile patrons, the military aristocracy, changed shape. As we have seen, the *sumō bugyō* of the middle ages gave way to the *jūjitsu* instructor, as did those wrestlers who plied their trade for the exclusive entertainment of a military patron. This did not mean, however, that the samurai class was content to give up this diversion without a struggle. At least as late as 1648, just shortly before it reversed its position and gave permission for a public sumō tournament to be held in Asakusa, the Tokugawa government was still trying to keep sumō in the martial arts camp, reserving it for the entertainment of the warrior class by forbidding all wrestling except that which took place privately inside daimyō residences.[22] Indeed the early history of wrestling in the Tokugawa period shows much evidence of official hostility and suspicion. *Shikona*, for example, the flamboyant aliases much in vogue among wrestlers, were officially forbidden as late as 1651 in a decree which read, 'All those employing the assumed names known as *shikona* are to be reported immediately; wrestlers have taken false names in the past, but these must not be employed in future.'[23] Street-corner wrestling, or *tsuji-zumō*, was regarded with as little enthusiasm as any other activity likely to attract a crowd. It was banned repeatedly – in 1648, 1687, 1694, 1703, 1707, 1719 and 1720[24] – and these bans were never relaxed; it would appear, however, that it continued, bans notwithstanding, since the 1694 prohibition refers to people 'gathering each night to wrestle at various streets within the city'.[25]

Formal wrestling tournaments, known as *kanjinzumō*, or fund-raising sumō (since they were ostensibly held to raise money for some worthy project, such as the repair of a religious structure), were also initially more often discouraged than not. 'There have long been decrees pertaining to *kanjinzumō* within city limits; their intent must be more strictly observed,' read a prohibition of 1661, 'it is not to take place in the city.'[26] There was a further prohibition in 1673,[27] and several petitions to hold tournaments were rejected during that decade[28] but there were no further formal prohibitions, and from 1684, when Honda Tadachika, the government minister in charge of religious affairs, gave permission for an eight-day tournament at Fukagawa Hachiman shrine, tournaments were held in Edo at regular intervals.[29] Yet there was obviously still a degree of residual unease in the official attitude, and, from the phrasing of two later laws, it is clear that that unease was related to considerations of social class. 'It has come to our notice,' read a decree of 1711, 'that commoners are employing wrestlers and holding displays in various quarters of the city ... This is inappropriate to commoners ... and it must henceforth come to an end.'[30] In 1743 the same decree was repeated, with a rider noting that the prohibition was still being ignored.[31]

Of course it was a losing battle. In this, as in virtually every other area of urban legislation, whether trying to prohibit gambling and brawling, or confine sideshows and prostitution to certain areas of the city, Japan's military aristocracy was obliged to come to terms with popular pressure. Sumō wrestling very quickly developed a momentum of its own, and as it did so began to attract aristocratic patronage in a new form. By the eighteenth century, instead of maintaining their own troupes of wrestlers, some twenty-seven daimyō around the country had begun to give stipends to the most noteworthy of the urban professional wrestlers.[32]

Initially those stipends could be quite large, as in the case of famous athletes like Shirayama Shinzaburō, who was receiving over 40 *koku* from the daimyō of Kii early in the eighteenth century.[33] A century later, however, the Tokugawa fiscal crisis having taken its toll, they became far more modest. Raiden Tamiemon, one of the great names of the 1790s — indeed, one of the most famous wrestlers of all time — received little more than 8 *koku* from his patron, the daimyō of Matsue, while Onomatsu, one of the two grand champions of the early 1830s, could attract no more than 50 *hyō* from the daimyō of Chōshū.[34] These stipends were, however, large enough to secure for the patron the privilege of having an athlete wear his livery — in this case the embroidered ceremonial apron, or *keshō-mawashi*, decorated with a distinctive pattern. As was the case with the British aristocracy in their patronage of jockeys in the nineteenth century and beyond, too, the sporting connection, tenuous though it was, was enough to confer upon the patron a degree of vicarious honour.

Official permission and individual patronage allowed sumō wrestling to survive and to develop over the Tokugawa period, but that development was far from unrestrained, and in the process was to take this most plebeian of sports and twist it into a far more complex shape. The Tokugawa government, in giving its sanction to tournaments, did its best to see that the development of sumō wrestling as a popular entertainment was as circumscribed as possible, hedging it about with a number of restrictions. When, gradually and reluctantly, sumō tournaments came to be allowed, it was only on condition that certain undertakings were given, and these undertakings were to end in total professionalization. In order to acquire accreditation, organizers of tournaments had to go through several steps. First of all it was necessary to establish the fiction that the tournament was not simply a commercial enterprise, but came rather under the *kanjin*, or fund-raising designation. With that done, promoters needed to obtain the consent of the Tokugawa government, represented in this case by the *Jisha bugyō*, a cabinet minister whose portfolio embraced the religious life of the nation and hence, by extension, all fund-raising activities on behalf of religious institutions. Before winning such consent, promoters were required to show themselves to be responsible persons, and to give specific undertakings that their tournaments would be trouble-free, with no fights in the crowd, no gambling, and no wounding. They were also obliged to furnish records of box-office receipts.[35] Further, to establish one's credentials as a responsible person, it became obligatory to be a *toshiyori*, or elder – someone with a reputation and influence in, as well as a familiarity with, the sumō wrestling business.[36] Effectively, by such regulations the government lent its support to the institutionalization of a hierarchy, with a certain limited number of promoters whose rights, privileges and authority would be officially guaranteed. It was a familiar system for the Tokugawa government, which always preferred to use self regulation wherever possible, and did so with qualified success in fields as diverse as oil-pressing and hair-cutting. Through the sumō *toshiyori*, who numbered thirty-eight by the 1790s and forty-five some fifty years later,[37] sumō became virtually closed to outsiders, and therefore distanced from its casual origins. It also became organized to a far greater extent than its haphazard seventeenth-century ancestor, with regular schedules, formal rankings, training facilities, and an accepted code of rules which, among other interdictions, forbade striking with the clenched fist, gouging, kicking, breaking fingers, and squeezing or tugging genitals.[38] There was no place now, nor had there ever been, for the kicks to chest and groin with which Nomi no Sukune had allegedly despatched Taima no Kehaya. Nor, for that matter, was there any claim to military effectiveness. On the contrary, before each tournament wrestlers were shown a diagram of the human body on which the prohibited areas – that is, those most dangerous to the contestants – were clearly marked.[39]

To survive, sumō had been obliged to submit to a degree of state regulation. This was not unusual. All the other popular arts and entertainments of the Tokugawa period followed the same path, and all, like sumō, were moulded to some degree by government pressure.[40] Towards the end of the eighteenth century, however, sumō was to achieve an eminence which no other popular entertainment could ever hope to rival. In 1791, in the Fukiage gardens of his castle at Edo, Tokugawa Ienari, the eighteen-year-old shōgun, viewed an exhibition mounted by 166 wrestlers for the diversion of himself and his senior vassals.[41] He was to view it a further four times, in 1794, 1802, 1823 and 1830, while his successor, Tokugawa Ieyoshi, saw it twice, in 1843 and 1849. No shōgun would ever have dreamt of dignifying the arts of the kabuki theatre, the puppet theatre, or the music-hall with his attention, but sumō – its equally raffish origins forgotten – was obviously different. Just why it was different – at least in the eyes of Japan's military aristocracy – is perhaps not too difficult to determine. It was, no matter how remotely, connected with military activity, and also, no matter how disingenuously, connected with fund-raising for religious purposes. Rather more significantly, it did not suffer from those features which brought so much suspicion upon popular theatre; wrestlers, like actors (although presumably for slightly different reasons), may have been the objects of sexual passion,[42] but at least what they did in public could never be construed as subversive in any way.

There was one further difference. All the evidence suggests that, if the authorities were content to give sumō their official blessing, the sumō promoters were in fact eager for it, and had spent much thought, much time and – in an age when, Matsudaira Sadanobu's reforms notwith-standing, bribery was still very much a factor in any new initiative – presumably money in bringing this about. This must surely be the explanation for the sudden emergence of the Yoshida Oikaze house into the world of commercial sumō. A line of hereditary Shintō ritualists, the Yoshida had been in service to the daimyō of Kumamoto since 1658, and their connections with sumō were, to put it mildly, tenuous.[43] They claimed descent from a line of priests in the Heian period, and a mid-seventeenth-century member of the family had written a work on sumō ceremonial, none of which was of the slightest relevance to the commercial wrestling of the big cities. Nevertheless, during the latter part of the eighteenth century the Yoshida Oikaze house suddenly emerged from the obscurity of central Kyūshū to assume a position of symbolic importance in the wrestling world, licensing its referees, formulating its rules and ceremonies and, as we shall see, crowning its champions. It seems more than likely that the sumō promoters, who, as men of affairs, would normally have been indifferent to claims based on tenuous connec-tions in the remote past, particularly if presented by a little-known country priest, were prepared to admit the pretensions of the Yoshida Oikaze house in return for the veneer of legitimacy such a patron could give. They continued to organize sumō, and to draw their profits from it, but with Yoshida Oikaze as their figurehead and lobbyist they would move closer to the world of political power, and in so doing render their investment that much more secure. It can hardly be a coincidence that the visit of Yoshida Oikaze to Makino Tadayuki, the *Jisha bugyō*, in 1790, a visit during which the matter of sumō was widely discussed, was followed sixteen months later by the special exhibition in front of the shōgun in 1791.[44]

Obviously Yoshida Oikaze had convinced the government that some form of official recognition was now appropriate. To do so, however, it seems likely that he had persuaded the promoters to strain towards a new kind of significance – a significance in which this popular sport laid claim to a fictitious lineage extending back to the *sechie-zumō* of the eighth century, the sumō of court ceremonial. It was in preparation for the shōgunal viewing of 1791 that the ceremonial trappings which have now become such an important part of sumō were introduced – some revived after centuries of disuse, others newly devised. This was obviously seen to be a matter of some importance, and it is easy enough to understand the reason. In a society where entertainment was regarded with very real suspicion by the authorities, and where physical activity of any kind had to be related to the achievement of some serious and worthy objective (such as service to one's lord, or service to the gods), the very notion of sport as entertainment, *asobi* (in itself a word overloaded with pejorative associations), was unpalatable, indeed reprehensible. Tradition, there-fore, no matter how spurious, had to be mobilized to give sumō respect-ability, and, particularly in the hands of a Shintō ritualist, what better kind of tradition than a religious one? For this reason religion, or more correctly religiosity, made its appearance in this mass spectator sport, heightening its appeal to those political authorities who were about to view it for the first time.

The element of religiosity had appeared first just three months before Yoshida Oikaze's visit to the *Jisha bugyō*, when a new rank was established. In addition to the ranks of (in descending order) *ōzeki, sekiwake, komusubi* and *maegashira*, already in regular use at least a century earlier, the rank of *yokozuna*, or grand champion, was created. Its first holders, Tanikaze Kajinosuke (whose waist measurement was 87 inches) and Onogawa Kisaburō, were given certificates to that effect by Yoshida Oikaze in 1789, and immediately became the focus of a new kind of sumō ceremonial. First, they were provided with attendants, one to carry the grand champion's sword (*tachimochi*), and the other to go before to clear the way (*tsuyuharai*). This was gentrification with a vengeance. The *yokozuna*'s accoutrements, too, were devised at the same time – a thick, white girdle of rope from which were suspended the white paper *gohei* which, to all Japanese, signified the presence of some supernatural phenomenon. In preparation for the shogunal viewing, too, both grand champions were specially coached in a new ceremony – the footstamping *dohyōiri*, in which the wrestler, slowly and deliberately, gave symbolic proof of his might, and his willingness to engage in unarmed combat. It is, and presumably was then, an impressive sight, but there is little doubt that it, and the whole *yokozuna* concept, was developed in preparation for the shogunal viewing of 1791. Certainly the demand for a *yokozuna* cannot have been generated from within the sumō profession itself, since between the retirement of Onogawa in 1798

(Tanikaze having already died three years earlier) and the appointment of the next *yokozuna* (Onomatsu Midorinosuke), a total of thirty years were to elapse.

The elaborate ceremonial attached to the rank of *yokozuna* was not the only change introduced for the shōgun's sake. The bow-twirling ceremony, resurrected from the *sechie–zumō* of the Nara-Heian periods, was introduced for this occasion, too, to be performed by the winner of the day's bouts. Similarly, Yoshida Oikaze – a Shintō ritualist himself, after all – devised a special ritual for consecrating the ring; during this rite priests first buried a variety of sacred substances in the centre of the raised earthen ring, and said prayers, together with the customary flourishing of branches of the sacred *sakaki* tree. All of these ceremonies, which have since become an integral part of the sumō spectacle, we owe to the shogunal viewing of 1791, to the public relations campaign of Yoshida Oikaze, and, beyond that, to the sumō world's quest for respectability and security.[45]

There can be no doubt that the quest was a successful one. By the nineteenth century sumō had acquired, as a token of respectability, the designation of *kokugi*, a term which, literally translated, means 'national skill', but which really means something closer to 'national sport'. More than that, it had come to acquire a degree of identification with the Japanese state which once would have been unthinkable. In 1854, at the time of Commodore Perry's second visit to Japan (a visit on which, in order to intimidate the Japanese into complying with his government's demands, Perry had assembled a squadron of seven ships, including three steamships) the Japanese government responded to the American offensive by mounting a display by a platoon of twenty-five wrestlers. The appearance of these men, 'enormously tall and immense in weight,' as the official narrative of the Perry expedition describes them, was obviously designed to demonstrate to the foreigners, if only at the psychological level, that Japan was not to be taken lightly. In this it failed: the official account leaves no doubt that the Americans were much more repelled than overawed by these 'twenty-five masses of fat'.[46] Nevertheless, the intention was there. Sumō had been mobilized in support of that Japanese state which had once done all it could to suppress it.

In another sense, too, sumō was to achieve respectability in the course of the Tokugawa period. To an extent which its plebeian origins and essentially homely nature would not lead one to anticipate, sumō came to attract the attention of least a certain section of Tokugawa Japan's intellectuals. Ueda Akinari's *Tandai shōshin roku*, written in 1808, displays quite a detailed knowledge of the wrestling world, including just which wrestlers have which daimyo for their patrons.[47] So too does

Okinagusa, the vast collection of essays written in the latter part of the eighteenth century by Kamizawa Sadamoto.[48] *Bukō nempyō*, too, a gossipy chronicle of Edo life written by Saitō Gesshin in the mid-nineteenth century, intersperses its news of scholars and poets, artists and doctors with bulletins from the sumō world.[49] Of course attention of this sort would have reflected, at least in part, the insecurity of sedentary scholars, half-afraid and half-admiring of more physical beings; a similar tendency is to be observed now whenever academics gather together to talk with nervous knowledgeability of football or cricket. But in the case of these Japanese scholars there was perhaps another yeast at work. Sumō for them was also a metaphor for decline, since a common theme is how inferior were the wrestlers of their day to those of earlier periods. Ueda Akinari, from his vantage point at the end of his life, looked back at the great wrestlers of the 1730s, forties and fifties – men like Maruyama and Genjiyama – against whom, in his view, later wrestlers were 'all as little fish swimming in a bowl'. Not even Tanikaze, the first yokozuna, met with Akinari's approval; he may have been cunning, goes the writer's appraisal, but he 'had the strength of a child ... and that is not sumō'.[50] *Okinagusa*, too, written perhaps thirty years earlier, sounds the same note. In the past, Kamizawa observes, 'unlike the present, good wrestlers were produced by every province – truly enormous. Today's top wrestlers are much inferior to the middle-ranking wrestlers of former days ... they have grown smaller over the past thirty years.'[51] The anonymous *Shimpan kasen sumō hyō rin*, written in 1756, is still more severe. 'The behavior of wrestlers these days is unbearably bad', claims the author, and if wrestlers continue to 'ignore the old ways and do as they please, then chaos will inevitably follow'. His standpoint, however, differs somewhat from that of later critics; like them he looked back to a golden age, but not to one where wrestlers were bigger, stronger or more skilled. To him the golden age was when 'most wrestlers were in the service of daimyō and hatamoto, and there was little wrestling among *rōnin* and commoners'. Now, he says, where once sumō was a 'samurai skill, studied so that it might be of use on the battlefield ... it has now come to be just a way of making a living'.[52] Yet, if he does not approve of the sumō of his day, the anonymous author is no less well-informed than Akinari or Kamizawa Sadamoto. He not only knows the names of the leading wrestlers, but is prepared to use those names as a springboard for a series of brief, punning poems. It should be added that the *shikona*, or ring-names employed by wrestlers, seem to have been well-suited for such a purpose; in the 1830s another anonymous author was composing yet another set of brief poems, this time with the express purpose of satirizing the government of the day.[53]

Other aspects of sumō were also to be incorporated into everyday intellectual life. The *banzuke*, or ranking list, format was the most immediately noticeable of these. From the mid-eighteenth century onwards, when, at the beginning of each tournament, it became customary to issue a large sheet of paper on which the wrestlers were ranked in order of seniority, the convenience of this *banzuke* format came to recommend itself to others in fields far removed from the gymnasium. In 1788, for example, appeared a *banzuke* which ranked the eminent Confucian scholars of the past eighty or ninety years, with Kumazawa Banzan as *ōzeki*, or champion, of the eastern side and Arai Hakuseki as *ōzeki* of the west. Below them, as *sekiwake*, were Ogyū Sorai (east) and Itō Jinsai (west).[54] Ten years later the same was done for *rangaku* scholars and, a little later still, for Osaka doctors.[55]

Sumō had acquired a veneer of gentility. Patronized by daimyō, approved by the Tokugawa government, taken up by Tokugawa Japan's urban intellectuals, it could hardly have been otherwise. Yet it must be emphasized that this was just a veneer. Patronage notwithstanding, sumō and its promoters won their following and their money from among the common people of the three major cities, who could not have cared less about gentility, noble patrons, or intellectual aficionados. To the ordinary citizen of Edo, or Osaka or Kyoto, the appeal of this sport was as earthy and as basic as it had always been. In Edo, tournaments were always held in the heart of the *shitamachi* area, where the bulk of the commoner population was centred, in the precincts of shrines and temples in Asakusa and Ryōgoku and, increasingly, at the Ekōin. This latter temple, built originally to house the remains of all those unidentifiable dead killed during the great fire of 1657, was to become Edo's Luna Park or Coney Island, an entertainment centre where freaks, sideshows, amusements and wrestling could be seen at various times throughout the year. Like the quarter of Edo in which it had been established, it was anything but genteel. So too were the wrestlers themselves, all country boys, and almost all of them sporting the emblem of their trade, the cauliflower ear. One has only to look at the image of wrestlers as it has survived in the popular culture of the Tokugawa period to see what these athletes meant to the general public. More than anything else they were the subjects of stories concerning their size, their strength and their capacity for food. Some wrestlers are known only by their physical statistics, like Akashi Shigano-suke, said to have wrestled in the early seventeenth century, and said to have stood 8 ft 3 inches high and weighed more than thirty stone. With others, like Ōzora Takezaemon, of whom more is known, the emphasis is still far removed from wrestling skills;[56] he was famous for his height (7 ft 2 inches), as the well-known print of him straddling an ox indicates, and for his eating habits (he would consume over six pints of rice a day).[57] With Ikezuki Geitazaemon, who was on the fringes of the sumō world in the 1840s, what was important was the allegation that he had 'the strength of eighteen men'.[58]

The awe inspired by men such as these, and the opportunity sumō offered to see such giants at close quarters as they went through their ritualized pantomime of combat, was quite enough, despite official restrictions, and despite necessary accommodations with the authorities, to keep sumō securely within the orbit of Japanese popular culture, exactly where it had always been.

NOTES

1 The quotation (with the spelling slightly modified) is, like the general account itself, taken from W. G. Aston, *Nihongi*, George Allen and Unwin, 1956, p. 173.

2 Yoshida Tōgo, *Dai Nihon chimei jisho* (Dictionary of Japanese place names), Revised edition, Tokyo, Fuzanbō, 1971, vol. 2, p. 335.

3 Funabashi Seiichi, *Sumō-ki* (Records of *Sumō*), Tokyo, Bēsubōru magajin-sha, 1982, pp. 26–8.

4 Wakamori Tarō chosakushū kankō iinkai (eds.), *Wakamori Tarō chosakushū* (The collected works of Wakamori Tarō), Tokyo, Shibundō, 1983, vol. 15, p. 5.

5 For example, Yokoyama Kendō, *Nihon sumō shi* (History of Japanese *sumō*), Tokyo, Fuzanbō, 1943; Wakamori Tarō, *Sumō ima mukashi* (Sumo past and present), Tokyo, Kawade shobō, 1963; P. L. Guyler, *Sumō: From Rite to Sport*, Tokyo, Weatherhill, 1979.

6 William H. McCullough and Helen Craig McCullough, *A Tale of Flowering Fortunes*, Stanford University Press, 1980, vol. 1, p. 392.

7 Harold Bolitho, 'Japanese kingship', in I. W. Mabbett (ed.), *Patterns of Kingship and Authority in Traditional Asia*, London, Croom Helm, 1985.

8 Wakamori, *Chosakushū*, vol. 15, p. 15.

9 Nishiyama Matsunosuke, *Iemoto sei no tenkai* (Evolution of the *Iemoto* system), Tokyo, Yoshikawa Kōbunkan, 1982, p. 29. Ariwara no Yukihira is the author of poems 23, 365, 922 and 962 of the *Kokinshū*.

10 McCullough, vol. 1, pp. 391–2.

11 *Utsubo monogatari* (*Nihon koten bungaku taikei*, 11), Tokyo, Iwanami Shoten, 1961, pp. 154 ff.

12 *Nobunaga kō ki* (Records of lord Nobunaga), Tokyo, Jinbutsu ōraisha, 1965, p. 100.

13 Nishiyama, pp. 31–2; Wakamori, 1963, p. 32.

14 For an example, see Shinji Yoshimoti, *Edo jidai buke no seikatsu* (Lives of samurai of the Edo period), Tokyo, Yūzankaku, 1963, p. 179.

15 Furukawa Miki, *Edo jidai ōzumō* (*Sumō* tournaments in the Edo period), Tokyo, Yūzankaku, 1968, p. 85.

16 Bunka-chō (eds.), *Nihon minzoku shiryō jiten* (Dictionary of Japanese folk materials), pp. 292–3.

17 Furukawa, p. 96.

18 Nishiyama, pp. 31.2.

19 Furukawa, p. 98.

20 *Ibid.*, p. 207; Cuyler, p. 69.

21 Furukawa, p. 114; Wakamori, 1963, p. 42.

22 Furukawa, p. 96.

23 Takayanagi Shinzō and Ishii Ryōsuke (eds.), *Ofuregaki Kanpō shūsei* (The Kanpō era collection of official proclamations), Tokyo, Iwanami shoten, 1958, Doc. 2658.

24 Furukawa, pp. 96, 113; *Kanpō shūsei*, Docs. 2664, 2665, 2666, 2669, 2671.
25 *Kanpō shūsei*, Doc. 2666.
26 Ishii Ryōsuke (ed.), *Tokugawa kinrei kō* (A study of Tokugawa prohibitions), Tokyo, Sōbunsha, 1960, vol. 5, Doc. 3416.
27 Furukawa, p. 101.
28 *Ibid.*, p. 103.
29 Wakamori, 1963, p. 45.
30 *Kanpō shūsei*, Doc. 2674.
31 Ishii, 1960, vol. 5, Doc. 3163.
32 Wakamori, 1963, p. 45.
33 Furukawa, p. 117.
34 Nishiyama Matsunosuke, *Kinsei bunka no kenkyū* (Studies in early modern culture), Tokyo, Yoshikawa Kobunkan, 1983, p. 451; Furukawa, pp. 198–9.
35 Wakamori, 1963, p. 43.
36 *Ibid.*
37 *Ibid.*, p. 47; Guyler, p. 89.
38 Furukawa, p. 155; Funabashi, pp. 26–8.
39 *Ibid.*
40 See for example the case of the kabuki theatre. Donald H. Shively, '*Bakufu* versus *kabuki*', *Harvard Journal of Asiatic Studies*, vol. 18, nos. 3–4.
41 Naitō Chisō, *Tokugawa jūgodaishi* (History of the fifteen Tokugawa Shogun), Tokyo, Shin jimbutsu ōraisha, 1969, vol. 5, p. 2,463.
42 Kamizawa Sadamoto, author of *Okinagusa*, a late eighteenth-century collection of essays, wrote of the wrestler Raiden Gompachi as a 'big, beautiful fellow' whom 'all the women would go to see'. *Nihon zuihitsu taikei* (Anthology of the Japanese essay), Third series, vol. 23, Tokyo, Tanchōsha, 1978, p. 151.
43 Nishiyama, 1982, p. 37.
44 *Ibid.*, p. 32.
45 Wakamori, 1963, p. 51; Yokoyama, pp. 56 ff.
46 Frances L. Hawks, *Narrative of the Expedition of an American Squadron to the China Seas and Japan*, abridged and edited by Sydney Wallach, New York, Coward-McCann, Inc., 1952, pp. 189–90.
47 *Ueda Akinari shū* (Collected works of Ueda Akinari), *Nihon koten bungaku taikei* (Anthology of classical Japanese literature), Tokyo, Iwanami shoten, 1963, pp. 341–3.
48 *Nihon zuihitsu taikei* Third series, vol. 23, pp. 151 ff.
49 Saitō Gesshin, *Bukō nenpyō* (A Chronological account of Edo, in the Province of Musashi), edited by Kaneko Mitsuharu, Tokyo, Heibonsha Tōyō Bunko series, 1968.
50 *Ueda Akinari shū*, pp. 341–2.
51 *Nihon zuihitsu taikei* (Third series), vol. 23, pp. 152, 153.
52 *Tokugawa bungei ruijū* (Anthology of Tokugawa period literature) vol. 12, Tokyo, Kokusho kankō kai, 1914, pp. 275–9.
53 *Nihon shomin seikatsu shiryō shūsei* (Sources illustrating the daily life of Japanese commoners), vol. 11, Tokyo, San'ichi shobō, 1969, p. 849.
54 *Tokugawa bungei ruijū*, vol. 12.
55 Nonaka Misao, *Ōsaka rangaku shiwa* (Tales of foreign studies in Osaka), Kyoto, Shibunkaku, 1979, Figs. 14, 10.
56 In fact he had no wrestling skills. Saitō, vol. 2, p. 77.
57 Furukawa, p. 198.
58 Saitō, 2, p. 104.

2 Osaka popular culture:
a down-to-earth appraisal

TADA MICHITARŌ

INTRODUCTION

The framework for the following analysis of Osaka popular culture has been borrowed from Roland Barthes (see figure). Following the Barthian model,[1] this discussion of Osaka popular culture in its social and historical context will begin at the level of least abstraction; that is, at the most basic and blunt area of popular culture, what I will call *food culture*. In looking at the place of food in the urban Osaka culture I came up with two phenomena to be explored later in this paper: tako-yaki, a foodstuff charged with symbolic significance, and a huge moving crab sign, an example of a visual symbol of a foodstuff ($S'e^1$). That which gives meaning to both of these is the food ideology in Osaka (Sa^1). This ideology will be looked at by considering the historical, social, and cultural development of Osaka, and especially emphasized will be the development of the area of the city known as *Minami*. (Part #1)

Once an understanding of the characteristics of Osaka's (and by extension, Japan's) popular culture is attained in this way, we will be able to analyze the relationship through which the verbal text known as *manzai* ($S'e^2$) or Japanese comic dialogue, receives its meaning from the general Osaka popular culture (Sa^2). From this newfound vantage point we should be able to make an adequate appraisal of Osaka's popular culture in the twentieth century. (Part #2)

The original inspiration for this paper came from an unusual source. When walking down the street in New York City many years ago I saw a huge advertisement for cigarettes. It was a big sign with a hydraulic moving arm which carried a cigarette to and from the lips of a smiling face towering high above my head. When I saw that sign (which may have vanished since then) I recalled a certain, equally *kitsch*[2] (or tacky) advertisement back in Osaka. The sign (of which I will say more later on) was of a huge crab with moving legs which advertised a seafood

33

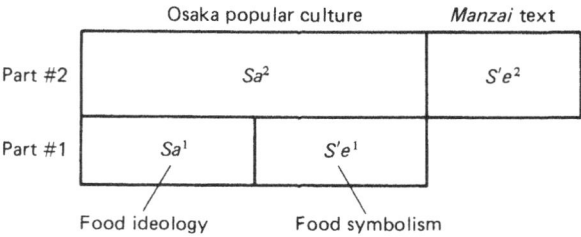

Figure 1 Conceptual layout

restaurant in Osaka's *Minami* area. Pondering the sign I asked myself, 'What is the cultural background from which such a gaudy example of kitsch culture has developed?' This giant red crab with its front legs crossing and uncrossing above me meant something if I only knew where to look for its meaning.

One of the first and most fundamental aspects of such a symbol to consider is its religious connotations. For it is important to note that the Japanese religious tradition related to the sea is both rich and complex. It is an essential component in present-day Japanese popular culture as well. While the relationships between this crab and other symbols of Japan's ocean-oriented religious culture will be discussed further on in this paper, I would like to consider now one particular aspect of that set of relationships.

The similarity between the crossing and uncrossing motion of the crab's legs and the gestures of Japanese people when they are caught up in the mood of the *matsuri* (a religious festival) is really quite striking. The participants in the festival, after only the slightest contact with the *mikoshi* (a sacred shrine carried through the city streets) enter into a trance-like state[3] and can be seen crossing and uncrossing their arms above their heads in a motion, much like that of the hydraulic crab's legs. This connection, although it may seem quite tenuous now, is actually very important when looking at the historical development of Osaka's popular culture. For as I will show, the *sakariba* (roughly, 'amusement areas') in Osaka (of which Minami, the area where the crab sign is located, is the most famous) developed on the borders (in the 'limen') between the sacred shrines and temples and the profane urban areas. These sakariba thus owed their existence to the special status of the shrines and temples in Japan. I believe that the *communitas* experienced through participation in ritual was extended into these more 'down to earth' pleasure areas. And so, from the beginning, the connection between religion and Japan's secular, often outrageous popular culture has been a fundamental one, a fact which should be borne in mind through the following discussion.

HISTORY

I was born at the beginning of the Shōwa period in Settsu; a coastside town near Kōbe. It was at a place near the Hankyū Okamoto station of today. I have heard that around the time of my birth octopuses used to crawl up onto the shore from out of the sea. Deep, deep in my memory that image lingers. Around Shōwa 10 (1935) I became obsessed with playing *pachinko* at night in the open-air stalls. Symbols taken from the sea were used on plates affixed to the winning slots. The octopus which had come up from the sea had been transformed into a symbol on a game machine. And then, after the war, we find a huge moving crab sign in Dōtombori. Shouldn't all of these things be included as part of the same cultural tradition which crawled out of the sea?

As a new popular culture began to form in Osaka during the feudal period, the land was growing. Thus the history of landfill (by which I mean to describe both natural and directed land reclamation) is tied to the history of Osaka's popular culture. We should think about this history of the landfill process in two parts. They are, of course, the *natural* and the *directed* (or mechanical) eras.

When we dive deep into the past we find that the city of Osaka was once two thirds under the sea. A large part of Osaka city was formed by the gradual build-up of silt from the Yodo River and mud from the ocean (a natural process) and was only later supplemented by mechanical land filling. Let us then call the mechanical stage in the landfill history of Osaka the 'second' stage. In this second stage a conscious and systematic large scale project was undertaken to fill in the ocean. The area thus created comprises as much as one third of present-day Osaka.

In those days Osaka was called *Naniwa* (originally meaning 'rapid currents'), and it was written and pronounced in many ways, such as 浪花 (Naniwa), 難波 (Nanba) or 難波潟(Naniwagata). The ocean tides used to collide with the many rivers coming down from the North in fantastic fashion. For this reason the currents around Osaka were said to be fast. The various forms of the word 'naniwa' fit the personality of the city of Osaka to a tee. (The people of Osaka are even today said to be especially hurried, and extremely fast and crazy drivers.)

The left radical in the *saka* of Osaka used to be written with the radical meaning 'slope' (坂). While this might have stood for the slopes of Osaka reaching down into the sea, the 'Naniwa' image of Osaka still seems more interesting. Osaka was created out of the energy of the collision of two distinct currents. Land was formed by the sand from the rivers and the mud carried in by the ocean. A long time ago, well, not all *that* long ago — let's say during the Asuka Period (seventh century) and the Nara Period (eighth century) – the ocean reached as far as Kusaka (日 下) at the foot of Ikoma Mountain (生 駒). To the North, it reached as far as the area which is now Hirakata city (枚 方).[4] This is the same Hirakata which is famous for the *kurawanka* boats, and the same city which I once lived in for ten years. At the time a place known as *Uemachi* was located on a plateau atop a promontory running south into the ocean like a beach-head. While the promontory was surrounded by water on both sides, on both the east and west sides land was being built up from silt accumulation. Before one's eyes, the land grew. People of the Ancient Period (*kodai*) and even Middle Ages (*chūsei*) saw the scene of Osaka's growth with wonder in their eyes.

In this wonderful way the land was created. There was nothing more strange or more worthy of the people's gratitude. This feeling of wonder created the deity *Ikutama-kami*, or in full, *Ikukuni-Sakikuni-Tamano-kami* (生国咲国魂神). Thus, we can assert that the *Ikutama-kami* deity is the landfill deity.[5]

The newly formed land to the east is today known to Osaka residents as Higashi Nari-ku while the land to the west is now Nishi Nari-ku. The name comes from the word *naru*, 'to become'; thus, Nishi Nari-ku meant 'land formed on the west', but this etymology is unknown to most Osaka residents today. The shape of the land to follow was influenced by the form of Nishi Nari-ku and Higashi Nari-ku. It was only natural that later the famous Toyotomi Hideyoshi should build Osaka castle on the Uemachi plateau which, as the reclamation of the city of Osaka progressed, assumed a strategic position at its base. The plateau ran southwards from Osaka castle to the very tip where we find *Sumiyoshi*, the dwelling place of the sea deities. Tennoji temple and the area around Sumiyoshi, including the especially important *Tennōji* complex (四 天 王 寺), are important areas in both the Buddhist and Shinto ocean deity traditions.[6]

When we look at the history of the landfill process of Osaka, we come to see that the source of the city's vitality is derived from its being a peninsula. While nowadays Osaka's Minami area is a vaguely defined place used in such phrases as '... the busy streets of Minami', historically the Minami area of Osaka has been the contact point with the ocean. Minami has become a proper noun which designates that area which has moved throughout the ages in order to stay in contact with the ocean.[7] There are some who go so far as to say that Minami *is* Osaka. This can be said because Minami is the most Osaka-ish place in the city.

Sumiyoshi used to be known as *Suminoe*. Around Suminoe fishing was an important industry. People would visit the shrines and temples in order to pray for safe sailing and for a good catch from the sea. And of course,

Suminoe was also the main reception area for cultural currents incoming from China and Korea. Suminoe was the point of cultural penetration into Japan from overseas. The base for all of this activity was Shitennōji temple.[8]

While even today the citizens of Osaka go to Shitennōji temple to pray, this tradition was actually started long ago by nobles from Kyoto who would go to pray to the setting sun. This practice is called *Nissokan* (日想観). From that point, the sun sets due West. Facing directly towards paradise in the heaven, they prayed. This Buddhist custom of praying to the West at the equinox is found in no other Buddhist country. When we think of Osaka then, we are compelled to remember the inflow of things and traditions from the West into Tennōji long ago.[9]

Even today place names related to the ocean or to promontories, dating as far back as the *Man'yōshū* era, can be found in Osaka: Tsumori (津守) 'point', Kohama (小浜) 'small beach', Suminoe (住ノ江) 'cove', and others, show us the influence of the ocean on the formation of the city. In the *Man'yōshū* text we find the line 'Yasoshima kakete' (八十島かけて) 'all that could be seen were eighty scattered islands'. In this context 'eighty' meant 'countless'. (Eighty was used in ancient Japan to express an extremely large number.) The following quotation by the modern Osaka poet Tomioka Taeko reveals the abundance of such place names even today:

And around the steel bridge there are still many place names which incorporate the Japanese word for island, *shima* (島): (四貫島) Shikanjima; (西島) Nishijima; (百島) Momojima; (中島) Nakajima; (出来島) Dekijima; (姫島) Hime-jima; (御幣島) Goheijima; (竹島) Takejima; (加島) Kajima; (初島) Hatsu-jima; (向島) Mukōjima; and (桜島) Sakurajima, etc. And when we look on the map we find many small islands which are literally referred to as the 'mouth of the river'. We can also see how the land was formed by such small islands in those areas and that the rivers and canals thread their way between them. And we also see that the closer these 'islands' are to what is now Osaka Bay, the smaller their population becomes.[10]

This is the 'real Osaka'. There are indeed places far inland containing the character for island that make one wonder how islands could have existed in such places. Even though the place known as Fukushima (福島) is quite far north, it seems that in the past even there it was mostly marshland. There is another spot known as *Kyū-jō-jima* which used to be written with the characters (崔壊島) meaning 'the island which wrecks the city'. As Kyū-jō-jima grew, the Yodo River became obstructed. Rather than let the town be flooded, a new river, the Aji River (安治川) was made. This river was named after the pioneer, Kawamura Yasuharu (河村安治), 'Aji' being an alternative reading of the characters in

'Yasuharu' (安治). Later the characters for Kyū-jō-jima were changed to the more benign ones (九条島) borrowed from Kyoto, which had a numerical street-naming system based on the Chinese model. This is just one example of the history of river improvement and land reclamation hidden in a place name. While one third of Osaka city land developed by natural accumulation, another third was the result of deliberate planning, with Toyotomi Hideyoshi playing a major role. He is the man who cleared away the greatest crisis in the history of Japanese civilization: he unified Japan after winning a series of civil wars against competing feudal lords. While it is true that places such as Kawachi, Izumi, and Settsu came to be centres of imported culture, originally the real political power was consolidated in the hinterlands upriver on the Yamato and Yodo waterways. From this base in the mountains, sword gripped in hand, the pioneers went on east and north to claim a still larger domain.

During the warring states era in the fifteenth century a furious war raged in the mountains. While concentrating on defending their territories, military leaders in the Eastern region such as Uesugi (上杉) and Takeda (武田) looked back West and waited. Once unified by Oda Nobunaga (信長), they began to gradually encroach on the West, that is, to head back towards the sea once again.[11]

Azuchi castle in Ōmi broke away from the convention of building mountain protected castles and started a new trend by opening up towards the ocean. And the man at the vanguard of this reversal of current trend was Hideyoshi. Why then didn't Hideyoshi build his castle in Nagoya, his hometown? While at that time Nagoya did have a remarkably high rice production output, it was nevertheless necessary to relocate to Osaka. This is because Osaka was the centre of commerce, and more importantly was the base for trade with the West. This connects up with Osaka's landfill history as well as with Hideyoshi's later invasion of Korea. Although Hideyoshi paid back his debt to East Asia with malice, he never realized his dream of *muso*, that is, his dream of establishing a Japanese capital on the mainland of Asia.

Hideyoshi's dream was to go to the Ningpō (寧波) area in China and to there establish a centre of power. Ningpō in particular caught his eye because, when cultural ambassadors left from *Naniwa* for China, they always docked at Ningpō, the location from which successive waves of enlightenment had flowed to Japan in the past. The famous song, 'Perhaps the moon is rising from Mikasa Mountain', was also written at Ningpō. This was a nostalgic song written about a man who, stranded on the other side of the sea from Japan, took solace in believing that perhaps the same moon had just risen over Mikasa Mountain near Osaka.

The vision of setting foot on mainland Asia, specially Ningpō, stimulated a precocious explosion of imperialism, a first blast of nationalism. While later on, the Tokugawa Bakufu was able to control this explosive nationalism, it never actually threw away the grand design of seizing part of the coastline on the continent proper. The wisdom of the Bakufu was to only gradually convert Hideyoshi's ideas into reality.

Even though the centre of political power had been moved to Edo in the East, which came to contain the central life force of the country, the pioneers in the West were not ever really abandoned. Over 300 years the Bakufu was able to recruit talented men to reclaim a large amount of land in Osaka. They also managed to have the landfill work done voluntarily. This was accomplished by lowering the workers' taxes and by not regulating profit-taking. So the Bakufu, while making virtually no national investment, was content to merely sit by and issue permits. It was a conscious policy of utilizing independent enterprise. In this way, the workers' activities were able to progress freely. This economic rule of increasing production by the use of independent enterprise quickly actualized the city of Osaka. This is the reason why Osaka overwhelmed the neighbouring port city of Sakai.[12]

It is indeed true that Sakai was the production centre for many remarkable firearms, and it was the centre of Oda Nobunaga's military strength. It also witnessed the emergence of a highly refined culture through its connections with Kyoto. Even though it was an open city which had achieved a high level of culture and technology, in the final analysis it was too much connected to the controlling powers. As a result, the strength of the common folk which had emerged from a mixture of various sub-cultures was repressed and finally abandoned. This was not the case in Osaka.

The land reclamation in Osaka during the 300 years of Tokugawa rule occurred in several major stages. Starting in the Keichō years (1596–1614), into the period of culture administration (circa 1800), and at the end of the Tokugawa Period (circa 1860), large, directed reclamation projects were undertaken. During these years many interesting episodes took place. For example, there was one large-scale landfill project which was repeated every forty years. Wooden stakes would be put out in the water and seaweed would be woven around them. Over the years the silt would build up and new land would emerge. It was approximately a forty year cycle. Throughout the Tokugawa Period this process of reclaiming the newly formed land and placing the stakes further out to sea was repeated over and over.

As a result of this process, the agriculture and fishing industries in Osaka became mixed together. Fishermen who wanted to start doing farming asked the Bakufu Administration for permits. A half-fishing, half-agriculture type of life style began to emerge.[13]

The areas where the stakes entwined with seaweed were placed became the best fishing places. Often areas which appeared to be inland were really marshes where fish could be caught. However, there were few large fish to be caught. Most common were sardines, silver carp, eel, clams and fresh-water clams which could all be found in abundance. It can be said that *dashijako*, a fish broth made from otherwise inedible fish, is the source of popular culture of Osaka. The reason, then, why Osaka's *udon* noodles are so delicious is because they use 'junk fish' instead of high class fish such as *tai* (red snapper). Unknown to those in Edo and Kyoto, this is one piece of Osaka's alimentary knowledge.

Thus the ordinary fishermen also contributed to the reclamation culture of Osaka. Later, it seems, the capitalists also got in on the action. The makers of Osaka's landfill culture came from all over: their native homes were in Nara (奈良), Ise (伊勢), Kawachi (河内), and Izumi (和泉). They all came from far and wide to mix together and form a new popular culture. And Osaka was more than just the spot for the mixture of native cultural traditions; it was the place where various native elements *as well as* everyday cultural influences from overseas merged and combined to form a unique, new, hybrid culture.[14]

As the window to the cultures of the West, Osaka could well have been the capital of Japan long ago. (It was for a brief time, in fact, the official government centre.) But the capitals were always placed out in the hinterlands: Asuka, Nara, Nagaoka, and Kyoto. Why?

Ever since the days of the *Man'yōshū*, Naniwa has been labelled as 'countryside', that is, rural and backward. It is indeed strange that the main location for the influx of foreign culture should be thought of as 'countryside'. Perhaps this was because of the Japanese consciousness that they were indeed backward in comparison with China, and that the Chinese never chose a peninsula as capital. Also there was the strange idea that it was necessary to leave a certain aesthetic distance from the active window to other cultures. *Osaka was worldly and vulgar*; the capital was sacred and aloof. In order to be sacred, an aesthetic refinement is necessary. But such a place lacks activity.

This pattern remained influential even into the twentieth century. And while it may be sidetracking a little, the reason why the capital was moved to the north and to the west from Asuka and finally to Kyoto is closely related to the progress of river improvement in the Kansai region. The capital was moved each time a new, larger river was made navigable. Finally, with the complete opening of the Yodo River, the spreading movement from east to west came to an end. This is also the history of the development of the hinterlands of Osaka as well. The set pattern of Osaka's peasants and fishermen supplying the capital's inland with the vast majority of their food did not change up until the Edo Period.

While it was noted earlier that the ocean stretched far up into what is now North Osaka and that of course fishing and agriculture were important industries, it is seldom noted that early on livestock breeding was also an important industry. Grazing of horses and cows took place around Osaka.[15] Thus the Osaka Plain was the first place where cows and horses were raised in Japan. While it is widely accepted that Japan has had a settled agricultural culture since the Yayoi Period, in Osaka cattle breeding came next after fishing. Next, farming was introduced. Then in the Tokugawa Period commerce grew in importance, to be followed by the rising smoke of industrial production which appeared in the Meiji Period.

Fishing, breeding, agriculture, commerce, industry – these many economic activities accumulated like the stratified earth and sand upon which the city of Osaka grew. But the older styles of life were not buried under the landfill. Although cattle breeding vanished quite early, fishing has hung on through the centuries. It can be said that the way in which various occupational groups all piled together to form a new hybrid culture is rare in the history of large city formation. I here speak of all of these various processes as making up what I call Osaka's 'landfill culture'.

I would now like to move on to discuss the artistic aspects of the landfill culture. The first problem to consider is why did a *sakariba* like Minami come to be built upon the newly reclaimed land in Osaka?

When one thinks of the Minami area in Osaka, the *Dōtonbori* section especially comes to mind. Though it is hard to say just where the centre of Minami is, the *Dōtonbori* (道頓堀) region seems to be very close to that elusive centre. It is marked by entertainment spots and showplaces. When consulting the historical record we find that the first attractions were displays of exotic curiosities. Unusual animals from all over the world such as elephants and giraffes roused the interest of the masses.[16] Why was it, then, that such spectacles were exhibited? Why were such shows necessary?

The answer is once again related to landfill. In these areas the reclaimed land was still soft and swampy and needed to be stamped down. It would be best if the work could somehow be done for nothing. The public had to be attracted so that the walking of the masses would harden the ground.

This strange logic created Osaka's *Minami*. It created Osaka. In order to gather people together, something amusing had to be found, something that was interesting and abandoned reason. That is the foundation of the popular culture of Minami, of Osaka. Hence all sorts of variety halls, attractions, and restaurants came into being. From the chaotic mixing together of businesses, convenience shops, restaurants, and amusments, the unique Dōtombori district came together. Throughout the world there are really no other *sakariba*s like those in Japan[17].

Osaka's Minami was the first of the *sakariba*. The first requirement was to stamp down the reclaimed land. Instead of machines, the general public was used. This is the Osaka logic.

FOOD CULTURE

It was noted early that the Japanese capitals were sacred places with a high level of aesthetic refinement; this becomes eminently clear when we compare the food culture of Kyoto with that of Osaka.

Kyoto was, and still is, the centre of Japanese first-class cuisine known as *kaiseki ryōri*. This ancient culinary tradition involves the selection of only the finest parts of the very best fish and vegetables; the remainder is discarded. These select (and usually few) ingredients are meticulously arranged and served with the utmost care and courtesy. The astoundingly high price of such a meal comes as quite a shock to many first-time diners, but many return again and again.

Osaka residents, on the other hand, view this 'ridiculous' tradition with contempt. For Osaka residents a good meal is both economical and nourishing. Using the 'garbage' thrown out by the elites in Kyoto, they make cheap and delicious dishes of their own. For example, while in Kyoto only the white meat of the sea-eel is eaten; in Osaka there is a delicious traditional dish called *hamo no kawa* (sea-eel skins) which can even today be enjoyed there in the summer.

Nowadays there is a popular inexpensive dish in Osaka prepared from a mixture of various cow entrails which is thought to be especially nourishing. For this reason it is called 'grilled hormones' (*horumon-yaki*) by the Osaka residents (the Japanese seem to associate the word 'hormones' with good health). And the name of this dish is even more interesting when one considers the word *horumono* in the Osaka dialect which literally means 'something which is thrown away'. Paradoxically, then, the Osaka folks eat their 'horumono' instead of throwing them into the garbage!

So, if any foreign visitors to Japan want to enjoy a delicious, inexpensive meal, by all means they should drop by Osaka (forget about Kyoto) for some 'horumon-yaki'. And if they want to dine even more cheaply, I recommend that delicious traditional seafood bargain, *tako-yaki*.[18] *Tako-yaki*, for those who haven't yet taken this alimentary adventure, is an extremely cheap dish (costing less than one U.S. dollar), consisting of a bit of octopus inside an omelette-like ball of fried egg and flour. It is sold by the dozen in convenient wooden boxes (called 'boats') on the streets in the *sakariba* of Osaka.

An urban anthropologist acquaintance of mine has gone so far as to assert that this one-dollar dish incorporates all of the various jumbled aspects of Osaka's urban culture: the octopus for fishing, the flour for agriculture, the oil for commerce and industry (since oil has only recently become an important industry in Osaka), and finally, even the small amount of dried seaweed sprinkled over the top to give it that special aroma can be (imaginatively) associated with the modern information-based society.

So when you eat *tako-yaki*, you are eating Osaka, all for less than one dollar! If you want to eat Kyoto, on the other hand, you can expect to pay well over one hundred dollars for the aesthetic experience!

No matter what the price, however, it is clear that the sea is an important element in the Japanese popular culture. This can also be seen in the nostalgic symbolism for the sea so prevalent in Osaka's popular kitsch culture. For example, if you go to the open-air stalls at Tennōji temple on festival days you can see the real relics and mementos of Osaka's ocean culture. Processed goods of the sea such as seaweed and dried cuttlefish as well as other culturally meaningful foods such as octopus and shrimp are massed together to show us how images from the sea are regenerated in the popular mind of the public.

While the bric-a-brac shops in Japan are full of all kinds of ocean artefacts, the most common by far are images of the 'Seven Deities of Happiness'. One of them in particular, *Ebisu*, is referred to as the deity who 'brings happiness from foreign lands'. This is one manifestation of the belief widely held in Japan that anything which crosses over the sea from foreign lands is naturally a good thing. Ebisu has at times been written with the unusual characters (蛭子) ('leech'). According to Kida Sadakichi's investigation into the subject, entitled *Fukujin* (福神),[19] 'Deities of Happiness', 'Ebisu' stood for 'a person unable to stand'. For this reason, legend has it, he was loaded onto a boat and set adrift at sea only to then come ashore on the Settsu Peninsula. Left over from all of this is the *Ebisu Nishinomiya* shrine in the district bearing the same name.

Long ago in Japan handicapped people were greatly worshipped. We find that in Peru and many other countries as well there is an abundance of images of deities and handicapped figures which are said to have come from the sea. This, then, is the tradition behind the Ebisu cult. The ancient belief in Ebisu as the deity of the sea, along with the modern association of

Ebisu with success in business, have combined to form the present-day Ebisu faith. Thus, he is more than simply the benevolent deity of trade; he is also, since long ago the deity who brought to Japan the riches of the sea.

Though it finally vanished in the Meiji Period, a notable attraction at Tennōji temple used to be the 'Octopus Dance' (*tako no odori*). A man dressed as a ridiculous octopus used to do the dance. There were also scenes of Ebisu fishing for *tai* (red snapper). These sights were all viewed through a kaleidescope. Through the hexagonal lenses the images of the sea would multiply and be enhanced. It seems that children greatly looked forward to viewing the octopus dance. In this way, at a very susceptible age, children were able to learn about the stratified landfill culture at the stalls of Tennōji temple and Sennichimae (they certainly didn't learn about such things at school).

Osaka's open-air shop culture carries important meaning for the study of popular culture. One example is *pachinko*. While *pachinko* is widely thought to have originated in Nagoya as a children's game, it actually began in Osaka in Taishō 12 (1923). It first came as a western game called 'smart ball'. Pinball was originally based on a Greek game where a man would have to find the right way out of a labyrinth of pillars. Thus, pinball machines lay horizontal. The machines came to Japan and were made to sit up vertically by the merchants in the stalls in Sennichimae. It was that simple. But it also had the effect of eliminating the face-to-face contact between the player and the spectators. When one asks why the machines were set up vertically the discussion returns once again to the subject of land – there simply wasn't enough space to spare for a horizontal machine. The cutting off of the participant from the spectator caused by the cramped conditions foreshadowed the individualistic and alienated aspect of modern society.[20]

As I observed earlier on, in the pre-war period symbols of the sea were affixed to the *pachinko* machines. Perhaps as I was trying to shoot the balls into the little holes decorated with octopuses I was also imagining the octopuses which used to crawl out of the sea onto the shore of my hometown when I was a toddler. Or, perhaps I was recalling the words to a song in the *Man'yōshū* much further back in antiquity – *Naniwa no Ashikani* ('The Reed-Crab in Naniwa'). Starting with 'Today, today I went to Asuka . . .' it is a suffering lament from a crab who was tormented after crawling up onto the table of a noble family.[21] Reed-crabs live in places where salt and fresh water mix. For this reason, they are perhaps apt symbols of the poor people who made up the landfill culture in Osaka. The recipe for cooking reed-crabs used in Osaka is, not surprisingly, exactly the same as the one used in areas of China.[22] This is another good example of the importation of food culture from China and Korea which took place in Osaka. This mixed-living area where the salt and fresh water meets was the base of the hybrid culture which developed on Osaka's reclaimed land.

MANZAI

Manzai is performed in small theatres along with a few other vaudeville-style acts; and the manzai can be found mostly in Osaka's Minami area — that is, in the *sakariba*. The present writing of the word *manzai* (漫才) is derived from the earlier writing (万歳) which meant 'long life and prosperity'. This betokens the fact that the term was originally used to describe gypsy-like vagabonds who used to perform music, dancing, and songs of felicitation in return for payment from the wealthy. This particular part of the Japanese popular culture can be traced back to the thirteenth century. The tradition was enriched throughout the 300-year Tokugawa regime, and it especially flourished in Edo, Kyūshū, and of course, Osaka.[23]

As time passed, the emphasis of the *manzai* routines gradually grew more comic and came to focus on themes such as the differences between the countryside culture, from where most of the *manzai* artists came, and the big-city culture. It should be remembered that the great cities in Japan did not really achieve modernity until after the Russo-Japanese War in the twentieth century.

As I pointed out in the last section, the Osaka culture was created from the mixture of different cultural traditions from all over, and this process was even more enhanced at the beginning of the twentieth century when unskilled workers (apprentices, messenger-boys, shopworkers), who were immigrants into Osaka, formed the major audience for the modern *manzai* which has persisted in much the same form to this day. The themes of the popular routines focussed critically upon the many problems in the hodge-podge of Osaka's urban culture: class differences, age differences, and cultural differences. It was largely for this reason that the characters used for writing the word *manzai* were changed from those meaning 'felicitous' to ones meaning 'comic'.[24]

The *manzai* was the cheapest form of entertainment in Osaka. Neither the performers nor the audience had any money to speak of. As opposed to the more traditional vaudeville entertainments such as *rakugo*, *manzai* had come in off the street and could only imitate the tradition and professionalism of the higher theatre. This cheap, modern *manzai* of the twentieth century had Osaka as its centre.

The mass culture which was built on the reclaimed land of Osaka was constantly changing and being recreated as people and cultures from all over Japan (and elements from overseas as well) gathered together on the newly formed land to live. Out of this mixture came the *manzai* which, through the use of metaphor and criticism, reflected and connected together the diverse cultures it was based in:

Manzai is an art which connects ... it blatantly connects things which have virtually no connection. When one thinks that there is no way that such completely unrelated things could be brought together, jokes and laughter are used as the omnipotent bond. Upon letting themselves laugh, all of their inner conflicts are cleared away and the audience feels ready to accept anything that comes later.[25]

The *manzai* is performed in dialogue form; there is the *boke* (meaning 'idiot' or 'stupid', the buffoon) and the *tsukkomi* (meaning 'interrogator', the 'straight man'). The comic element in *manzai* is that the *boke*, in trying to understand and come to grips with cultural elements on a higher level, *always* misinterprets them and unwittingly drags them down to his own level. This pattern is clear when we look at a sample of a *manzai* text from the 1930s, when a flood of cultural influences were streaming into Japan from Europe and America. The following dialogue, 'Crossed Sukis' from 1936 is a representative example.

MANZAI TEXT SUKI[26]

A	Hey you, what do you think of *suki* (skiing)?
B	*Suki*? (Sukiyaki?) I often treat myself to it when it's cold.
A	I often go (to ski) to Ibuki Mountain, how about you?
B	You go a long way don't you! I always go downtown.
A	There are places to suki in town?
B	When I'm cold, I just go to the ordinary places (to eat).
A	Well, when I went there was snow piled up over fifteen centimetres.
B	You look at the snow when doing it?
A	You see, suki depends on the snow ...
B	Yeah, well, I guess it can be done when it snows too ...
A	When I do it, I grip tightly with both hands.
B	Oh yeah? I see ...
A	And when I do it, I dig in with both hands, 'splurch! splurch!'
B	You're really weird ... I just hold on lightly with one hand.
A	You do it with one hand? How strange!
B	I probe lightly ... here it's soft, here it's hard.
A	Whether it's hard or soft, I just go for it!
B	I go for the soft places.

A	The soft places? That's dangerous, ya' know ...
B	No, actually the hard places are more dangerous ...
A	What are you saying? The soft places go 'mush, mush' when you hit them, don't they?
B	Nah, they go 'slush, slush'!
A	That's dangerous!
B	Really?
A	When I went last time, I hit one place that was melting ...
B	Yeah, yeah, the really soft places. That sound is indescribably happy to me ...
A	I know what you mean.
B	But the steam is a bit of a pain ...
A	It doesn't steam – though it can splash once in a while ...
B	You use a white apron, don't you?
A	Nah, I don't wear everyday (Western) clothes.
B	*I* always go in a kimono.
A	You, you go to suki (ski) in a kimono? Isn't that hard to do?
B	Nah, not really; I'm used to it.
A	My speciality (read 'ju-hachiban') is paralleling.
B	Ohhh ... throwing food into your mouth, huh?
A	In a slightly bent over, squatting position, it's really hard to use all your power to guide you on your way.
B	What? Doing suki in a squat? How unusual! I do it with my legs crossed
A	What? Suki (ing) with your legs crossed?
B	And, if the other guests are important people, I do it while sitting properly ...
A	Huh? And you don't fall over?
B	Well, I guess if you do it for too long you could fall over, but it's not really a problem. Anyway, let's go together sometime!
A	I'd like to, but there isn't any snow yet ...
B	Even if there's no snow, let's do it while watching a river.
A	A river? How can you do it in a river? That's a completely different type of activity!
B	And I bet you don't like Konnyaku (Devil's tongue) noodles either, do you?
A	Not very much.
B	You're that type – I thought so ...
A	When I go, you see, I bring rice balls and pickles and ...
B	What? When you go to suki you bring food along?
A	Well, it's cheap and delicious.
B	Don't the waitresses get angry?

A   Well, where I go there aren't any people like waitresses. There are a lot of workers on the mountain, though.

B   Mountain workers? Well, I don't know the place at Ibuki Mountain that you went to, but at most restaurants there *are* waitresses.

A   Yes, of course a restaurant would have waitresses.

B   You, you don't like meat?

A   I can eat it, but if I eat too much it gets too overpowering.

B   I see, you're leaning more towards the chicken type, huh?

A   What are you talking about?

B   Chicken suki, y'know?

A   Who's talking about chicken suki (sukiyaki)?

B   Didn't you say that you go to eat suki when it gets cold?

A   Have you been listening? I'm talking about the suki you do on top of the snow!

B   Really? Is there really such a thing?

A   What the hell have you been talking about?

B   I thought you meant *suki* as in *sukiyaki*.

A   I'm talking about the sport called *suki*-ing.

B   Ohhhh ... and I *thought* there was a 'gap' (read *suki*) in our conversation!

This text is a classic example of Osaka *manzai*. The humorous quality of the dialogue comes from the crossing of lines of communication. While the correct line speaks of skiing, it is totally misunderstood by the wrong line, represented by the buffoon. While in Japan today over ten million people enjoy skiing, fifty years ago the figure was closer to one thousand. The buffoon, then, represents many of the common folk who could only think of the food *suki*yaki when hearing the word 'ski' (pronounced *suki* in Japanese). This attempt to interpret high leisure culture from the more base level of food ideology in the context of their own food culture is an important pattern in modern manzai.

In this text, A is the straight man while B is the buffoon. Neither A nor B use the verbs 'to eat' or 'to slide' in this script; instead they use the vague and ambiguous 'to do' (*yaru*). Due to the particularly ambiguous Japanese word for 'do', both lines can extend out to hilarious lengths. In this way the *boke* interprets everything in the context of food and the two lines remain crossed until the end of the routine, when they reach a long-belated mutual understanding only to deliver the final joke known as the *ochi*.

B   I thought you were talking about the suki in sukiyaki!

A   I'm talking about the exercise type of suki.

B   Ohhh ... I *thought* there was a gap (suki) in our conversation!

This final, surprising introduction of yet a third meaning for suki, 'gap', is the climactic joke which is not only a good example of a classic *ochi* ending, but is also significant for us in that it ends with the word 'gap'. It has been noted earlier that an important role of *manzai* is to bridge the gaps between various cultural and social elements.

I don't believe that the Japanese *manzai* is unique. The history of two-man stand-up comedy dates back approximately 2,000 years in China. It also flourished in America (Laurel and Hardy) and in England (Punch and Judy). An important characteristic of Japanese *manzai*, that is Osaka *manzai*, is the attempt to interpret everything through metaphors and symbols from the food culture. This food culture is not that of the refined first-class Japanese cuisine, but instead it comes from the lowest and cheapest forms of nourishment.

Let's give one more example of this food-centred mentality: the 1930 routine entitled 'The Baseball Coach'. When baseball first came to Japan it was widely misunderstood because of its English-language terminology. The Japanese saw foreign elements as being high-brow and inaccessible due to a mental complex (which remains to this day) towards the English language, that it is some aggressive thing to be feared. In this *manzai* routine almost all the Western sports terms are misconstrued and confused with food names and symbols. For example, 'umpire' is confused with *anpan*. *Anpan* is a popular snack in Japan in which traditional Japanese sweet-bean paste is surrounded by a Western-style bread exterior. This foodstuff is a good example of the Japanese tendency towards syncretism.

The use of everyday (but symbolically rich) *anpan* in the comedy routine is an example of the process by which *manzai* interprets and connects originally foreign cultural elements with the indigenous style of life. One characteristic of the Japanese modernization often noted by historians and social scientists about the Japanese is their seemingly uncanny ability to adopt and modify foreign ideas and goods into a particularly Japanese product, often one which is superior in many ways to the original. Little is said, however, about the actual *process* by which this takes place. The following example is meant to give one illustration of that process.

When walking along the *sakariba* in Osaka, I noticed the many plastic models of food which are displayed in front of almost every restaurant. They are realistic reproductions of sushi, tempura, spaghetti, omelettes, and any other food you would want to imagine. More so than the Japanese education system's attempts to spread knowledge of the West and its culture, these delicious-looking models did far more to promote interest in other countries. It was largely because of these models that the common folk, not only the elite, were able to be internationalized in such a fundamental area of everyday life.

The first wax model of a foodstuff was created in Osaka in 1932 by a man named Iwasaki Takizō. At the time, he was working in a shop which manufactured replicas of human body parts in wax for use in physiology instruction in schools. He was a young apprentice of thirteen or fourteen years and he had no interest at all in science; what he was interested in was (of course) food. Thinking of how great it would be to make a wax model of an omelette, which he had seen somewhere, he went to work. Thus, the first wax replica of a foodstuff was of an exotic Western dish: the omelette. The omelette was the first item to provide a window to Western food culture, and it was followed quickly by others such as croquettes and curried rice. Indeed the only restriction on what was to be introduced was the degree of difficulty required in making the moulds.

The aesthetic qualities of these wax models, their shapes and colours, were vulgar and definitely kitsch. This symbolic expression of a base desire, with the aid of newly arrived technology, produced a new naive and unadorned cultural product. Wax (and later plastic) models of food are representative of the kitsch culture in Japan in the 1930s.[27] Along with the *manzai* tradition, they have persisted until this day as incarnations of Osaka's popular culture.

Another important point to be mentioned is the fact that Iwasaki had come from Tokushima (in the countryside) to Osaka to become an apprentice. In his childhood he would often play with wax by a stream, which, when it fell into the water, would spread out and form exquisite shapes just like Japanese water-flowers. Upon seeing this, the young boy thought to himself, 'How pretty! I would like to be able to make beautiful things myself.' This aesthetic motivation built up inside him and was later combined with the Osaka food mentality to create a new urban artefact of popular culture. Thus the old aesthetic sensibility from the rural area was transformed into a typically urban form of kitsch culture.

It is apparent that the plastic replicas and the *manzai* in Osaka are both elements of the modern Japanese popular culture which were born from the 'gaps' between rural sensibilities and urban kitsch, Japanese and Western things, young and old.

In order to analyse and understand the modern popular culture in Osaka today it has been necessary to descend to its roots. We saw how basic food culture is in the *manzai* tradition and how essential it is as part of the way of life of the common folk in Osaka. This use of food as a symbolic tool in understanding the foreign and 'higher' cultures leads to an extremely 'low-level' type of universalism: 'in the long run people are all the same'. This ideology is founded in the Jōdo-Kyō sect of Buddhism.[28] Unlike the Zen Buddhist philosophy, here the attempt to rise up to salvation is abandoned and salvation is instead sought at a more 'down-to-earth' level, sprawling in the depths of society along with one's lowly peers.

NOTES

1 Roland Barthes, 'Element in semiologie', *Communication*, 1964.

2 Cf. Gillo Dorfles, *Le Kitsche*, Presse Universitaire de France, 1978.

3 See Yanagita Kunio, *Nihon no matsuri* (Japanese festivals), Kōbunsha, 1942.

4 Takigawa Masajirō, *Nihon shakai keizai-shi ronko* (A study on Japanese social economic history), revised edition, Tokyo, Meicho fukyūkai, 1983, pp. 507–8.

5 *Ibid.*, pp. 535–6.

6 After the Heian period, worshippers for the Kimano Shrine made it a rule to land at Watanabe first and then to proceed to the Tennōji temple. From there, they proceeded to Kumano via Abeno Ōji and Sakai Ōji.

7 Throughout the Tokugawa period, land was reclaimed in Edo (present-day Tokyo) as well as in the Inland Sea area. But the magnitude of landfill was largest in Osaka. 'The reclaimed area in the present-day City of Osaka includes the western parts of Sumiyoshi-ku and Nichinari-ku; all of Taisho-kū, Minato-ku and Konohana-ku; and the large western area of Nishiyodogawa-ku. In other words, approximately one third of the entire area of the city is the area which was reclaimed from the sea and developed as a field' (Tamaki Toyojirō, *Osaka kensetsu-shi yawa* [Stories of construction history of Osaka], 1980, p. 88). Dr. Tamaki estimated the scale of reclaimed land and newly developed rice field on the basis of the topographical map of original survey which the Land Survey Department of the Army General Staff office drew up in 1885. According to his estimate, which is the most reliable available, Minami is located at the centre of reclaimed land.

8 'Tennōji, which is on the route for worshippers who visited the Kumano shrine, was an old temple originally established by Prince Shōtoku, and a centre of worship offered by both the Imperial Household and by commoners. Since the ending phase of the Heian period, both retired emperors and imperial princes have strengthened their faith at this temple with its large precincts being equipped with halls, pagodas and cathedrals. Since the reign of Empress Shinko, the Sumiyoshi Shrine has constituted another centre of devout worship offered by people in all walks of life, as the residence of the god of navigation, war and *waka* (31-syllable Japanese poem). There was no end to people of all ranks who came from Kyoto and paid reverence at the shrine (Miura Shūgyō 'Chūsei no Osaka' [Osaka in the medieval ages], *Osaka to Sakai* [Osaka and Sakai], Iwanami, 1925).

9 *Nissōkan* is at times called *rakujitsu shonen* (chanting a prayer to the setting sun). According to *Nanba kagami* published in 1680, 'These stone shrine archways (*torii*) were believed to constitute the centre of the eastern gate of the Heaven, through which the sun was said to go at noon at the equinox. For this reason, people used to come around at this time to pray to the sun, but the practice had ceased to exist.' In the middle of the Tokugawa period, religious devotion apparently changed to secular orientation. 'Worshipping at the shrine during the equinoctial week apparently turned into the occasion for competition for better dressing, for better figure and for more spending. The entire precinct of Tennōji reportedly became the location for a grand garden party, as it were, which was held by priests, nuns and laymen and women' (Kitani Hogin, 'Higan no Tennōji to jidai-sō' [Tennōji at the equinox and the changing faces of the age]), *Kyodo shumi Osaka-jin* (Local hobbies and Osakaites), September 1929. The tradition of the 'grand garden party' was carried over to the Lunar Park of the Shinsekai area during the Taisho period. This transformation deserves particular attention as a radical example of alteration from religion to popular pastime.

10 Tomioka Taeko, *Umi kara mita tochi* (Land seen from the sea), Bungei shunjū, 1979. Bunshun paperback edition re-titled *Samazama na uta* (Various songs), pp. 61–2.

11 The original of the Osaka Castle as we know today is the Osaka Buddhist Lodge opened by the Saint Priest Rennyo in 1496. At the time, Chinese ships were said to come into the area right below the Lodge, indicating that this area was the hub of transport. Both the Yoshizaki Buddhist Lodge and the Yamanashi Buddhist Lodge which were opened by Rennyo were located at the centre of transport facing the Japan Sea and the Biwako Lake respectively. Thanks to his geo-political ingenuity, Rennyo established his strategic point in the Osaka Kamimachi tableland during the civil war period. The Ikko Sect of Buddhism, a new religion at the time, was the religion of the lower class, including tradespeople (such as oilmen, charcoal dealers, miniature shrine dealers, and fan makers), immigrants from Yamashina, itinerant harlots and gardeners (Miura, op. cit. 1925).

12 Sakai, famous for its municipal autonomy, is similar to Osaka in several respects. First, Sakai ground filled in from the sea. Secondly, both places were engaged primarily in fishery. Thirdly, both were first-class ports, and Sakai sometimes outperformed Osaka in this regard. Fourthly, both established municipal autonomy without any connection with religious forces (Miura, 1925).

13 The second characteristic of reclaimed land in Osaka is that the landfill extended as a shoaling beach which formed a low, swampy place. As a result, the area was not inhabitable and was cultivated as a newly reclaimed rice field (Tamaki, 1980, p. 93f).

14 In the Tokugawa period, the cultivation of such non-cereal goods as tobacco, spring onion and indigo plant was prohibited all over the nation, except in Osaka where it was permitted tacitly and sold along with various kinds of vegetables in the market. The Minami area of Osaka was characterized by mixed agriculture during this period (*Nishinari kushi* [A history of Nishinari Ward], 1968, pp. 18–19).

15 See Takigawa, 1983, pp. 275–6.

16 Dōtonbori was named after a construction engineer, Yasui Dōton who took charge of the construction of the Umenokawa River in 1612 under the order of Toyotomi, the then ruler of the nation. The labourers who were used for the task were peasants of Kuhōji Village. After the fall of Osaka Castle, Dōto, younger brother of Dōton, took over the work and in 1615 opened a channel to the Totsugawa River. Matsudaira Tadaaki who entered the Osaka Castle named the area Dōtonbori and authorized the construction of an amusement centre and playhouses there. In the middle of the Meiji period, Maruman, the noodle shop in the area, became famous and was often referred to in *rakugo* (comic stories). The area is now noted for good eating places (Mita Jun'ichi, *Dōtonbori*, 1975, pp. 16–17).

17 A chronology of amusement shows in the Dotonbori area is shown in Okubo Atsumaro, 'Kamigato no misemono (Shows in Osaka)', *Kamigata shumi*, June 1918.

18 *Tako-yaki* was an imitation of a red-stone ball which was in turn produced as an imitation of a red coral ball during the Tenpo years (1830–43). *Tako-yaki*, therefore, was a replica of a replica. It appeared for the first time in 1902 at a night fair of the Ishiya Shrine of Akashi, Hyogo Prefecture. See Kumagai Mana, 'Bunka to shite no takoyaki (Tako-yaki as culture)', *Gendai fūzoku '85*.

19 Kita Sadakichi, 'Isaburō kō (On Isaburō), *Fukushin*, Hōbunkan, 1976.

20 See Tada Michitarō, *Asobi to nihonjin* (Play and the Japanese), Chikuma shobō, 1975.

21 'Hogai-bito no yomeru uta (a poem by a beggar)', *Manyōshū*, vol. 16, 3888.

22 'Crabs in ditch reeds in Naniwa' became famous owing to the practice by which residents in this area captured and donated them to the Imperial House. 'Two poems on ditch reeds in Naniwa are found in *Ogura hyakunin isshu* though the *Manyōshū* poem cited in footnote 21 is best known. The fishing village near Fukōnoike described in Kaibara Ekken's *Shokoku meguri* (A travelogue of various provinces) is the retains of the vestiges of the Naniwa Inlet. Crabs sold at the riverside fish market of Kyōbashi, as illustrated in *Settsu meisho ezu taisei* (A pictorial compilation of noted places in the Osaka and Kobe area), are perhaps crabs in Naniwa reeds.' (Takigawa, 1983, pp. 508–9).

23 See Maeda Isamu, *Kamigata manzai happyaku-nen-shi* (A eight-hundred year history of Osaka vaudeville), Sugimoto shoten, 1975.
24 See Akita Minori, *Osaka Shōwashi* (A history of Osaka comic stories), Henshū kubō Noa, 1984.
25 Tsurumi Shunsuke, *Tayū Saizō-den* (A biography of Tayū Saizō), Heibonsha, 1979, p. 211.
26 Originally entitled *Konsen Sukii*, performed for the first time in 1936 by Ashinoya Gangyoku and Hayashida Jūrō. *Akita Minoru meisaku manzai senshū* (A selected collection of masterpieces of vaudeville dialogue by Akita Minoru), vol. 1, Nihon jitsugyō shuppansha, 1948, pp. 151–5.
27 *Shinsekai* (New world) deserves special attention as the heavily technologically oriented amusement quarters of Minami during this period. It came into existence before Disneyland. See Tokuyano Arinari, *Shinsekai ryūseishi* (A history of the rise of *Shinsekai*), 1934 (private printing), which describes how *Shinsekai* was planned and constructed.
28 See note 11. The backbone of the popular culture of Osaka which is said to be 'bonelessly flexible' is Jōdo-Kyō, especially Jōdo Shinshū.

3   New trends in Japanese popular culture

KOGAWA TETSUO

Before dealing with my main subject, I would like briefly to examine the concept of 'popular culture'. 'Popular culture' has two Japanese translations: *taishū bunka* and *minshū bunka*. While *bunka* covers the whole semantic extent of 'culture', 'popular' has the two different meanings of *taishū* and *minshū*. *Taishū* means a large number (tai) of population or groups (*shū*). *Minshū* means a good deal (*shū*) of ordinary people (*min*). For instance *min-shushugi* means democracy; *min-kan hōso* private broadcasting; *min-pō* the civil law; *min-yō* folk song; *min-zoku gaku* ethnology. Thus, *minshū bunka* is a more faithful translation of 'popular culture' than *taishū bunka*. Yet, the expression *minshū bunka* is not as popular as *taishū bunka* anymore in Japan today, so that 'popular culture' is usually translated into *taishū bunka*. This means that in the Japanese context 'popular culture' is absorbed by 'mass culture'. When the difference is intentionally expressed, the term of *minshū bunka* is recycled.

Conceptual complications of this kind are not a mere language problem but derive from the very process by which *minshū bunka* becomes *taishū bunka*. In the English cultural-linguistic context, too, popular culture is now nothing but mass culture in the sense that traditional oral cultures are dying under the domination of overpowering mass media.[1] Thus, I would like to use 'mass culture' as the expression of the popular culture that mass media create or circulate.

As mass media permeate into every facet of our life-world, very few phenomena are independent of mass media. Therefore, the formation of mass culture means the transformation of popular culture. While popular culture presupposes various types of people and groups, mass culture transforms the original character of such entities. When the mass media create and circulate homogeneous information, it creates and cultivates the homogeneous media audience who uniformly encapsulate concrete people and folk in the real world.

This has happened in Japan since the 1940s when the American type of mass media, especially television, was introduced. Television has homogenized Japan's diverse local cultures and traditional folk cultures. In order to conceive how effectively this occurred, one has only to recall the fact that in the few years after Okinawa was retroceded from the United States in 1972 and Japanese broadcasting was introduced, young people soon became accustomed to the standard language that radio and television announcers circulate and thus suffered less from 'culture shock' when they came to Tokyo. Even in Tokyo, radio and television played a major role in erasing dialect differences between *yamanote kotoba* (South-West Tokyo dialect) and *shitamachi kotoba* (North-East Tokyo dialect). It is impossible to guess a person's profession according to accent anymore. Differences between female and male language among urban young people have also diminished.[2]

Given this indisputable situation, I would like to begin with an analysis of the popular culture that manifests itself as mass culture. The 'mass' of mass culture connotes a coherent, homogeneously integrated matter, rather than a 'molecular'[3] ensemble. Modern technology is indispensable for cohering and integration. Technological development from printing to electronics has greatly intensified the extent and scale of this function. Electronics can organize not only nation-wide but also space-wide audiences. In this condition, individuals, folks and people who would otherwise maintain their 'autonomy', are fused into an electronic mass. So, I would call the present stage of mass culture 'electronic mass culture'. This culture has been formed through the pervasion of electric and electronic devices like refrigerators, washing machines, vacuum cleaners, stereo phonographs, radios, telephones, radio-cassette tape-recorders, electronic calculators, computer game machines, vending machines for food, tickets and banking, 'Walkman', *karaoke* and FM mini transmitters.

Today's Japanese popular culture manifests itself through the medium of these electric or electronic devices. They have changed the social relations of individuals, the communication environment, social and aesthetic values, and even political consciousness.[4] Among these devices, information devices like television and radio have an effect strong enough to change socio-cultural conditions, but even electric tools have done much for the change. For instance, electric refrigerators (89 percent of households in 1970), washing machines (96 percent in 1982), and vacuum cleaners (91.2 percent in 1975) drastically changed the housewife's conditions and traditional life-style and food habits. Also, vending machines totally transformed the shopping patterns and attitudes of the consumer. In 1971 the total number of vending machines for soft drinks, food, cigarettes, train tickets, and so on, was 1,391,470; in 1975 doubled; in 1982 it is 4,861,140 – 4.1 per 100 persons.[5]

The most powerful device of electronic mass culture is television. The first television broadcasting in black and white began with NHK and NTV (Nihon TV, a private station) in 1953. When thirty-four stations throughout Japan were licensed by the Ministry of Posts and Tele-communications in 1957, only 5.1 percent of national households owned television sets. In 1964 when the Games of the 18th Olympiad were held in Tokyo, 90.3 percent of households had a black and white television set. Colourcasting began in 1960 (experimentally in 1957), and over 90 percent of households owned a colour television set as early as 1975. In 1969, over 90 percent of households had either a black and white or colour television set.

In this changing media environment, people more and more spent time watching television. In 1960 people over the age of ten watched television for 56 minutes every weekday; in 1975 for 3 hours and 19 minutes. Inversely, sleeping hours decreased from 8 hours and 13 minutes to 7 hours and 52 minutes. (*Kokumin seikatsu jikan chōsa* by NHK hoso seron chosajo.) In this way, television has become a strong apparatus to transform everything into popular images – mass culture. As television has become like one's extended 'body', one does not necessarily recognize what is happening on television. At a level beyond one's consciousness, happenings on television directly become a popular culture. In the earlier stage, television was a medium between reality and its image. When it conveyed a programme of *proresu* (American-style wrestling), one of the most popular television programmes in the 1950s, for instance, people knew that the real show existed and the television represented it: the real world was considered the final referent. However people tend now not to consider the reality of the referent: the question is whether the video image is real or not. I would call this 'reality' *simulated reality*.[6]

The most typical and explicit example of this simulated reality that constitutes today's mass culture of Japan is illustrated by the recent 'Suspicious Bullet Incident'. This saga began in 1984: the 26th January issue of the weekly popular magazine *Shūkan Bunshun* contained an article accusing Mr. Miura Kazuyoshi, an upper middle-class citizen, of killing his wife in Los Angeles. In 1981, Miura and his wife were mugged on a street in Los Angeles, and his wife was shot and eventually died. At that time, Mr. Miura repeatedly appeared on television in Japan, and expressed his anger and sadness. Many viewers sympathized with him. It became a daytime television melodrama. However, soon after his wife was killed he got married again. So, people had the feeling that there might have been something unclear in the case. Needless to say, weekly magazines like *Shūkan Bunshun* dealt with this story, but it was not such big news. All of a sudden, three years after the mugging, *Shūkan Bunshun* reminded people of it, skilfully exploiting this vague aspect of the case and accusing Mr. Miura as if he had killed his wife. However, I think this 'Suspicious Bullet Incident' – the front line of the magazine's article – would have become a bigger topic if television reporters had rushed to Mr. Miura's house or he had been forced to appear on television. So far, neither in Japan nor in the U.S. have the police given any comment on Miura's alleged suspicion.[7] Nonetheless, the 'Suspicious Bullet Incident' was a big topic throughout the whole of 1984 and the Miura family had to escape from Japan to Europe.

This incident suggests that in Japan civil rights are diminished and ordinary people bear resentment against a new type of middle class like Mr Miura who is not an ordinary salaried worker but an independent dealer of precious metals and stones, and whose life-style is 'luxurious' and chic in the sense of Yuppie. This incident indicates the changing class structure of Japanese society. In fact, the economic and cultural difference is so obvious that there is a popular phenomenon by which people playfully discriminate one from another as either *marukin* or *marubi* (encircled – abbreviated – 'rich' or 'poor'). These terms derive from a recent best-seller *Kinkonkan*[8] by Watanabe Kazuhiro and Tarakopuroda-kushon, which is a kind of Japanese counterpart of *The Yuppie Handbook*.[9] From the perspective of popular culture, this phenomenon is more interesting than the change in class structure. People talk about the informational difference between 'rich' and 'poor', not the substantial difference. Given the informational atmosphere of 'rich' and 'poor' that *Kinkonkan* illustrates, one could, referring to this book, play at being 'rich' even if one is lower middle class. This is a game with simulated reality.

As the 'Suspicious Bullet Incident' suggests, simulated reality needs no authority outside reality. Everything is decided within the media, especially the video display, and is immediately accessed/forgotten by ON and OFF. Television and computer games are very effective at cultivating this ON and OFF electronic mass culture. Advertising spots for commodities on television have been a fundamental apparatus. They interrupt the stream of consciousness. Bertold Brecht once referred to the 'V-effect' ('Verfremundung-Effekt') when a continuing performance is suddenly interrupted by means of an unusual performance or action.[10] While Brecht used this technique in order to waken his audience when it was inclined to be passive during the performance, 'spots' on television are designed to relativize or bracket the very programme that 'V-effect' was supposed to activate. This pseudo-'V-effect' of 'spots' has been accelerated as they have become more and more 'artistic'. As early as the mid-seventies, the better 'spots' were a kind of 'short-short' with sophisticated film techniques and original ideas. The budget is usually much higher for a twenty-second 'spot' than for thirty minutes of regular soap opera. Some feature (mostly male) international actors – such as Alain Delon, Marcello Mastroianni, Orson Wells, Sean Connery, Paul Newman, Woody Allen and Harrison Ford. Therefore, the spots are sometimes more interesting than the regular programmes, and they enable the viewer to endure the interval of advertising. Spots on television, therefore, have a very strong influence on mass culture and have changed viewers' time consciousness.

In addition to this, *Jōhō-shi*, guide magazines to entertainment and leisure activities like the late *Cue* or *New York Magazine* began to organize new information channels. The most influential is *Pia* which started to publish in 1973. This magazine is now so influential as to be able to control the world of show business. Given the popular impression that *Pia* covers all the information concerning movies, performances and other entertainments in metropolitan cities, people become indifferent to shows not listed in it, and which people used to learn about through their oral information network among friends (*kuchikomi*). This type of magazine transforms the total reality of city culture into fragments of information that one can assemble according to one's preference (*konomi*). And it goes to an extreme in which informational reality is substituted for the substantial reality that everybody used to experience. For instance, many readers of *jōhō-shi* use the magazine for the fun of assembling-information play (*jōhō-asobi*), not for a guide to theatre going. This extremity is now taken over by computer games where everything is simulacra.

Throughout the seventies, the popular culture of television and magazines was based on various sorts of parody techniques in language and visual images. One typical example was a television show in which *sokkurisan* (human imitations) of popular stars and figures, even Marilyn Monroe and J. F. Kennedy, appeared. Fake aesthetics was dominant in this period, and fake feeling was preferred to 'real' touch. Spots on television clearly reflected this popular aesthetics, and many paronomasia (word play) and visual image quotations from famous films – especially American movies – were well-known advertising techniques. This fake aesthetics still had a referential real world from which the fake image derived its 'reality'. Electronic mass culture disregards such a referent.

The substance of comic strips (*manga*) changed from the fake aesthetics to the simulated electronic mass culture. The late sixties and early seventies were a golden age of Japanese comic strips, in which American popular culture, Japanese regional folk culture, and new waves of art were unified. These comic strips were influenced by television, too. The main streams emerged: *gekiga* and *shōjo manga*. Literally, *gekiga* means a theatrical picture and *shōjo manga* means young girl's comics. The former stress the action of character and expressionistic visual image; the latter tends to develop romantic images. Both seem to borrow the viewpoint or the image of settings of the Western world and to dream of Westernized culture. Although *gekiga* deals with old native conventional worlds, the touch of exaggerated drawings is as if American movies dealt with Japanese. In the beginning, there may have been such a dream to some extent, but I think it is getting less in today's *manga*.[11]

The myth of Western culture and civilization has been losing effect in the last decade as the Japanese have developed an awareness of an 'advanced industrial country ranked with the U.S.' and people have had plenty of opportunities to actually observe the Western world in tourism. Nowadays, there is a popular saying, '*Datsu-ou nyū-a*': that is a total reversal of *Datsu-a nyū-ou* (to disassociate from Asia, then to join Europe) of the nineteenth century.[12] These trends are compatible with the change in Japanese National Railways campaigns from 'Discover Japan' in the seventies to 'Exotic Japan' in the eighties. In the period of 'Discover Japan', there were diverse and regional cultures in every part of the country, especially in northern Japan. The tour companies could find differences and interesting places. This was a kind of exploitation of the diversified culture resources which existed. But in the process, these differences were flattened out. Today, one can hardly find differences even in the northern parts of the country like Iwate or Yamagata. In every place one can find vending machines, paved roads, supermarkets, the same commodities, and of course television. This is a problem for tourist companies. Thus, they have to create artificial differences, a simulated culture of Japan. This is the basic intention of the campaign called 'Exotic Japan'. The countryside *must* be exotic, but it *is* not.

Given that even exoticism is simulated, it would be one-sided just to point out a romantic fantasy of Western culture in the exotic image of *manga*. One must pay attention more to the social aesthetics side of this phenomenon. The typical reader of *manga* does not mind the imitating aspects but is fascinated by the constant distance from the real world. Blue eyes, blond hair, and long and slender legs – stereotyped images of *gaijin* – of the characters of *shōjo manga* aren't intended to refer to the real existing Western world but, on the contrary, to alienate the reader from any kind of real world. Therefore, *manga* especially *shōjo manga* succeed in creating science fiction fantasy and may be combined with animated cartoon movies or video. A recent hit movie, *Kazenotanino Naushika* directed by Komatsubara Kazuo, is based on *manga* by Miyazaki Shun.

Karaoke and Walkman belong to the electronic mass culture, too.

Karaoke is an electronic device with tape-recorder, amplifier, effecter, speaker and microphone, by which one can sing a song to the taped accompaniment of a professional orchestra or band. In a party with *karaoke*, people sing songs one after another. Even at a conventional party, people used to sing songs, so it might seem that the appearance of *karaoke* has not changed the form of conventional Japanese collectivity at all. However, there is a big difference. Careful observation of the relationship between the performer and the audience reveals that they are united only through the help of electronic media, only *on* and *off*. *Karaoke* performs the function of unifying people who have lost their ability to communicate with each other by their oral media. The popularity of *karaoke* means that Japanese collectivity becomes more and more temporal, rather than a continuous collectivity based on oral language, race, religion, taste or folkways.

Walkman,[13] a cassette tape-recorder with a headphone is for personal use only, so that the users are isolated from each other even if they listen to the same music source. Each user is united to the extent that he or she maintains and belongs to the same mass culture: electronic mass culture. Outdoor users of Walkman express their unconscious desire to be absorbed in the complete electronic culture. People attach themselves to the electronic environment on the street where otherwise they would be separated from it. From a different viewpoint, one gains protection from the more and more electronics-oriented city environment with blaring speakers, noisy traffic and flickering shop windows.

In this environment in which television, *karaoke* and Walkman were gradually establishing the electronic mass culture, 'Mini FM' has appeared. 'Mini FM' is a very low powered license-free FM radio station. In Japan, there has been no public access to the regular broadcasting spectrum except on *bijaku-denpa* (very low powered airwaves). According to the Japanese Radio Law (Article 4), 'a station whose broadcasting wave is in a very low power needs no license'. Also, the Article 6 of the Enforcement Regulation defines this 'very low power' to mean that the wave must be 'below 15 microvolts per meter at the distance of 100 meters from the transmitter.'[14] These regulations were established in 1957 when electronic technology was gradually brought into the market. Thus, the *bijaku-denpa* was not intended to be a stingy 'public access' measure but a pragmatic measure to promote more electronic devices like wireless microphones, television remote control devices, garage-door openers, model airplanes and other wireless transmitters. At that time, *bijaku-denpa* was useless for broadcasting except for the childish fun of playing at wireless communication over a limited distance, playing at professional radio announcing or music show. However, the situation has changed as the performance of the receivers has improved and the cost of radio sets has decreased. Today, most people have at least one or two transistor radio sets at home, which can catch signals at the distance of 500 metres from transmitters submitting to the regulation of *bijaku-denpa*. This means that the legally permitted airwave could cover a one kilometre radius of the city – which in a dense population area contains 20,000 residents, all potential listeners. 'Mini FM' found this unknown 'public access'.

In the late 1970s, there were some individuals and groups who tried this method in order to open up a free radio station, stimulated by free radio fever in Europe or CB mania in the U.S. However, enthusiasm for 'Mini FM' did not begin until August 1982 when a station called KIDS in Aoyama, Tokyo was opened and newspapers, popular magazines and television reported on it. In the first month after their opening, they had at least thirty interviews with the mass media, and then numbers of advertising agencies and service industries got in touch with them. During the next few months, hundreds of imitators opened their stations in campuses, coffee shops, private spaces, stores and gathering places. Given the excitable nature of Japanese mass media, daily newspapers and weekly magazines began to report about what was happening in 'Mini FM' and on how to open up a 'Mini FM' station. Consequently, other unknown 'Mini FM' stations had a chance to become public along with the imitators of KIDS. By April 1983, almost all types of 'Mini FM' station appeared before the public.[15] How enthusiastic they were could be illustrated by the fact that in March 1983 the Ministry of Posts and Telecommunications referred the matter of growing 'Mini FM' to their consultative body out of an anxiety for the future consequences.[16] So far, they have not proclaimed any change except in the way of measuring the *bijaku-denpa*. In future, the regulation is to be changed to the following: 'Below 500 microvolts per meter at the distance of 3 meters from the transmitter.' This does not mean that the power allowance is to be increased. The substantial power to be allowed is almost the same as before. The difference is for the authority to be able to measure the power more exactly for the measurement is done at a very close distance – three metres – to the transmitter. This could be a subtle control of 'Mini FM', but the activity will, for some time, be undamaged. 'Mini FM' is celebrating its fifth anniversary in 1987.[17]

This 'quiet' interim has something to do with the present unstable situation of government radio policy. The government has just begun to discuss revision of the broadcasting law that was basically established in the 1950s and is apparently inconvenient for promoting new media. In comparison with the radical change of FM radio policy in the advanced industrial countries, the Japanese government has been very reluctant to open the VHF airwaves to FM broadcasting, although this is indispensable for their new policy, stressing localism, cultural diversity and new media, to proceed.

A short history of Japanese radio policy shows how long the government has monopolized the people's air spectrum. Regular FM radio broadcasting was authorized in 1969 with 170 public stations (in the whole country) of NHK, most of which had been experimentarily licensed, and with only one private station – FM *Aichi*. As far as Tokyo is concerned, NHK FM was the only authorized station on FM dial and the private FM Tokyo (FM Tokai at that time) didn't receive a regular licence until 1970 although they had had approval to broadcast experimentally since 1961. Thereafter, the government approved more licences, but there were only four private FM radio stations in all (FM Tokyo, FM Aichi, FM Osaka, and FM Fukuoka) until 1982 when five additional stations (FM Ehime, FM Hokkaidō, Hiroshima FM, FM Nagasaki and FM Sendai) were licensed. Even today, Tokyo has only two FM radio stations in all. This is unbelievable in comparison with other countries. In Italy, although an extreme case, each major city has at least a hundred stations on the FM dial: there are over 200 stations in the city of Rome for instance.

As economic activity grew, a strong need for more FM stations developed among private institutions. Already in 1964, the Nippon Hōsō – a private broadcasting company on AM dial – declared the management policy of 'Audience segmentation' which was to accord with the coming consumption-oriented society, comparable to the U.S. In the following years, this form of management became prevalent in major broadcasting companies on AM dial. However, given the Japanese broadcasting law basically obliges stations to operate nationally, not on a local basis, the audience segmentation policy didn't work. Only segmentation by age, sex and profession was introduced. Yet it is beyond doubt that segmentation of local needs is most appropriate to make best use of FM broadcasting. In Japan, FM broadcasting has been different from AM broadcasting in its quality of sound.

Although hundreds of institutions – including advertising agencies and political and religious organizations – have continued to apply for licences to the Ministry of Posts and Telecommunications since 1945, no additional AM or FM stations have been approved in Tokyo, for instance, during the past fifteen years, a period of increasing cultural diversity in the context of economic development, in spite of the availability of the VHF spectrum. This abnormal situation is partly because the government, headed by the Liberal Democratic Party, and those private stations already operating are in collusion. While the government controls them through the restriction of licences and through *amakudari-jinji* (appointment to the executives from ministries of the government), private stations are willing to submit to governmental supervision in order to monopolize the market and avoid competition with newcomers. The government has balked at dismantling this intervention in the private sector.[18]

The abnormal situation did not lead to illegal broadcasting as in France in the late seventies because of the governmental direct and indirect control of the airwaves.[19] In addition to the severe supervision, a common idea that airwaves belong to the government has been deeply implanted in the people's minds. Therefore, people's discontent about broadcasting is to be found in various unconscious phenomena. One of the most interesting examples is the growing number of the listeners to FEN, Far East Network – a special broadcast service for U.S. troops stationed in East Asia. FEN's English programmes, which stress current American popular music, are so welcomed by Japanese young listeners that a special quarterly and more than ten books on how to listen to FEN are in circulation.[20] This means that Japanese radio broadcasting does not satisfy young people.

It was in this situation that 'Mini FM' came out and immediately became a popular sub-culture. How it immediately caught people's interests is not entirely clear, yet the following motives may be hypothesized: (1) public desire for more diverse radio programmes, (2) influence from free radio in Italy and France, (3) preference for fake culture, especially American pop culture, (4) an accumulation of a kind of techno-youth culture (playing with cassette tape-recorders, stereos, wireless microphones, walkie talkies, model airplanes and electronic game machines), (5) availability and reducing cost of electronic devices like tiny FM transmitters, and (6) desire for new communication with others. All the motives above are interrelated with each other and each of them has something to do with what is going on in Japanese society today.

While thousands of 'Mini FM' stations rose and fell in three years, at least a hundred survive in Tokyo today. They could be classified into three categories. The first is most interested in a kind of radio performance with talk and music shows by disc jockeys in an open studio. KIDS was the founder of this type and has been busy in sending out their staff and equipment to university campuses, festival scenes, department stores and supermarkets. There are many coffee shops where an open 'Mini FM' studio is set up. They are less interested in communicating through radio media. 'Mini FM' is a kind of performance show to catch one's attention or to entertain customers. Therefore this type of station is a paying business. The second one is interested in establishing an alternative radio station on FM dial. They broadcast programmes of music, news, talk shows and discussions and intend to communicate with local listeners. AD AD in Nerima, Tokyo has established a quite large network with a relay system and broadcasts mainly music programmes every day. JOGG FM Kōenji broadcasts every weekend and publishes a complete programme every two weeks. They have many talented D.J.s who have created various kinds of discourse that cannot be heard on the dial of NHK or FM Tokyo.

The third one has found a radically different function of 'Mini FM', and I think this represents one of the most viable sides of 'Mini FM' and the electronic mass culture. Setagaya MaMa's studio is housed in a small shack serving both as a gathering place for neighbours and as an alternative retail store carrying natural foods and other daily necessities. This station is radically free from professional programming and has been open to anyone who wants to talk by radio. Even babblings, clatters, and slams in this store-station are on the air. Once, when several people began talking about community politics, some listeners rushed to the station and joined in the discussion. Radio Komedia Suginami have established their station in a coffee shop for a totally different purpose from the first category. They have a programme, but it is so flexible that anybody in the coffee shop can join in the discussion. Some listeners are inclined to visit the station when they are at home and eventually come to the coffee shop to take the microphone. Radio Home Run tries to be more conscious of this line – being *centrifugal*, rather than *centripetal* in function. Their station defines itself as a gathering place with a transmitting device. The programme consists of discussions. For five days a week, various kinds of discussion are broadcast. But more important is the very process of discussion and gathering, rather than broadcasting. Workers, political activists, feminists, artists, performers, journalists, students and jobless people get together in order to talk on the air and maintain a temporary collective. Needless to say, visiting listeners are allowed to join in the discussion. There are always guiding individuals or groups who prepare the theme of discussion and organize the temporal collective every time. Sometimes a temporal collective becomes permanent and begins to work outside the station, being involved in artistic or political activities. So, Radio Home Run is a catalystic space to create new groups of art, alternative magazines, social activities, theatre and music.

The collectivity through radio, i.e. the electronic collectivity, brings together individuals on the air. This provisional unfetteredness can alienate individuals in an isolated electronic capsule of mass media, but at the same time can reactivate individual desire for free, spontaneous *communion*. The electronic collectivity either integrates individuals into a homogeneous 'fused' collectivity or emancipates every *singularity* in the collective. Electronic media and the mass culture they cultivate have this dual potentiality. The present situation of Japanese popular culture is oscillating between the poles of this dual potentiality. The question is whether popular culture will remain in the homogeneous capsule that was established with the modern emperor system (*tennō-sei*) and has been reorganized with the electronic mass media, or will it overcome it.

NOTES

1 Among hundreds of critiques on contemporary popular culture, Theodor W. Adorno's criticism is most penetrating. He called the change of culture as 'all post-Auschwitz culture, including its urgent critique, is garbage' (*Negative Dialectics*, translated by E. B. Ashton, New York, The Seabury Press, 1979, p. 367). Given his definition, today's popular culture must be found among 'garbages', that is mass culture. As for the homogenizing process of culture from popular culture to mass culture, see my 'Adorno's "strategy of hibernation" ', *Telos*, no. 46, Winter 1980–81, pp. 147–53.

2 These are the consequence of 'Americanization' that is much more visible in Japan than in the U.S. See my 'Japan as a manipulated society', *Telos*, no. 49, Fall 1981, pp. 138–40.

3 Felix Guattari, *Molecular Revolution – Psychiatry and Politics*, London, Penguin Books Ltd., 1984.

4 See my 'Electronic individualism', *Canadian Journal of Political and Social Theory*, vol. 8, no. 3, Fall 1984, pp. 15–20.

5 *Jihaikiseihin sōgōkatarogu* (IAM '84 buyers guide green book), Jidō hanbai kenkyūjo, 1983.

6 Myron W. Krueger's terminology 'artificial reality' belongs to the same situation, but it lacks the critique upon the status quo itself. *Artificial Reality*, Reading, Mass., Addison-Wesley, 1983.

7 On 11th September 1985, Mr. Miura was eventually arrested by the police on suspicion of another murder case and is still detained in November 1985.

8 Watanabe Kazuhiro *et al.*, *Kinkonkan* (Money spirits book), Tokyo, Shufunotomosha, 1984.

9 Marissa Piesman and Marilee Hartley, *The Yuppie Handbook*, New York, Pocket Books (Simon & Schuster), 1984.

10 Bert Brecht, *Brecht on Theatre*, edited and translated by John Willett, New York, 1964, pp. 143 f.

11 On the basic trends of *manga* in the sixties Ishiko Junzō provides many brilliant accounts. See his *Gendai manga no shisō* (Thoughts in contemporary *manga*), Tokyo, Taihei shuppansha, 1976. Also, Tsurumi Shunsuke writes a concise note on the history of *manga* in his *A Cultural History of Postwar Japan, 1945–1980*, London, Kegan Paul international, 1987, chapter 3.

12 Fukuzawa Yukichi wrote his 'Datsua-ron' (An exhortation beyond Asia), *Jiji Shinpō*, 16th March 1885. (*Fukuzawa Yukichi zenshū*, bd. 10, Tokyo, Iwanami shoten, 1970, pp. 239 ff).

13 This is a trade name of Sony's product. 'Walkman' is generally called 'headphone-stereo' in Japanese mass media, too.

14 Denpa kenkyūkai (ed.), *Denpa hōreishū* (Radio wave statute book), Tokyo, Denpa shinkōkai, 1983, p. 2 and p. 236.

15 On the prehistory of 'Mini FM', see my *Korega jiyū-rajioda* (This is free radio), Tokyo, Shōbunsha, 1983; and my 'Free radio in Japan' in Douglas Kahn and Diane Neumaier (eds.), *Cultures in Contention*, Seattle, The Real Comet Press, 1985.

16 Strangely enough, this news was reported by only one newspaper, *Suponichi*, 26th March 1983.

17 Six months later after this draft was written, the authority made a move to change the situation. On 4th September 1985, a 'Mini FM' radio station in Mita, Tokyo, was

suddenly attacked by the police detectives. The station owner and disc jockey were arrested, accused of broadcasting with illegal 'strong' transmitting power. This station was one of the community-oriented 'Mini FM' radio stations which functioned as a people's voice. Young rock 'n' roll fans would gather there at night. The 4th September incident, then, was a case of communal activity being repressed, rather than a matter of technological policy. The Ministry of Posts and Telecommunications gave no warning to the station of this allegedly 'strong' transmitting power. The incident has ended in a 20,000 yen fine against the owner. Although the authority apparently expected that many 'Mini FM' stations would get anxious about the authority's surveillance, very few people hesitated to continue broadcasting and mass media didn't support the authority. Numbers of criticisms against this control appeared. See the feature articles on this case in *Asahi Journal*, 8th November 1985, pp. 84–91, for instance.

18 As for the complicated relationship between the government and the applicants for the FM licence, see Ando Tatsuo, 'FM kakujū o meguru 10 yonen no ōshū', *Sōgō jānarizumu kenkyū* no. 102, Autumn 1982, pp. 38–43; Denpa Shinbun (ed.), *FM menkyo, asudewa ososugiru* (FM broadcasting licence – don't put off till tomorrow), Tokyo, Denpa shibunsha, 1978. Terebi bunka kenkyūkai, *Terebi fushoku kenshō* (Examination of the eclipsing television), Tokyo, Chōbunsha, 1980, has the very critical chapter, 'Denpa media no gendai shinwa (Modern myth of airwave media)' on the Japanese radio policy pp. 237–60.

19 As early as January 1979 when Italian free radio 'fever' was spreading in France, a pirate FM radio station 'FM Nishi Tokyo' began to broadcast and created many enthusiastic listeners. However, it was shut down by the police after three months. See *Asahi Shinbun*, 11th April 1979, p. 23.

20 An electric company sells even a special radio set which can only receive FEN dial.

II Popular movements: alternative visions of 'modernization'

4 Popular movements in modern Japanese history

IROKAWA DAIKICHI

I propose to examine the common characteristics of popular movements in Japan over approximately the last hundred years. Of course the popular movements over this period were of many different kinds, there were many of them, historical conditions differed, and it is difficult to generalize. However, no matter which of them one considers, the objective of the movement was not properly realized; most of them were blocked along the way, or defeated.

What can this mean? Before clarifying the reasons for it there are certain preliminaries to be dealt with. First, the popular movements that are dealt with here are confined to independent movements started by ordinary people in accordance with their own social needs or the requirements of their livelihood. Movements of intellectuals or movements led by professional politicians and the like are not included.

However, since popular movements occur as the summation of social contradictions, a purely independent popular movement cannot exist in fact; in some form or other all are subject to the influence of state ideology or the thinking of intellectuals.

For example, the *yonaoshi* ('World Renewing') movement, which was the largest popular movement of the 1860s and included *uchikowashi* (smashings) and the wild mass dancing of *eejyanaika* 'we'll do as we bloody-well like', was stimulated by the 'restoration' ideology of the samurai of the anti-Bakufu faction who became the central force of the Meiji Restoration; the Liberty and Peoples' Rights movement and the armed peasants' revolt at Chichibu, which involved millions of people in the 1880s, were influenced by the idea of the natural rights of man and by Liberty and Peoples' Rights thinking; the rise of the universal suffrage movement and of the *buraku* liberation movement (through the establishment of the *suiheisha* or Levellers' Society) in the 1910s and 1920s was reinforced by Taisho Democracy, the main trend of thinking at the time, and by socialist ideology; the popular movements of the period since the

1960s were started off under the overwhelming influence of the occupation period reforms and of post-war democracy. Nevertheless, these movements embody also quite distinctive ideas and demands, and, in the sense that they did not develop just out of the ideas of the times, they fall within the category of 'popular movements' as I have defined it here.

Let us divide these popular movements of the last hundred odd years into three periods and consider them in sequence.

1 1860–1890 From Divided Feudal States (the Tokugawa *Bakuhan* system) *to the Establishment of Central Authority* (the imperial absolutism of modern *tennōsei*). This is the period of the Meiji Restoration in the broad sense. The largest popular movements of this period were the 'World Renewing' movement and the Liberty and Peoples' Rights movement.

(i) Discontent and world renewing aspiration. The rebellions, 'smashings' and movements against the new Meiji government that went under the collective name of 'World Renewing' were fought under conditions in which the people had no political rights and so in the end degenerated into 'riots' and were suppressed by the authorities. However, the *Bakuhan* system was severely shaken in 1866 as the 'World-Renewing' movement became generalized; and in 1876 the Ise riots won from the new government the concession of a substantial reduction in the land tax.

Popular longing for 'World Renewal' spread through Eastern Japan from the eighteenth century, together with the cult of Amitabha (the 'future' Bodhisattva). The combination of 'World Renewal' movements with the longing for the advent of Amitabha the saviour became one of the bases through which poor people came together. This Amitabha faith was a kind of Buddhist messianism, but it did not possess in Japan the considerable power to spread and become organized as the Taiping movement had in China. Those people who felt they were living in a 'last age', because of the inflation and social disturbances that followed the arrival of Commodore Perry, linked up with others in society who felt a desire for the transformation and reordering of society. They lacked a consciousness of themselves as a subjective force for social change but they generated a number of elements of reformist thinking. Japan did not produce intellectuals able to systematize this experience and these ideological glimmerings as a modern ideology. Instead it became the arena of the popular new religions of 'World Renewal' such as Tenrikyō (founder: Nakayama Miki), Maruyamakyō (founder: Itō Rokurobei) and Omotokyō (founder: Deguchi Nao).[1]

(ii) The 1880s. In the later phase, the 1880s, the Liberty and Peoples' Rights movement, which developed under the leadership of rich farmers as a nation-wide movement for 'people's rights', was a popular movement independent of samurai intellectual leadership. Its members formed groups in the local town or village they resided in and involved themselves consciously and deliberately in study movements, movements to petition for the establishment of a National Assembly, or movements for the drafting of a constitution. This demonstrated the passion for political participation on the part of those who had never had any rights. In one mountain village in the Kantō region which I have studied, a group of local youth who possessed a radical consciousness allied themselves with various wandering people's rights youth from other villages, organized study groups on political activity within their village, and accomplished the task of drawing up their own draft constitution (the Itsukaichi draft constitution). They involved themselves in the education of local people through running regular lecture meetings and organized their own hygiene movement to protect their villages from contagious diseases. These 'People's Rights Societies' took different forms in different localities, but at their peak they numbered probably over one thousand.[2]

The People's Rights movement was therefore a large self-education movement of the people, possessing epochal significance as a grass-roots political and social movement. In 1981 the 100th anniversary of this movement was celebrated throughout Japan – 7,000 people took part in the central 'Peoples' Rights 100th Anniversary Meeting' – and many new historical materials were discovered and introduced. The fusion of modern Western Enlightenment thinking with the Confucianism that constituted the traditional culture of the samurai and rich peasant elements and with the 'world renewing' ideas of the lower classes can be seen in this popular movement.

Although the People's Rights movement reached an impasse in 1884 as a result of the fierce repression and divisive strategems of the central government, and although the Liberal Party, which had been the guiding political party, dissolved iself, the 'Chichibu Uprising', which had the character of an armed uprising, was launched immediately afterwards by farmers in the several counties surrounding and including Chichibu. This was an unprecedented affair: the organization was carried out for more than a year mainly by the Chichibu Party of the Poor (*Konmintō*) and military units (troops or *Shōtai*) were set up in each village to make a clean sweep of the local authorities. At its peak the rebellion comprised a force of 10,000 men (three battalions) and clashed in several places with the army sent by the central government to suppress it. The demands of the movement were for the reduction of high interest loans, reduction in taxes, and an end to new policies. There were some among the local Liberal Party members in the leadership who were demanding change in the political system. Among the voices of the people who rebelled some could be heard proclaiming 'Mr. Itagaki's World Renewal' and 'Join forces with Marquis Itagaki's army and establish a new government'. Itagaki was Itagaki Taisuke, who had been president of the Liberal Party and who exercised great charismatic force.

Samurai intellectuals did not participate in this popular movement, and in recent years the view that the ideology of the rebellion was a unique popular aspiration for social equality, going beyond 'world renewal' to 'world levelling', has become generally accepted, although it is a fact that it could not have taken place but for the influence of the Liberty and People's Rights Movement. The history of the Chichibu uprising was handed down among the descendants of those villages whose members took part in it en masse; folklore transformed Tashiro Eisuke and other executed leaders into martyrs, and the proud popular tradition was revived in 1984 at the 100th anniversary celebrations of the peak of the Chichibu incident.

2 1890–1945: Popular Movements Under the Meiji Constitutional System

(i) The beginnings of constitutional government – the period of expansion through foreign wars. In this period the movements on the part of small sections of people with Christian connections whose demands included the right of participation in politics for women and the abolition of prostitution are worthy of note, but they were blocked by the current of familism which was particularly strong among conservative men and could not expand as popular movements. But on the other hand the movement of the sufferers from Ashio mine pollution developed into a large movement embracing people living along the Watarase River who demanded that the Ashio mine, which was protected by state militarization policies ('Enrich the Country, Strengthen the Army' policies), cease operations and clean up its poison effluent.

In the forefront of this movement stood a Liberty and People's Rights activist and son of a local village headman called Tanaka Shōzō. Although he struggled for twenty years, right up until the day he died, Tanaka was unable to secure a fundamental solution (an end to the pollution). Yanaka village, which resisted to the end together with Tanaka, was flooded and destroyed by a government-built reservoir. This tragic struggle earned the humanitarian sympathies of intellectuals and students in Tokyo and for a time moved general public opinion, but the government's clever efforts at changing the nature of the problem – through flood control works designed to ameliorate the damage – destroyed the local unity and the movement was split through the exploitation of patriotic sentiments stirred up by the Russo-Japanese War. However, this earliest anti-pollution struggle in Japan was reevaluated more than half a century later, in the 1970s, during defensive struggles by local farmers over the Minamata disease and at Sanrizuka against the construction of the New Tokyo International Airport at Narita. Its lessons have been passed on to the present-day.

Upon his death Tanaka Shōzō came to be revered as a charismatic leader, and the victims of pollution built shrines to him and for long afterwards praised his virtue, just as the farmers of the Edo period made pilgrimages to the Sōgo shrine built in honour of Sakura Sōgo who fell as a victim in their rising. The pure and passionate character of Tanaka Shōzō served to engrave the Meiji anti pollution movement deeply in the hearts of the Japanese people.

(ii) Taisho Democracy – the emergence of political party cabinets. The universal suffrage movement and the self-liberation movement of the people of the lowest strata of society were the key movements in this period, but the leadership role of political parties, scholars and journalists was very pronounced in the suffrage movement, so that it can scarcely be considered an independent popular movement.

Likewise the 1918 Rice Riots were a large scale affair in which the energies of a million people, enraged over the sudden rise in the price of rice and the government's resourcelessness, exploded, but this movement falls outside the scope of popular movements as considered in this paper because it was sudden and impulsive, lacking in plan or organization.

I am inclined to think rather that in this period it is the *buraku* liberation movement which constitutes the model of a movement which emerges from the depths of Japanese society to indict the society. From the Edo period Japan had at least a million low-caste people who were beneath the 'samurai, farmer, artisan, merchant' status levels and were discriminated against as pariahs and outcasts ('*eta hinin*'). After the Restoration this term was abolished under the 'Liberation Edict' and the people were entered in family registers as 'new commoners', but in practice discrimination continued unchanged. When these people engaged upon a *buraku* improvement movement in an effort to eliminate the causes of discrimination from within their own communities they found that however much they exerted themselves in self-improvement the discriminatory attitudes of society at large and the injustices of the bureaucracy did not change. The *buraku* masses, deeply disappointed, gradually turned a sharp eye upon the contradictions of society, whereupon the government, fearful of a radicalization of the movement, quickly set about joint official-private *buraku* improvement projects, propagandized that the 'Liberation Edict' was a fruit of the Meiji emperor's benevolence, and organized 'assimilation associations' one after another.

When the mass movement erupted at the time of the Rice Riots, socialism revived, and the *buraku* people were influenced as a result; they saw through the deceit of the *buraku* improvement policies, issued their own declaration of human rights, and in 1922 set up the National Levellers' Society (*suiheisha*), thus commencing, by their own efforts, a fundamental liberation movement. This movement spread like a prairie fire as it repeatedly issued fierce denunciations of discrimination. Within the space of just over a year over 300 branches were set up throughout the country. However the movement, because of its closed character, the violence of its tactics, and its inclination to class struggle, invited the intervention of the authorities and gradually became distanced both from the people in general and from the *buraku* masses.

In 1925 the government set up a 'Central Assimilation Enterprise Society' (president: Hiranuma Kiichirō, later Prime Minister) in an effort to unite all the assimilation organizations throughout the country. As these soft policies permeated the people of the *buraku* they came in due course to be mobilized within the war support system. In 1937 one of these, the Levellers' Society, came out in support of the aggressive war on China.

It was because of the weakness of the National Levellers' Society towards *tennōsei* (imperial absolutism) and nationalism that this happened. Even though it was possible to oppose the external discriminatory structures of *tennōsei*, the movement was unable to recognize or to overcome *tennōsei*'s internal sham of succour and relief. *Tennōsei* was a total structure in which external and internal aspects of control and 'succour' complemented each other. It was this internal structure of 'universal benevolence' and 'limitless embrace' in particular that enjoyed the support of the people. Those at the bottom of Japanese society and beyond the reach of any helping hand, in the extremity of their despair not surprisingly came to rest their illusions of succour upon the benevolent and charismatic emperor who was located at the pinnacle of Japanese society. When the government pursued its assimilation policies in the name of the emperor, the Levellers' Society leadership should have recognized that the desire for succour on the part of the people at the bottom of society would be swayed. But not only the *suiheisha* failed to grasp this point; the same is true of the Marxists in the 1930s. Neither was able to achieve an overall grasp of *tennōsei* as a psychological structure. This was one of the causes of the frustration and the reversals suffered by the *buraku* liberation movement during the fifteen years' war.

(iii) 1931–1945. During the fifteen years' war period neither the popular movements of the left – the anti-military and anti-war movements and the various proletarian movements associated with them – nor the movements of the right – the Showa Restoration movement – developed into broadly based popular movements. They did not mobilize the masses like the Nazi party did, and eventually they were absorbed in the Imperial Rule Assistance Association in 1940. Needless to say, during World War Two even the most trivial of popular movements was forbidden or suppressed under the Peace Preservation Law. (Immediately after defeat in the war approximately 3,000 people who had been imprisoned under this law were released.)

3 1945–1985: the period of post-war democracy. The pre-war absolutist emperor system and militarist structures were completely transformed with the introduction of post-war democratic reforms as Japan experienced defeat and occupation by the allies. Popular movements were for the first time able to develop freely through the whole of Japanese society under the new constitution which pledged respect for basic human rights, popular sovereignty, rejection of war, and international reconciliation. The post-war abolition of repressive laws, educational reform, land reform, and the promotion of the labour movement, stimulated various popular movements in response. However, dominance by the (American) GHQ and by the political party leadership was strong in the early stages, and it was only in the late 1950s that really independent popular movements emerged.

The largest of these was the signature movement for the banning of atomic and hydrogen bombs which was started by the grass-roots popular movement after the hydrogen bomb tests at Bikini atoll. Next came the movement of opposition to the Japan–U.S. Security Treaty, the *Ampo* struggle, which reached a peak in June 1960. Both of these became extremely complex in character and developed into long-lasting movements involving millions and even tens of millions of people and became linked with political parties, labour unions, religious organizations, women's organizations, and various other people's movement organizations. Consequently it seems not appropriate to describe them just as popular movements.

Here I should like to summarize the characteristics of contemporary Japanese popular movements by taking up the cases of *Beheiren* for the peace movement and the Minamata struggle for the residents' movement.

Apart from these, the movement in the 1970s for environmental protection and the reformist local self-government movement, as well as the movement of the Buraku Liberation League which revived after the war, the new Womens' Movement and the Consumers' Movement should not be neglected, but there is not enough space to deal with them here.

Beheiren, the Citizens' League for Peace in Vietnam, was born in 1965. This was an epochal event, marking a change in the character of the Japanese citizens' movement. The struggle over Minamata disease got properly established in 1968. This too was of great significance in opening new horizons for local residents' movements in Japan. During the time of high economic growth both of these movements developed a sharp critique of the deformed structures of Japanese society and of the original sins of 'modern' war and 'modern' industry.

From its inception *Beheiren* was not a selfish Japanese movement. It was not just an anti-war movement of protest against the horrors of war from the side of the victims, but a spontaneous movement in which, recognizing the role of the Japanese state (and their own role) in support of the Vietnam war, people chose to try to stop the war. In this sense it transcended the level of thinking of the movement for the banning of atomic and hydrogen bombs which had been undertaken by the Japanese people as nuclear bomb victims. It was able to adopt the perspective of the people of the Third World and to look back at Japan from it. As an autonomous citizens' organization completely independent of established political parties and democratic organizations, rich in an internationalism suited to the information society, and generating various new styles of movement, *Beheiren* exercised a considerable influence over other popular movements. As its symbolic leader, the young radical liberal, Oda Makoto, spelt out the Beheiren position in easy-to-understand terms. A free citizen of the world, with absolutely no local connections, he was able to draw to the movement Japanese youth who were in the same position. Oda provides a nice contrast with Ishimure Michiko, the local Minamata novelist who emerged as chronicler of the Minamata disease. Exercising a different kind of attraction, she too was able to draw youth from all over Japan to Minamata.

The movement in protest against Minamata disease was another completely independent organization which kept free of control by established political parties and the like and, by sticking closely to the Minamata-based movement of the sufferers from the disease and persisting in their support, like *Beheiren* was able to generate dozens of groups with the same name and with over 10,000 members all over the country. It played a considerable role in influencing public opinion.

In the Minamata struggle, this support organization developed jointly with the movement in which the sufferers themselves demanded compensation. However, since the principle of sufferers as masters, supporters as followers, was adhered to strictly, the Minamata movement was able to develop along proper lines of democratic autonomy. There were some groups which reversed this relationship, calling themselves a 'vanguard', but they did not effect any substantive improvement in the sufferers' movement.[3]

While Beheiren grieved over the thousands or millions of dead and wounded in the battlefields of Vietnam, and developed strong international links, the Minamata movement, while exerting itself to the utmost for the succour of the Japanese sufferers from the disease, whose numbers rose from 2,000 to 20,000, also shared their concern with the victims suffering from the same methyl mercury poisoning among the Canadian Indians or in South-East Asia, and exchanged delegations with them. Both movements, through establishing a way of thinking by which the people themselves stretched out helping hands to each other at a level which transcended states, broke through one of the barriers that has been constraining Japanese popular movements to this time.

Furthermore both movements, stressing the activity of learning but distrusting the establishment's specialist scholars, took the initiative in setting about specialist studies themselves. In the case of Minamata, members of the organization themselves set up in the local city of Kumamoto the Minamata Disease Research Society, a high level research organization which over the past twenty years has gathered an enormous quantity of data and carried out continuous research. Also the '*jishu kōza*' ('independent lectures') group which produced 'Basic Theory of *Kōgai*', under the leadership of Ui Jun, an assistant in the Engineering Department of Tokyo University, performed an important role as information and communication centre for the anti-pollution movement throughout the country for fifteen years from 1970 to 1985. These various efforts produced an advantageous result from the many pollution-related court cases and caused the Diet and the Government to adopt a series of pollution-related laws.

I have referred elsewhere to the difference in the character of these various movements.[4] Popular movements like Beheiren which are made up mainly of liberal urban people and which I call 'citizens' movements by the few who are (relatively) strong' are different in character and function from those popular movements which take up local environmental problems such as Minamata, which I call 'residents' movements by the many who are (relatively) weak'. In my view mutual cooperation between them is now necessary to construct a new movement theory. I warned that if the residents' movements failed to construct a nation-wide network of cooperation and action within a few years they would almost certainly be rolled back by the government, their policies preempted, and the movements themselves brought to the brink of collapse. Since then the situation has developed as I predicted. In the early 1970s the residents' movement was flourishing and there were some three thousand such movements throughout Japan; now only a few dozen survive and the movement is in the doldrums. This is partly due to the responses that have been developed by business and government, but it is also a fact that the weaknesses and disadvantages of a minority movement have been responsible. In 1982 there was a brief burst of activity on the part of Japan's anti-nuclear movement on the occasion of the United Nations special session on disarmament, but this was not necessarily to be seen as a grass-roots movement. After the advent of the Nakasone government there was a lull. Recently, however, deep concern in the citizens' movement over the extraordinary advances of militarization and the rightward shift of society has led to a recovery; spearheaded by the local government anti-nuclear declaration movement, there is a process of internal reorganization of the movement going on, and there are signs of a new peak approaching.

From the above summary of modern Japanese popular movements what characteristics can be drawn?

Of course if one looks at these popular movements in detail there are differences in the problems they faced and in their objectives, in the social circumstances surrounding them, and in the character of the authorities they had to resist. It would be quite unscholarly to ignore these differences of objective conditions and stress only the similarities. However I believe it is meaningful to take a broad view of how the people have responded, reacted to or resisted the government-led process of modernization since the Meiji Restoration.

These popular movements were, according to their period, influenced by the ideas of the Liberty and Peoples' Rights movement, democracy (*minpon-shugi*), socialism, fascism, and post-war democracy, but were not inspired just by such influences. They had their own ideas as popular movements and their own basis of legitimacy. This leads to the question: 'What are the intrinsic ideas of Japanese popular movements?'

1   By comparison with other neighbouring Asian countries, religion is a very minor characteristic in Japanese movements. Compared with the Taiping rebellion in China (1851–1864) or the Korean Tonghak rebellion (1894), the Japanese 'world-renewing' or Chichibu affairs lacked the power to unite the broad masses of the people in a movement tran-

scending their direct material interests because they had no religious force. Of course it is clear, if we look at the *Ikkō* or Shimabara rebellions of the sixteenth and seventeenth centuries and see how the *Ikkō* sect and the Christians of the pre-modern period were able to unite large numbers of people in stubborn resistance to the military authorities, that this was not always the case.

From the Edo period such religions were severely proscribed, and in more than 3,000 peasant rebellions and world renewing disturbances no religious element capable of serving as an ideological catalyst to develop a nation-wide peasant war can be detected. Some would see this as a positive element, and say that it is precisely this absence of religion which is evidence of Japan's modernization; others see it as a negative factor, saying that this prevented the Japanese people from developing a world view and blocked the expansion of the democratic movement.

2   Japanese people's movements generated their leaders from within, and the ultimate martyrdom of those leaders made possible the continuity of their tradition. This charismatic leader emerges as the movement develops from origins through high point, to decline and end. The resentments, aspirations, and various emotions of the people are entrusted to this leader, who combines them all as the symbolic embodiment of the struggle. When the people of the village were faced with suppression and forced to retreat, they would call upon their leaders to die as martyrs, in order to save the lives of the rest of the people. But the care of the surviving family of such leaders and care for their graves then becomes a common task. Those who are sacrificed enter into folklore as popular heroes and this makes it possible for the movement to be revived in a later age.

This is a characteristic which can be seen through the peasant uprisings, world renewal and Chichibu movements, and the Ashio anti-pollution movement. This popular logic is well understood on the part of the suppressors too. The pattern of severe punishment of the charismatic leader only, combined with leniency to the participating masses, is well established. The leader too follows the traditional way, taking his stand at the head of the movement determined on such a fate from the beginning, with his posthumous name emblazoned on his headband.

This is the dramatic ritual, and also the aesthetic, of the popular movement. The classic examples range from Amakusa Shirō of the Shimabara rebellion, the hundreds of martyrs of the peasants' rebellions headed by Sakura Sōgo, the leaders of the Chichibu affair Tashiro Eisuke and Sakamoto Sōsaku, to Tanaka Shōzō of the Ashio mine pollution case. Itagaki Taisuke, president of the Liberal Party, is famous for having cried out, when attacked by an assassin in 1882: 'Though Itagaki die liberty shall not die.' Had he then obliged by dying he might have been endowed with charisma as a 'god of liberty' and the Liberty and Peoples' Rights movement have reached new heights.

Although the notion of charisma has been weakened, this tradition has been handed down to the present day. Those who died in various struggles – Kanba Michiko of the Security Treaty struggle in 1960, Kamimura Tomoko of Minamata struggle, Ōki Yone of Sanrizuka struggle – are all turned into martyrs and live on as a source of inspiration for the participating people. The study of popular movements as folklore has only recently begun.

3   It is a particular kind of relationship between local radicals and radicals who drift into the local community from outside that sparks off rebellion. When once the structure of contact between permanent residents and outsiders is established at a certain *ba* or place within the locality, an earth-shattering political energy and cultural creativity is produced. I noticed this in my investigations of the mountain village of Mitama, which a hundred years ago gave birth to a popular constitutional draft of extraordinarily democratic content.[5] When, under the impact of the Meiji Restoration and the People's Rights movement, the radical energy which had been built up locally in that mountain village came in contact with new information and new radical energy as radicals were admitted from other regions, an equilibrium was reached between the two forces and a creative force generated; other examples of the same phenomenon have been confirmed. (What is referred to in studies of nuclear physics as a magnetic field is hinted at in this theoretical model.)

Many other examples of this phenomenon can be observed in the Chichibu affair and the Ashio mine pollution affair. In the villages which constituted the core of these movements there are many cases where the feelings of local residents were brought into contact with and then fused with those of a floating population. Village people only stirred themselves to rebel after some peasant leaders, possessed of a traditional morality, left the village for a while, became outlaws, and then returned and were incorporated within bands of gamblers and ruffians.

In present-day terms this can be seen in the form of the close relationship, which is evident at Sanrizuka, Minamata, and in a majority of the anti-pollution movements between local residents who are involved, who have suffered in one way or another and who constitute the 'subjectivity of the movement', and the volunteer groups who came from outside as 'supporters'. The fusion between the League of Minamata Disease Sufferers and the Society for the Prosecution of the Minamata

Disease is one example. Even now the popular movement advances through the relationship of tension and cooperation between two different groups.

4   It is characteristic of Japanese popular movements that they give primacy to morality over ideology; they are strong when their fight is rooted in local solidarity, and when it is not they are prone to split, to defections, or to total collapse. When the authorities belittle the movement's conviction that what counts above all is that morality of self-identity which is rooted in the relationships of the local community, and when they belittle the movement's sense of compassion and justice, then popular resentment deepens and becomes a powerful force. Whenever the people proclaim their legitimacy to the state or to business leaders it is always this morality rather than theory that they stress, because they have staked their livelihood to defend this self-identity and they believe in its universality.

The best example of this is the direct negotiations, in 1973, between the sufferers of Minamata disease and the president of Chisso Corporation.[6] The sufferers made repeated efforts to get the president as a human being to declare his commitment to righteousness and compassion, and to apologize for having transgressed against them. The demand for economic recompense was secondary. Compensation was merely the partial outward expression of this moralistic apology. The ordinary people of Minamata decided that this should be so.

The uniqueness of the theory adhered to by popular movements in Japan can be seen in this attitude. Such a theory is quite distinct from the theory of modern bourgeois society or capitalist society. Furthermore, when combined with a transcendent notion such as 'It is heaven's will that the world be renewed (or that inequality be eliminated from the world)', or, with the idea that 'more than country, more than anything, it is the people itself that matters', such a theory becomes the basic principle sustaining determined action. Without it, frontal armed uprisings against such a powerful *tennō* state would have been impossible.

When Tanaka Shōzō proclaimed that the country that could flood and obliterate the village of Yanaka was already 'finished as a country' and the Japanese people 'the people of a ruined country', he was referring to the state-transcending idea that the people themselves are the foundation of government and the source of worth. His words must have deeply moved the hearts of the tens of thousands of farmers living along the Watarase river who had dropped out of the movement. They felt thereafter a sense of guilt for having abandoned Shōzō, and when Shōzō died while travelling around on behalf of the movement tens of thousands lined up in front of his remains. Fifty years later Tanaka Shōzō's words 'I declare there to be injustice in Kanto', inscribed in large letters on straw mat banners, were unfurled over the plains of Sanrizuka.

Furthermore, while it might seem obvious to point it out, popular movements are strong when they are able to fight on the basis of local community solidarity, and when they cannot they are easily divided and defeated. This can be seen in pre-modern times too. Of course what constituted this community may have been a fiction, having already more-or-less lost its economic reality; still it had power over local residents. This was well understood by the establishment, which skilfully exploited the weak point of the popular movement.

5   Japanese popular movements are weak in the face of *tennōsei* and nationalism, have little sense of internationalism, and have failed to develop their thinking so as to transcend the notion of the state, by the idea of a place of refuge abroad (in times of extreme adversity) for example. The weakness of Japanese popular movements towards *tennōsei* and nationalism was mentioned earlier in the discussion of the pre-war *buraku* liberation movement. This tendency to lapse into national chauvinism is partly due to the fact that the Japanese people, as the 'single race' of an island country, having experienced a long period of isolation from the world, find it difficult to look at themselves objectively. Furthermore, the long survival of this consciousness of community solidarity made it possible for the Japanese state to be seen as a single family entity under the emperor as overall household head (the family state view) and for people to have the feeling of 'common destiny'.

For this reason, when popular movements encountered the frenzy of public opinion over an aggressive war or an imperial rescript, their weakness was exposed and they suddenly lost strength. This is what happened with the Liberty and Peoples' Rights movement and the Ashio anti-pollution movement.[7] A similar course was followed by the anti-war movement at the time of the Russo-Japanese and Sino-Japanese wars. And the way in which the leadership of the communist and labour movement, which in the 1930s and 1940s was struggling fiercely against state power and being vociferously anti-imperialist, was converted en masse and absorbed within the structures of cooperation during the development of the Pacific War, demonstrates the point well. 'Country', 'home village', 'state', 'people', and 'homeland' are all mixed up and difficult for the Japanese popular movement to distinguish clearly; consequently it has had great difficulty in developing a notion of refuge or an ideology that transcends the state. For this reason pre-war popular movements virtually never developed international connections.

6   Even though the popular movement suffers complete political and military defeat, so long as it does not suffer moral or ideological defeat its tradition will revive and be born again. It is obvious that the conditions for victory by the people's forces, against the combined weight of the forces of a well-equipped centralized state bent on their suppression, were virtually non-existent. To the extent that this is so the frustration and defeat of the popular movement was unavoidable. But the political, economic, military (violent) defeats of the people's struggle did not amount to a real defeat.

A really decisive defeat of the popular battle would only occur when pride and belief in their legitimacy was lost in relation to the very matters for which the people were battling, in other words, in the event of a moral or ideological defeat. If the movement were to end in such total defeat the tradition itself would be buried beyond hope of revival. The task of breathing new life back into it could only be accomplished through ideological awakening and a process of disinterment by the people of a later age, with the cooperation of historians.

It is partly because there was no place where the experience of popular struggle could be preserved after such defeats, where it could be grasped in terms of ideas and located in the tradition, that so many popular movements in early modern Japan were buried in this way.

For example, had the experience of armed uprising of the Chichibu farmers been embraced and preserved by the left-wing peoples' rights ideologists of the time like Nakae Chōmin, Ōi Kentarō, or Ueki Emori, and had that experience been grasped in terms of ideas, the Chichibu affair would have become tradition, and the subsequent movements of the Japanese people been so much richer.

It is worth noting that the post-war *Buraku* Liberation League, perhaps because of reflection on this point, supported thorough research and scholarship into the history of discrimination, poured a considerable amount of money and effort into theorizing the movement, maintained a fine *Buraku* Liberation Institute, whose reports it published, and held regular annual congresses. In the anti-pollution movement too, it is important that specialists from various fields – such as the executive committee of *Jishu Kōza* (Autonomous Lectures) – have acted as receptacle for the movement and have been active in the task of theorizing its experience and providing and disseminating both within the country and abroad the latest information from their accumulated researches.

7   A perspective on how to overcome the weak points of the popular movement to date has thus been developed within the contemporary movement, and new possibilities have been opened.

One evidence of this is the effort that the movement itself has devoted to the creation of a new fundamental principle in place of the religious character which in the past served to create spiritual interrelatedness on a broad scale among the people. (Thus at its inception in 1980 the ' "Japan – Is-This-How-We-Want-It-To-Be?" Citizens' League' issued a draft 'New Citizens' Declaration'.) And in the place of charismatic leaders, new leaders have begun to emerge within the movement who combine in their character a broad and correct international consciousness, scientific rationality and popular morality.

It goes without saying that the subject and principal of the popular movement is, and must always be, the people. In order for this to be the case support groups must adhere to the role of 'disciples'. Only when this rule is followed can a creative, cooperative relationship, based on the indigenous sentiment of the people, be developed between permanent residents and the drifters, and the boundless radical energy of the people be tapped.

The task that the vanguard popular movement of today faces is to overcome ideological weaknesses such as in relation to *tennōsei*, nationalism, dogmatism, pseudo-community, familism, the tendency to toady to the great and powerful and to despise the masses, and discrimination. Since the 1960s a real possibility of achieving these goals has been opened up. In other words, I believe that, sustained by its long historical experience, and despite the many weaknesses outlined above, the Japanese people's movement is now in the process of creating the possibility of a breakthrough to the future.

It is true that, as of 1985, we are far from the peaks of the people's movement that were reached around 1970. This fact is closely related to the attitudes to life that we have developed in recent years since Japan came to monopolize so much of the wealth of Asia. So many of us Japanese do not realize that our enjoyment of a materially rich livelihood creates many rejects at the bottom of society and is based on the sacrifices of more than half the world's people that live in poor and developing countries. The Japanese people, and their leaders, lack a correct understanding or response to the deceptions and tricks by which this happens. So long as Japan as a whole exists in the form of a 'Great Japan Company Limited' which stands on the 'false bottom' of the weak, and so long as we try to maintain our own prosperity only, it is obvious that no movement for radical change of the status quo can emerge. The people's movement cannot develop in future without a global consciousness and a perspective of self-denial, together with an appreciation of indigenous sentiment.

Although only a tiny force, we have a citizens' movement whose consciousness and activity is global in scope. There are various movements with global perspective, such as the Asian Environmental Conference (president: Ui Jun) Pacific-Asia Resource Centre (PARC) which continues to publish the quarterly journal *Sekai kara* (Mutō Ichiyō and others), the 'Community Movement for a Nuclear-free, War-free World'; and the movement to develop local government nuclear-free declarations on a world-wide scale. The 'Japan-Is-This-How-We-Want-It-To-Be? Citizens' League', of which I am a member, is one of these movements.

These citizens' movements, with the exception of the Hiroshima-based movements calling for an end to nuclear weapons, do not yet enjoy the support of large numbers of the people, but there is no doubt that they are preparing for a new upsurge, and that they are ready for it.

It is the people who make history, but not the people by themselves, because the people possess a naturally conservative bent. The decisive role in moving history is played by the minority of low level leaders and groups which self-consciously are one step ahead of the people but do not separate themselves from the people. The people – the principal force in the creation of history – will never be able to emerge as a giant in the realm of politics unless the minority movement persists in preparing for radical change.

NOTES

1 Best known of the popular movements of this period of approx. 1840–60 are the 'peasant rebellions', the 'world renewing' disturbances, which were more profound in character and developed on a larger scale, and the popular new religions. On these, see Katsumata Shizuo, *Ikki* (Rebellions), Iwanami shinsho, 1979; Sasaki Junnosuke, *Yonaoshi* (World renewing), Iwanami shinsho, 1979; Yasumaru Yoshio, *Nihon no kindaika to minshū shisō* (Popular thought and Japan's modernization), Aoki shoten, 1973, and *Kamigami no Meiji ishin*, (The Gods' Meiji Restoration), Iwanami shinsho, 1979; Murakami Shigeyoshi, *Kindai minshū shūkyōshi no kenkyū* (Studies in the history of modern popular religions), Hōzōkan, 1963.

2 In 1968 I was one of a group that discovered an outstandingly democratic draft constitution, drawn up in 1881 by the people themselves, at the village of Fukasawa, in the mountains of the Tama region, about 50 kilometres from Tokyo. Academic circles were astounded at this example of grassroots popular movement when we published an introduction and analysis of this constitution (Irokawa, *et al.*, *Minshū kenpō no sōzo* (The creation of a popular constitution), Hyōronsha, 1970). The complete set of relevant materials has been published in Irokawa Daikichi (ed.), *Santama jiyū minken shiryōshū* (Collected materials on liberty and people's rights in Santama), 2 vols. Yamato shobō, 1978.

3 Together with Hiroshima, Minamata is known to the world for the calamity that befell it, but the nature of the movement that developed there is not well known. There are many reports on the Minamata struggle, of which some of the most easily accessible, comprehensive, and best are: Harada Masazumi, *Minamata-byō ni manabu tabi* (Journey of learning from Minamata disease), Nihon hyoronsha, 1985, Irokawa Daikichi (ed.), *Minamata no keiji* (Minamata inspiration), 2 vols. Chikuma shobō, 1983. On the reality of the actual Minamata disease there is nothing to match the collection of photographs by the American cameraman, Eugene Smith: W. E. Smith, *Minamata*, New York, 1975.

4 See my article in *Ushio*, September 1973.

5 The Chichibu Incident was an armed uprising which occurred in 1884 when several thousand poor farmers from the Chichibu region in Saitama prefecture linked up with some of the Liberal Party's grass roots elements under the leadership of Tashiro Eisuke. Their first demands had been for a reduction of debts and taxes, but these petitions were repeatedly denied. As a last resort they organized the villages and rebelled, captured the whole Chichibu district and clashed with the army and police. After ten days they were routed. Many are the tales of heroes of the people which stem from this period. The question of whether this incident should be regarded as the brightest flower blooming at the end of the Liberty and People's Rights Movement, or as a final culmination of present rebellions, is currently debated. The most readily available and best book on it is Inoue Kōji, *Chichibu jiken* (The Chichibu affair), revised edition, Chūkō shinsho, 1968. The first three volumes of a six volume set of documents edited by Inoue and others, *Chichibu jiken shiryō shūsei* (Compilation of historical materials on the Chichibu affair), have so far been published by Nigensha.

6 See Ishimure Michiko, *Ame no sakana* (Fish in the heaven), Tokyo, Kōdansha, 1980.

7 From the 1880s poisonous effluent from the Ashio copper mine, one of Japan's most important mines, polluted the Watarase River and brought misery to 200,000 people who lived along the banks of the river. Protest has continued, if intermittently, through the hundred years since then. The leader of this anti-pollution movement was Tanaka Shōzō, a local member of parliament from Ibaragi prefecture. Tanaka had originally been an advocate of Liberty and People's Rights, who firmly believed that the welfare of the people was the first principle of politics. He threw himself totally into the struggle against the government, and died while active on behalf of the movement. He was seen as a modern Sakura Sōgo, and revered as a charismatic figure by Japanese farmers. The best study of him is Hayashi Takeji, *Tanaka Shōzō no shōgai* (The life of Tanaka Shōzō), Kodan shinsho, 1976, and on the sufferings of the farming people see Tamura Norio, *Kōdoku nōmin no monogatari* (Tales of the farming people sufferers from mining pollution), Asahi shinbunsha, 1975. *Tanaka Shōzō zenshū* (The complete works of Tanaka Shōzō) has been published by Iwanami shoten. [Trans note: In English see also Kenneth Strong, *Ox Against the Storm: A Biography of Tanaka Shōzō, Japan's Conservationist Pioneer*, London, 1977.])

5 For self and society: Seno'o Girō and Buddhist socialism in the post-war Japanese peace movement

STEPHEN S. LARGE

Yukichi, in Osaragi Jirō's novel, *Homecoming*, says this about his hopes for the future after Japan's defeat in World War Two: 'Now we've been relieved of the burden of history. If we can make up our minds to march forward completely empty-handed, unhindered by anything from the past, something really new will come to life here in Japan.'[1] Many Japanese shared these expectations of a 'new Japan' after the war, but they soon found that the 'burden of history' remained and that in significant ways, the 'new Japan' was not so new. This was the experience of the Nichiren priest, pacifist, and Buddhist socialist, Seno'o Girō (1889–1961). Seno'o, along with others in the post-war Japanese peace movement, discovered that his earlier struggles against pre-war fascism and militarism had to be sustained if peace and democracy were to prevail in Japan after 1945.

Seno'o is seldom mentioned in Western scholarship on modern Japan.[2] However, much has been written about him in Japanese. Tokoro Shigemoto, for example, locates Seno'o in a long tradition, going back to Meiji, of an intellectual and political confluence of Buddhism and socialism comparable in many ways to that of Christian socialism. In Meiji, this tradition was epitomized by the zen priest Uchiyama Gudō, who was executed with Kōtoku Shūsui in the High Treason Case of 1911. For Tokoro, Seno'o embodied the same tradition of Buddhist socialism in Taishō and Shōwa.[3]

In particular, Seno'o figures prominently in Japanese accounts of the unsuccessful pre-war popular front (*jinmin sensen*) movement of the mid-1930s when Seno'o cooperated significantly with Katō Kanjū, Suzuki Mosaburō, and Takano Minoru in efforts to unify socialist and communist protest against fascism and militarism.[4] The mixture of religious and political ideas underlying Seno'o's radicalism in the popular front and his participation later in the post-war peace movement has intrigued many Japanese writers. Historians such as Ienaga Saburō and Shimane Kioyoshi see him as an exceptional and original thinker in the intellectual history of modern Japanese socialism due to his attempts, dating from early Shōwa, to combine Buddhism and Marxism into a potent religious and secular ideology of liberation. Shimane calls Seno'o a 'revolutionary thinker' who emphasized the social revolutionary character of classical Buddhism and its synthesis with modern Marxism.[5]

This portrait of Seno'o Girō in the post-war Japanese peace movement has two broad purposes. The first is to provide perspective on the intriguing but little-known tradition of Buddhist socialism in modern Japanese social protest. Seno'o's career was governed by a fundamental question of importance to many politically engaged Japanese Buddhists: what is the relationship between the private quest for the ideal self in Buddhist religious terms and the public quest for the ideal society in socialist political terms? His approach to this problem, which demanded clarification of the subtle and intricate linkages between individualism and the politics of collective protest, was that of the 'prophet-reformer' who believed that the achievement of peace and democracy was not just a political undertaking but also involved a profound moral transformation of self and society based on universal ethical values.

What this transformation entailed for Seno'o is suggested by Fujita Shōzō's general observation about social movements. Fujita writes that as a social movement engenders universal values under a 'prophetic leadership',

these values impinge on the inner man and alter his attitudes. The movement thus develops on the basis of its ability to 'convert' people to its values and thus produces a new inner order. In seeking to express itself in concrete external form, a renovated inner order in turn creates a new social order . . . In this way, there is a fundamental rejuvenation of the society from its deepest, most inner point: man's heart.[6]

How Seno'o endeavoured to apply his religious ideals for a moral transformation of self and society to the political realities of the post-war peace movement, and with what consequences, are the principal concerns of this study.

The second purpose is to investigate the ways in which their experiences as members of pre-war resistance movements against fascism and militarism shaped the thought and behaviour of people like Seno'o once they became active in the post-war peace movement. Seno'o is an interesting case study for this inquiry because he left behind a sensitively written diary, recently published in seven volumes, which reveals much about his psychological and political disposition as he made the difficult passage from pre-war to post-war Japan. Accordingly, the diary is used extensively in the paper at hand.

After the war, Seno'o was optimistic about the prospects for peace and democracy in Japan. The left, including labour and proletarian party elements, was emerging as a strong force for peace and democracy in the fairly tolerant climate of the early occupation. As well, Japan's defeat had vindicated the anti-fascist and anti-militarist goals of the pre-war popular front in which he had been a major participant and the way now seemed clear, with the wartime political structure in ruins, for Japan to embark on a new, pacifist, course. These hopes conformed with Seno'o's Buddhist socialist outlook which continued to shape his idealism as it had in the 1930s.

However, there was another, more personal, impetus behind Seno'o's hopes for peace. This was an urgent determination to atone for his role in the failure of the pre-war popular front movement, by helping to re-create and sustain a new popular front after the war which would have as its purpose the preservation of peace. In seeking atonement, the 'burden of history' weighed heavily on Seno'o. To appreciate why this was so, it is necessary to briefly examine religion and politics in Seno'o's career before the war.[7]

A chronic pulmonary ailment, which had cut short his early education, prompted Seno'o's conversion to the Nichiren faith in 1911 and his decision in 1915 to become a Nichiren priest. Nichiren Buddhism attracted him as a means of overcoming physical weakness which he associated with weakness of moral character and a selfish nature. He believed that through the attainment of *muga-ai* (selfless love) within the heart, one's physical well-being and the moral quality of one's *jinkaku* (personality) could be strengthened and that this self-strengthening would ensure a capacity, characteristic of the bodhisattva ideal, to nurture *muga-ai* in the lives of other individuals and in society, generally.

Accordingly, Seno'o practised austerities to strengthen his moral character and physical health, including the chanting of sutras before dawn and late at night, abstaining from sexual intercourse for long periods, and cleaning the toilets of his home to cultivate humility and service to others. With 'Butsuda o seoite gaitō e' (Bear the Buddha into the streets) as his motto, Seno'o felt compelled to communicate his faith to the Buddhist world. During the 1920s, his ability to do so in simple eloquence and his erudite command of Buddhist scriptures made him a well-known evangelist in the Buddhist student movement. Some of his admirers called him 'today's Nichiren'.

However, 'Bear the Buddha into the streets' soon acquired a new, political, thrust in the late 1920s when Seno'o, increasingly conscious of the great disparities under Japanese capitalism between the privileged and the poor, found himself attracted to Marxism after reading Kawakami

Hajime's *Bimbō monogatari*, or Tale of the Poor, and other critical works
on capitalism by Yamakawa Hitoshi, Takabatake Motoyuki, and similar
writers. Before long, Seno'o became convinced that for Buddhism to assist
the poor and powerless, it had to be combined with Marxism. The result
was his synthesis of Buddhism and Marxism, which will be discussed
shortly. The point here is that by the early 1930s, Seno'o Girō had coupled
his earlier religious faith with a political commitment to proletarian
liberation.

Believing that 'The value of one's faith must be seen in what one does'[8]
politically, in 1931 he founded the Shinkō Bukkyō Seinen Dōmei (New
Buddhist Youth Federation) which attracted progressive young Buddhists
from all sects who shared Seno'o's desire to apply Buddhism to proletarian liberation in the darkening context of the Manchurian Incident,
terrorism within Japan, and the growth of Japanese military power.
Besides criticizing the established Buddhist Church for remaining com-
placent about these developments and for generally supporting Emperor
and Holy War in Asia, Seno'o and the Seinen Dōmei became active in
protests against Japanese militarism and fascism. For example, Seno'o
took part in many anti-war labour strikes, joined the Han-Nachisu Fassho
Dōmei (The Anti-Nazi League to Crush Fascism), and as editor of *Rōdō
Zasshi* (Labour Magazine) was instrumental in the political mobilization
of the Japanese popular front movement. For these endeavours, Seno'o
was arrested and charged with treason in December, 1936, whereupon he
was faced with precisely the test of moral character for which he had
prepared himself earlier as a devotee of Nichiren.

At first, Seno'o denied false police accusations that he had plotted the
revolutionary destruction of the Emperor system and capitalism. He
insisted that socialist reform of capitalism, resistance to fascism, and
above all to militarism and war, had been his goals and those of the Seinen
Dōmei. But under great harassment from his interrogators, Seno'o then
suffered a collapse of will and ended up confessing that the charges against
him and the Seinen Dōmei were all valid and that henceforth he would
fully support Emperor and nation. His *tenkō* (conversion) was then used
in 1937 to justify police repression of the Seinen Dōmei, as part of an
escalating crackdown on the popular front movement as a whole. Seno'o
himself was sentenced to three years in jail, his confession notwithstanding. After his release, he spent the duration of World War Two in quiet
seclusion.

Seno'o had been aware of the risks of taking part in leftist protest
movements. But he had been confident of having sufficient faith and
strength of character to pass any tests that might eventuate. In 1934, while
reflecting on Sano Manabu's famous *tenkō* the preceding year, he vowed

in his diary, 'I shall never tenkō.'[9] And in early 1936, he had written, 'There is no righteousness without adversity and through adversity our faith is strengthened.'[10] Given these sentiments, it is scarcely surprising that in 1939 he would condemn his own *tenkō* as 'the inexcusable act' of a 'crystal man of weak character' who had completely betrayed both his own moral integrity and the entire popular front movement out of a craven desire to save himself from persecution.[11] Thus, the guilt he felt over the moral and political failure represented by his *tenkō* constituted Seno'o's personal 'burden of history' which he carried into the post-war period when he looked to the peace movement for expiation. Since Buddhist socialism continued to be central to his own self-recreation as a seeker (*kyūdōsha*) of *muga-ai*, and to the pacifist reconstruction of post-war Japan, it is well to consider briefly his ideas concerning war, peace, and social change which were grounded in his adaptation of Buddhist and Marxist concepts.[12]

Seno'o traced the root causes of war and all other irrational forms of exploitation, repression, and suffering to the individual and collective selfishness of the ego in the spiritual realm of mankind and to inequalities of wealth and power in man's material circumstances. These spiritual and material roots of war had existed throughout history but they were especially characteristic of modern capitalism, a system which sanctified the selfish pursuit of wealth and power.

From this analysis, it followed that the problem of war required both spiritual or religious and secular or political solutions. Specifically, he held that a lasting peace depended upon victory over selfishness in human nature and the replacement of capitalism with socialism, the only system wherein man could achieve the ideal self and society of *muga-ai*. In combination, Buddhism and Marxism offered complementary spiritual and material means to these greater ends, with Buddhism operating as a rational, religious science of human nature and Marxism as a rational, secular science of social and political reality.

As a Nichiren priest, it was natural that Seno'o would wish to emulate Nichiren's prophetic defiance of secular authorities in proclaiming the message of *risshō ankoku* (establish the truth and pacify the country) with an emphasis on returning to correct spiritual doctrines. However, his understanding of the latter transcended Nichiren Buddhism to embrace the classical Buddhism of ancient India which Seno'o believed was universally applicable to the problems of modern society, especially to the penultimate problem of war. Of special importance to Seno'o was the classical Buddhist doctrine of *sankirei*, the 'three-fold refuge'; *kiebutsu* (reliance on the Buddha), *kiesō* (reliance on the sangha), and *kiehō* (reliance on the dharma or Buddha Law).

To Seno'o, *kiebutsu* meant reliance on the historical personage of the Buddha Shakamuni who, possessing the perfect *jinkaku* pervaded by selfless love, was the peerless model for all mankind in the universal struggle to overcome evil within the heart. Shakamuni had shown the way to salvation from sin through the doctrine of the Eight-Fold Path which enabled one to achieve the ideal Buddhist state of 'emptiness' (*kū; śūnya* in Sanskrit), signifying the absence of greed and lust emanating from the ego. In Shakamuni's teachings, and countless sutras – especially the *Muryōgikyō* (Sutra of Infinite Meaning) – Seno'o discerned the means to achieve the perfect *jinkaku* or authentic Buddhist selfhood: the rational logic of negation (*hitei no ronri* in Japanese), a spiritual dialectical process ending with the negation of selfishness. He likewise honoured Shakamuni's compassionate commandment against the taking of life (*fusesshō-kai*) as the greatest of all the Buddha's teachings. For Seno'o, to be a bodhisattva following the example of the Buddha meant being a pacifist.

Seno'o's emphasis on *kiesō* sprang from the conviction that the sangha, the classical community of monks, was the ideal model for the uncompromising application of the Buddha's teachings in social life. The sangha's chief attributes as the perfect *kyōdō shakai* (corporate society) included absolute moral selflessness, social interdependence, and a communal ownership and sharing of property for the good of all. The philosophy and world view of the sangha was the 'third refuge', *kiehō*. This encompassed the immutable laws of dependent origination, cause and effect, and impermanence and change. Together, these impelled Seno'o to believe that the wheel of historical change turned on the unfolding potential of man's capacity to save himself from sin while striving for peace and justice.

Clearly, Seno'o advocated the revival of classical Indian Buddhism to accomplish the moral revolution of man's nature, beginning with the individual and spreading throughout society, which he saw as essential to a world without war. If the Buddhist sects could unify around this reclaimed legacy of classical Buddhism and proclaim it to the world, then Buddhism could become a powerful transformative religious force for peace in contemporary Japan. However, Seno'o believed that a moral revolution by itself would mean little to the goal of peace unless Buddhism was linked with the Marxist promise of political change, through class struggle by the proletariat, leading to socialism. He was optimistic for several reasons about combining Buddhism and Marxism.

First, unlike Christianity and other theistic religions which Marxists rejected as irrational and therefore useless to liberation movements, Buddhism was atheistic and, in its logic of negation, inherently rational. Accordingly, it was possible to conjoin Buddhist and Marxist dialectics as a way of understanding and facilitating the parallel processes of spiritual and secular revolution. Secondly, like Marxism, the Buddhist concept of *busshin ichinyo* (union of mind and matter) recognized the inter-dependent relationship of the material and spiritual in human life. Here, too, was a basis for reconciling Buddhism and Marxism.[13]

Thirdly, Seno'o emphasized the similarity between the sangha and the Marxist ideal of a socialist society characterized by the absence of private property and all of the associated inequalities and injustices that spawned conflict and war. Finally, Seno'o's synthesis of Buddhism and Marxism owed much to the shared humanism which he admired in the lives of Buddhists and Marxists, including the Buddha, Nichiren, Marx, and Kawakami Hajime. All of these men were alike in seeking a just, peaceful, society. Kawakami was espcially important to Seno'o as a modern Japanese interpreter of Marxism who had demonstrated the compatibility of dialectical materialism and the Buddhist dialects of negation, as propounded in the *Muryōgikyō*.[14]

Now, it may be questioned whether intellectually Seno'o's synthesis was altogether successful in terms of Buddhist and Marxist categories of thought.[15] But he intended it to provide Buddhists with a way to relate their faith to problems in secular society, just as he hoped that his ideas would deepen the moral consciousness of Marxists engaged in class struggle. Nor was his purpose academic in nature. To reiterate, there was an unbreakable connection between his quest for authentic Buddhist selfhood and his participation in social movements on behalf of peace and justice. Only through service to an effective post-war peace movement could Seno'o hope to expiate the guilt caused by his pre-war *tenkō* and thereby prove to himself that he had indeed overcome selfishness and weakness of will.

After the war, Seno'o first plunged into the movement within Buddhism to democratize the administration of the sects and expose the complicity of some sect leaders with the aggressive policies of the wartime order.[16] But soon he turned to politics, advocating the formation of a Buddhist socialist party to implement his vision of the future. When that plan proved imprac-tical in 1946 he formed a successor to the pre-war Seinen Dōmei. This was the Buddhist Socialist Federation (Bukkyō Shakaishugi Dōmei) which in fact included many people from the Seinen Dōmei in its fold plus others who were attracted to Seno'o's ideas. The new Federation, under Seno'o's direction, quickly declared its support of the Japanese Socialist Party which Seno'o himself joined in 1949. He also joined Sōhyō (General Council of Trade Unions) the next year. The reason was straightforward: he had come to believe that the JSP and Sōhyō were potentially the backbone of a des-perately needed new popular front movement.[17]

The necessity of a popular front reminiscent of the pre-war front arose from his perception by the late 1940s that his earlier optimism about the prospects for peace had been unfounded and that political developments had now become a serious threat to peace. He was especially disturbed by the mounting repression of the left, epitomized by Supreme Commander Allied Powers' prohibition of the 1947 General Strike, the general rehabilitation of Japanese capitalism, and the rising probability of Japanese rearmament. These developments and others, including renewed respect in society for the Emperor, persuaded Seno'o that it was American policy and that of the collaborating Japanese bureaucracy and so-called 'liberals', like Yoshida Shigeru, to convert Japan into an American satellite under the same formation of capitalism, militarism, and the Emperor system which had led the country into the Pacific War.[18]

The general objective of the popular front, as Seno'o saw it in these circumstances, was to oppose militarism, capitalism, the emperor system, and Japan's cooperation with American imperialism while working for a neutral foreign policy of friendship with the People's Republic of China, Russia, and Japan's other Asian neighbours. The popular front was also his solution to the problem of maintaining cohesion in what had already become a highly ramified and ideologically diverse peace movement where it was often difficult to agree on a definition of 'peace', much less the means to achieve it. Seno'o defined his own mission in the popular front as the mobilization of religious support for the political and labour arms of the new *jinmin sensen*.

Accordingly, in April 1949 Seno'o and the Buddhist Socialist Federation were instrumental in founding and influencing the Zenkoku Bukkyō Kakushin Renmei (All-Nation Buddhist Renovation League). With Seno'o as its chairman and peace as its first priority, the Renmei was vigorous in its opposition to Japanese rearmament and the Emperor system which it regarded as the linchpin of capitalism and militarism. Besides the Buddhist Socialist Federation, this organization included over 200 groups from many sects. Among them were the Jōdoshū Minshuka Dōmei (Pure Land Democratization Federation), the Tendaishū Kakushin Dōmei (Tendai Sect Renovation Federation), the Nichirenshū Kakushin Dōmei (Nichiren Sect Renovation Federation) and so forth.[19]

The emergence of the Bukkyō Kakushin Renmei and other Buddhist groups in 1949 and thereafter indicated, as Murakami Shigeyoshi points out, that in this period 'progressive Buddhists became a major force in the peace movement'.[20] However, their impact was greatly magnified through cooperation with other religious groups and in this regard, too, Seno'o was significant. For instance, he was among the founders of the all-inclusive Nihon Shūkyōsha Heiwa Kyōgikai (Japan Religionists' Peace Conference), organized in June 1951.[21]

Notable for the fact that 'for the first time it brought about the cooperation of labour unions and political parties' and that 'it was responsible for the development of the international peace movement in Japan', the Heiwa Kyōgikai, assisted by Sōhyō, created the even larger religious peace front, the Nihon Heiwa Suishin Kaigi (Japan Peace Promotion Congress) in July 1951.[22]

Seno'o quickly emerged as a leading light in Heiwa Suishin, primarily because he could exploit close contacts in Sōhyō's unions, and he addressed the annual conventions of Sōhyō and the socialist party movement for the benefit of Heiwa Suishin. For instance, he enjoyed the personal backing of Takano Minoru, Sōhyō's chairman, with whom he had worked in the pre-war popular front. Takano became a patron of Heiwa Suishin and used his influence to have Seno'o elected as *jimukyo-kuchō* (Director) of Heiwa Suishin in 1952. This, and the fact that Seno'o sat on Sōhyō's Political Committee, ensured Sōhyō's support for Heiwa Suishin in the peace movement.[23] Similarly, he had the backing of Suzuki Mosaburō in the socialist party movement. Recalling their days together in the pre-war popular front, Suzuki consulted Seno'o often on matters pertaining to the peace movement. As well, Seno'o was appointed to a special committee in the JSP's leftwing faction to coordinate relations with Heiwa Suishin.[24]

With this type of outside, leftwing, support, Heiwa Suishin became widely known for its large-scale peace rallies, its opposition to American military bases in Japan, and similar activities in the burgeoning peace movement. Despite his advancing age as a man in his sixties, and chronic ill-health, Seno'o worked hard to strengthen cooperation between Heiwa Suishin, Sōhyō, and the socialist party ranks. He spoke at countless rallies staged by Heiwa Suishin, lectured on peace issues at many factories through the sponsorship of Sōhyō and the socialist party movement as a respected proponent of peace. He also campaigned vigorously for JSP candidates in local and national elections and on several occasions was asked to run for election to the House of Councillors, although he declined for reasons of uncertain health.[25] Throughout all of these activities, Seno'o emphasized the theme of peace and the importance of a popular front to the cause of peace.

In addition, his pacifist commitment to international goodwill led Seno'o to become active for several years as *rijichō* (Director) of two organizations which advocated full and friendly relations with Korea and China, respectively. These were the Nitchū Yūkō Kyōkai (Japan-China Friendship Association) and the Nikkan Yūkō Kyōkai (Japan-Korea

Friendship Association). Of particular importance here is his participation in two exercises of people's diplomacy undertaken by the Japan-China Friendship Association in 1953.

The first occurred in the spring when Seno'o visited China on one of several ships organized by the Association ostensibly to repatriate Japanese in China but more fundamentally to express political goodwill in the context of the Korean War. He visited China again in July on another expedition arranged by the Association, this time to return to China the ashes of 560 Chinese 'martyrs' who had perished in Japanese mines during the war. Seno'o explained to Chinese officials that the mission was a symbolic gesture of goodwill and contrition for the cruelties these people had suffered as forced labourers. He also expressed the solidarity with the Chinese of the Japanese peace movement in seeking victory over the 'reactionary forces of war', a reference to American imperialism and Japan's cooperation with it.[26]

Seno'o's trips to China in 1953 climaxed a period of intense involvement in the peace movement. After this, he relinquished his positions of leadership and withdrew increasingly to the sidelines. One reason for this was declining health. Another, more important, reason was mounting despair over the poor progress of the popular front. Seno'o had tried to assist the front by using his considerable influence in Heiwa Suishin, Sōhyō, the socialist party movement and other pro-peace organizations to weld them together into a unified front. But here, he encountered difficulties leading to great personal frustration.

Some of these difficulties may appear insignificant. Yet they upset Seno'o because they reflected the actual weakness of the front he was attempting to build. For instance, only with much reluctance did Heiwa Suishin's governing committee approve of his appointment as *rijichō* of the Japan-China Friendship Association on the grounds that the Association was too close politically to the Japanese Communist Party. Seno'o was thus alerted to, and became very anxious about, anti-communist feelings in Heiwa Suishin and their negative import for socialist-communist cooperation in the peace movement.[27]

Rather more serious in this vein was the growing tension between Sōhyō and Heiwa Suishin in 1953. Once again, the problem of Heiwa Suishin's anti-communism was the cause. Sōhyō itself was divided over whether, or to what degree, to work with the JCP. But it tended to be more open to the idea than Heiwa Suishin which, until now, Sōhyō had backed in the peace movement. The issue climaxed when Takano Minoru moved to have the office of Heiwa Suishin relocated physically to Sōhyō's headquarters where he could increase his leverage over Heiwa Suishin. When Heiwa Suishin resisted, Sōhyō's December 1953 national convention called for it to be closed down. Seno'o tried assiduously to effect a compromise, but to no avail. Deprived of Sōhyō's financial and political support, Heiwa Suishin soon fell apart, to Seno'o's disappointment.[28]

But he was the most discouraged by the worsening problem of disunity in the socialist party movement. In 1951, the JSP had split into two parties – the JSP Left and the JSP Right. Seno'o had immediately sided with the JSP Left, led by Suzuki Mosaburō's faction, because it advocated cooperation with the communists where possible, something the JSP Right, led by Nishio Suehiro, opposed. In 1953, Seno'o became a major spokesman for the JSP Left on this issue and was elected vice-chairman of the JSP Left in Nagano Prefecture as a result.[29] He was furious with the JSP Right for its rigid anti-communism, which he felt was detrimental to his vision of a popular front for peace. He later likened the JSP Right to the socialists of Weimar Germany who had refused to close ranks with the communists to resist Hitler.[30] As it turned out, the two socialist streams came together again in 1955 but by then Seno'o had no illusions about the durability of their cooperation and indeed, he had become quite disenchanted with the spectacle of socialist in-fighting. Socialist reconciliation proved short-lived, in any event, for in 1959 Nishio and the JSP Right broke away to form the rival Democratic Socialist Party. The continuing problem of relations with the communists was a major catalyst for this final separation, as one might have expected.[31]

These divisions on the left caused Seno'o much distress because they occurred at a time when a strong popular front was needed more than ever before to resist what he saw as Japan's steady drift towards full-scale remilitarization and fascism during the 1950s. He repeatedly attacked the cabinets of this period for reviving the Japanese military, as in the creation of the Self-Defence Forces in 1954. He also accused the government of making Japan a 'catspaw' of American imperialism and American anti-communism under the 1951 United States–Japan Mutual Security Treaty and the 1954 Mutual Security Agreement. Later, that Japan stood on the brink of fascism reminiscent of early Shōwa was illustrated for Seno'o by new terrorist incidents such as the assassination of the socialist, Asanuma Inejirō, in October 1960.[32] Well before then, Seno'o had concluded that radical new responses were needed if the peace movement was to survive with any hope of realizing its objectives of peace and democracy.

Politically, the radical new response Seno'o himself undertook was to embrace communism and join the JCP in January, 1960, hoping that, with the socialists in disarray, the JCP could somehow keep alive his fading dream of a popular front as Japan headed for a titanic struggle that year over the renewal of Ampo, an issue which symbolized for Seno'o the great horror of nuclear war.

It may be true as some writers claim that Seno'o's conversion to communism did not mean that he had completely abandoned his Buddhism.[33] Certainly the humanitarian, pacifist, component of his Buddhist outlook remained. Nevertheless, his diary reveals that he had become disillusioned with Buddhism as a force in the peace movement. On 2nd June 1957, he wrote that Marxism was the only effective ideology for analyzing the world and governing society in the nuclear age and that accordingly, 'I have no choice but to change my Buddhist world view.'[34] On 15th August 1957, he stated, 'I no longer see Buddhism as my ideology. There is no choice but for me to serve my brothers and sisters as a Marxist in the years left to me.'[35]

In part, Seno'o's disillusionment with Buddhism was due to the failure to mobilize the established Church behind the peace movement. Many of the so-called Buddhist 'new religions', such as Nihonzan Myōhōji, associated with the Nichiren sect, were active in the peace movement. But this scarcely compensated for the general 'silence' of the Church, as Seno'o characterized it, concerning peace issues. By 1960, he concluded that despite the presence of progressive minorities in the Church, the Church as a whole was devoid of any capacity to emulate the Buddha, revive the sangha ideal, or communicate the Buddha Law to modern Japan. The Church was a gross perversion of authentic, classical, Buddhism.[36]

But there was more to Seno'o's disillusionment than that. To put it simply, by 1960 Seno'o was consumed by an overwhelming sense of personal failure for not being able to live up to the bodhisattva ideal he had set for himself in the peace movement. It will be recalled that he had believed 'The value of one's faith must be seen in what one does' and that a man of strong faith and outstanding moral integrity would be efficacious as a modern bodhisatvva in the realm of public service. The reverse side of this peception, however, was that the value of what one does necessarily reflects the moral quality of one's *jinkaku* and that a man of weak faith and flawed moral character is identifiable by his failures in the field of public service. As Seno'o had defined it after the war, the litmus test of his personal triumph over the selfishness and weakness of moral will which had triggered his pre-war *tenkō* was success in building a strong popular front for peace in post-war Japan. This indeed was how he had sought to expiate the guilt he felt over his pre-war *tenkō*. But in his last years, he saw all too clearly that political divisions in the peace movement had sabotaged his vision of a popular front, which is why he wrote, 'All the work to which I have devoted myself has failed.'[37]

Thus did Seno'o, near the end of his life, interpret the inability of the popular front to meet his very high expectations as an undeniable sign of a flawed faith and a weak moral will on his own part, precisely because he had so personalized the search for an ethical basis of protest in the peace movement. And, the more he judged himself harshly (in retrospect, too harshly) in this way, the more overpowering his sense of unexpiated guilt became. He had always regretted his *tenkō*. But now his self-condemnation was merciless.

One entry in his diary is especially revealing. It was written in September 1957 after an interview in August with the historian Shimane Kiyoshi who had visited Seno'o while researching an essay about him that would later appear in Tsurumi Shunsuke's edited series on *tenkō*. This entry begins with a reference to Kawakami Hajime who had refused to recant his beliefs when subjected to the same kind of police harassment which Seno'o had suffered. Then, Seno'o wrote,

When I think of his unshakable convictions when he was in prison ... I am ashamed of my own *tenkō*. I should not have refused to die in prison ... My cowardice and meanness were pitiful. For that, I have lost my eternal soul. My existence is wretched.[38]

Thus, weighed down by guilt, Seno'o succumbed to a spiritual fatigue that could only be relieved by setting aside, as he did, his own 'burden of history', the now insupportable burden of 'bearing the Buddha' not only to the streets but in his inner life as well. Once unemcumbered, he achieved a new clarity of purpose in the peace movement, proclaiming with renewed optimism, 'pacifism and communism are the roads to victory'.[39] This new peace of mind, which had eluded him earlier as a seeker who could not find the Buddha Way, lasted until he died in 1961, in Matsumoto, at the age of seventy-two.

Beneath the distinction between 'pre-war' and 'post-war' Japan there are notable continuities bridging these periods. Seno'o's career discloses that this was certainly true in the case of Japanese social protest movements. The post-war peace movement needed symbols of continuity with the past in order to increase its legitimacy and enhance its public appeal. Seno'o Girō, whom Inagaki Masami aptly calls a 'populist priest', was one of those symbols because of his advocacy of a popular front to facilitate resistance to militarism and war.

In fact, Seno'o's most significant practical contribution to the post-war peace movement – the mobilization of support for a pro-peace popular front – grew directly out of his similar endeavours on behalf of the pre-war popular front. Their recognition of this continuity explains why the socialists and later the communists welcomed him to their respective banners in the movement. This was somewhat ironical. Although they professed allegiance to the idea of a popular front, their political sectarianism, which Seno'o opposed, helped to undermine the realization of this goal during the late 1940s and 1950s.

Seno'o's commitment to finding a universally applicable basis for protest in the peace movement also made him a symbol of continuity with other pre-war social movement activists who, as Tetsuo Najita observes, 'sought an intellectual faith' as a way of 'finding a principled basis upon which to act out moral convictions'. Their 'insistence on religious commitment' and 'reliance on faith in a traditional belief' in this respect was exemplified by Seno'o before and after the Pacific War.[40]

The distinctive core of Seno'o's intellectual faith was his synthesis of Buddhism and Marxism. And here, Seno'o's ideological significance in the peace movement lay in his prophetic reinterpretation of classical Buddhism to relate the fulfilment of the individual to collective social protest by linking the personal pursuit of the compassionate 'true self' as defined in Buddhism to political pursuit of an ideal secular order as expressed in Marxism. Given that Japanese Buddhism was religiously and politically conservative and largely remote from social problems, this was a compelling affirmation that within Buddhism there existed progressive potentials which offered the modern Japanese, in association with Marxism, a means to individual and social liberation.

In many ways, Buddhism and Marxism would appear to be antithetical, but not to Seno'o. Robert Tucker remarks, 'Marx sees the Communist revolution as a revolution of self-change and communism itself as a new state of the generic human self ... It is man's "regaining of self"'; Tucker adds that in Marxism, 'The enemy of human self-realization is egotistic need; the drive to own and possess things'.[41] Seno'o believed that this was also the perception of Buddhism. That Buddhism and Marxism shared this concern for the 'regaining of self' constituted for him the common ground where they could be combined in social movements. In terms of broader comparisons, Seno'o reminds one of thoughtful Buddhists in other countries – notably Burma, for example – who have likewise integrated Buddhism and Marxism in their world view.

A central feature of this world view, which is readily apparent in Seno'o's idealism, is a comprehensive concept of man wherein the compassionate potential of the individual is actualized for achieving righteousness and justice in human affairs. The importance of a balance between compassion and righteousness in any social movement, especially a peace movement the goals of which invariably encompass morality and politics, is underscored by Hidaka Rokurō, who writes:

When righteousness and compassion combine, people are moved. When compassion is absent under the flag of justice, people keep their distance, at best feeling some respect ... It is impossible to understand man in his entirety unless one understands both his capacity for righteousness and for compassion. And without understanding man in his entirety, it is difficult to find any clue to his total liberation.[42]

As a prophet of 'total liberation', Seno'o sought to prod the polarized and politically particularistic peace movement to return to the fundamental universal values inherent in its collective enterprise, but in this role he was to be disappointed. Only his closest followers in such groups as the Buddhist Socialist Federation seem to have subscribed to his views. Perhaps Seno'o's public discourses on such concepts as *sankirei* and their relevance to the politics of the peace movement were too abstract to exert much appeal to his audiences. More to the point, Seno'o stood intellectually at the then-unexplored boundary between Buddhism and Marxism and hence was peripheral to the mainstream of both currents of thought in post-war Japan. Many Buddhists may have ignored Seno'o because he was too Marxist. Many Marxists may have ignored him because he was too Buddhist.

However, if in his own day Seno'o's ideas registered only a limited impact, they have not been forgotten. It is interesting to see that since he died, Seno'o has been invoked often by Buddhists and Marxists alike who have continued the discourse he began concerning the possible conjoining of Buddhism and Marxism to help create a peaceable and just society.[43] An outstanding example of this recent discourse is found in the writings of the distinguished Buddhist scholar and philosopher, Ichikawa Hakugen. Ichikawa's 'Sūnya-Anarchism-Communism', as he characterizes it, differs from Seno'o's Buddhist socialism in some respects, but not in its fundamentals. His many references to Seno'o indicate how much Seno'o inspired Ichikawa to take up where Seno'o left off in exploring the application of Buddhism and socialism to liberation movements in contemporary Japan.[44] Although the parallels should not be exaggerated, it is possible to discern in the ideas of Seno'o, Ichikawa, and similar thinkers some of the same incipient concerns of 'liberation theology' in Third World Catholicism today, in the sense that Japanese proponents of Buddhist socialism have also reinterpreted a religious tradition to adapt it to Marxist concepts of liberation.

Finally, this study has considered, through Seno'o Girō, the personal experience inside the post-war peace movement of Japanese whose commitment to peace was in significant measure forged in the crucible of guilt for having failed in one way or another to prevent Japanese aggression in the Pacific War. Clearly, this guilt was a positive catalyst in motivating participation in the post-war peace movement, as Seno'o's story illustrates.

Yet the same guilt exerted powerful negative pressures on the lives of people like Seno'o who were fated to carry into post-war Japan overwhelming memories of failure in pre-war Japan. Nothing Seno'o could do as a post-war peacemaker was sufficient to expiate the guilt he felt over his pre-war *tenkō*. By defining the terms of expiation so impossibly, including nothing less than building the elusive popular front, Seno'o in effect sentenced himself to perpetual despair. Because his guilt was both political and what Robert Lifton in another context calls 'existential guilt' – which 'reflects the inner gap between the man one is and the man one could be'[45] – guilt also led to religious disillusionment once Seno'o concluded that the realization of the bodhisattva ideal lay beyond his reach. Ultimately, Seno'o comes across as a man who always regretted not having died in prison many years ago as a martyr for the pre-war popular front. He was the archetypal survivor who condemned himself for having survived.

All this is perhaps a reminder that while it is important to stand outside the peace movement and study its organizations, ideologies, programmes of action, demonstrations, slogans and so forth, it is also important to try and penetrate inside the movement to understand more precisely why its individual members took part, what their participation meant to them at the personal level of experience, and what it may mean to Japan today.

NOTES

1 Osaragi Jirō, *Homecoming*, Tokyo, Charles E. Tuttle, 1966, p. 263.
2 An exception: Whalen Lai, 'Seno'o Girō and the dilemma of modern Japanese Buddhism: Leftist prophet of the Lotus Sutra', *Japanese Journal of Religious Studies*, vol. 11, no. 1, March 1984, pp. 7–42.
3 Tokoro Shigemoto, *Nichiren kyōgaku no shisōshiteki kenkyū*, (Studies in the history of ideas of Nichiren teaching), Tokyo, Fusanbō, 1976, p. 356.
4 Cf. Matsune Takashi, *Seno'o Girō to Shinkō Bukkyō Seinen Dōmei* (Seno'o Girō and the New Buddhist Youth Federation), Tokyo, San'ichi shobō, 1975; Inagaki Masami, *Butsuda o seoite gaitō e: Seno'o Girō to Shinkō Bukkyo Seinen Dōmei* (Bear the Buddha into the streets: Seno'o Girō and the New Buddhist Youth Federation), Tokyo, Iwanami shoten, 1974; and Shimane Kiyoshi, 'Jinmin sensen: Seno'o Girō to Nakai Shōichi' (The Popular Front: Seno'o Girō and Nakai Shoichi), *Gendai no me*, February 1973, pp. 152–65.
5 Shimane Kiyoshi, 'Shinkō Bukkyō Seinen Dōmei: Seno'o Girō' (The New Buddhist Youth Federation: Seno'o Girō), in Shisō no kagaku kenkyūkai (ed.), *Tenkō: kyōdō kenkyū* (Joint research on tenkō), I, Tokyo, Heibonsha, 1967, p. 349. For Ienaga on Seno'o, cf. 'Zadankai' (Roundtable discussion) in Seno'o Tetsutarō and Inagaki Masami (eds.), *Seno'o Girō nikki* (Diary of Seno'o Girō), vol. 7, Tokyo, 1975, pp. 501–18, *passim*, Hereafter the *nikki* is cited as SGN.
6 Fujita Shōzō, 'The spirit of the Meiji Restoration', *Japan Interpreter*, vol. 6, no. 1, Spring 1970, p. 78.
7 The following account of Seno'o's pre war career is based chiefly on the books by Matsune and Inagaki, cited in footnote 4, and on Hayashi Reihō, *Seno'o Girō to Shinkō*

Bukkyō Seinen Dōmei: shakaishugi to bukkyō no tachiba (Seno'o Girō and the New Buddhist Youth Federation: Socialism and the Buddhist position), Tokyo, *Hyakka-en*, 1976. Note that Seno'o was born and raised in the village of Tōjō, in Hiroshima Prefecture. His family's business was sake brewing, with little commercial success. In fact, Seno'o experienced great poverty as a youth. Seno'o was educated through the first year at the First Higher School where he came under the humanitarian influence of its Head, Notobe Inazō. In 1919, Seno'o formed the Dai Nihon Nichiren-shugi Seinendan (Greater Japan Nichiren Youth Corps), which he served as president, with the support of the Nichiren philosopher Honda Nisshō, until 1930. He also edited this organization's publication, *Wakōdo* (Youth), which was well-read in Buddhist circles. For an autobiographical account, cf. Seno'o Girō, 'Yonjū-sannen no bukkyō taiken' (Forty-three years experience of Buddhism), Seno'o Girō kinenkai (ed.), *Seno'o sensei o shinonde* (Remembering Seno'o), Tokyo, Seno'o Girō kinenkai, 1974, pp. 115–41.

 8 Entry for 1st January 1925, *SGN*, vol. 3, p. 3.
 9 31st December 1934, *SGN*, 4, p. 236.
10 19th January 1936, *SGN*, 4, p. 322.
11 30th May 1939, *SGN*, 5, p. 45.
12 This discussion of Seno'o's ideas is based on his major prewar and postwar writings compiled in Inagaki Masami (ed.), *Seno'o Girō shūkyō ronshū* (Collected religious writings of Seno'o Girō), Tokyo, Daizō shuppansha, 1975.
13 However, Seno'o questioned the Marxist view that the material determined the spiritual; he claimed that in this world of constant flux, they were mutually determined.
14 24th April 1929, *SGN*, 3, p. 365. Gandhi was another of Seno'o's heroes, for his humanism and pacifism, although Gandhi was neither Buddhist nor Marxist.
15 For criticism of Seno'o's ideas, cf. Tokoro, pp. 361–2.
16 Cf. Murakami Shigeyoshi, *Japanese Religion in the Modern Century*, Tokyo, University of Tokyo Press, 1980, p. 126.
17 Seno'o first called for a post-war popular front in 1946; cf. 27th January 1946, *SGN*, 6, p. 10. A good survey on the Bukkyō Shakaishugi Dōmei is found in Nakano Kyōtoku, *Kindai Nihon no shūkyō to seiji* (Modern Japanese religion and politics), Tokyo, Aporonsha, 1968, pp. 212–22. For a general history of the peace movement, see Heiwa undō sanjūnen kinen iinkai (eds.), *Sensō to heiwa no Nihon kindaishi* (Japanese modern history of war and peace), Tokyo, 1979.
18 Cf. 1st January 1950, *SGN*, 6, p. 357 and 10th November 1952, *SGN*, 6, p. 434.
19 Nakano, p. 223.
20 Murakami, p. 133.
21 Nakano Kyōtoku, 'Natsukashii ronsō no omoide', (Sweet memories of disputation) in *Seno'o sensei o shinonde*, p. 220.
22 Murakami, p. 133. Heiwa Suishin, like the Heiwa Kyōgikai, included Buddhist, Christian, Shintō groups and some of the 'new religions'.
23 Matsune, p. 122.
24 18th February 1952, *SGN*, 6, p. 365.
25 Cf. 1st April 1950, *SGN*, 6, p. 304; 11th May 1950, *SGN*, 6, p. 305; and 11th November 1950, *SGN*, 6, p. 343.
26 14th July 1953, *SGN*, 6, p. 499.
27 4th October, 1952, *SGN*, 6, p. 424.
28 *Ibid* and 31st December 1953, *SGN*, 6, p. 531.
29 19th February 1953, *SGN*, 6, p. 454.
30 18th January 1956, *SGN*, 7, p. 120.
31 These events are discussed in J. A. A. Stockwin, *The Japanese Socialist Party and Neutralism: A Study of a Political Party and its Foreign Policy*, Melbourne, Melbourne University Press, 1968, pp. 39–48 and pp. 78–81.
32 12th October 1960, *SGN*, 7, p. 445.
33 Mibu Shōjun, 'Zadankai' (Roundtable discussion), *SGN*, 7, p. 513.
34 2nd June 1957, *SGN*, 7, p. 214.
35 15th August 1957, *SGN*, 7, p. 231.

36 8th April 1960, *SGN*, 7, p. 413.
37 16th August 1957, *SGN*, 7, p. 231.
38 10th September 1957, *SGN*, 7, p. 237.
39 26th September 1959, *SGN*, 7, p. 373.
40 Tetsuo Najita, *Japan: The Intellectual Foundations of Modern Japanese Politics*, Chicago, University of Chicago Press, 1974, p. 141, and p. 129.
41 Quoted from Trevor Ling, *Buddha, Marx and God: Some aspects of Religion in the Modern World*, New York, St. Martin's Press, 1979, p. 170.
42 Rokurō Hidaka, *The Price of Affluence: Dilemmas of Contemporary Japan*, Melbourne, Penguin Books Australia, 1985, p. 60.
43 Cf., for example, the essays in Inaba Noburō (ed.), *Bukkyō to marukishizumu* (Buddhism and Marxism), Tokyo, 1976.
44 Cf. Ichikawa Hakugen, 'The problem of Buddhist socialism in Japan', *Japanese Religions*, vol. 6, no. 3, August 1970, pp. 15–37 and his book *Bukkyōsha no sensō sekinin* (The war responsibility of Buddhists), Tokyo, Shunjū shuppansha, 1970.
45 Robert Lifton, *et al.* (eds.), *Six Lives, Six Deaths: Portraits From Modern Japan*, New Haven, Yale Univesity Press, 1979, p. 109.

III Uneven development and its discontents

6   The other side of Meiji: conflict and conflict management

IAN INKSTER

INTRODUCTION

The political and intellectual conflicts evident in the early years of Meiji were not just ineffectual, populist responses to industrial modernization and institutional change. Alternative political and industrial programmes were developed by groups who extended in influence from the dissidents who were within or ex-members of the strategic elite, through both metropolitan and provincial intellectuals who remained outside the immediate vortex of the political executive, to popular societies, newspapers and discussion groups.

Conflict, especially at its most 'popular' level, was often successfully contained by the direct, interventionist measures of the State. *Intellectual* opposition was contained by the process of economic change itself. This was possible due to the nature of both the ascribed and achieved social configurations of the opposition intellectuals. This hypothesis requires a fairly complex argument.

The politics of containment generates economic wastage. The size of that waste during a period of profound industrial development depends to a great extent on the nature of *technology transfer, adoption* and *diffusion.*

THE OTHER SIDE OF MEIJI

Many writers simply ignore the extent of intellectual and political threat during the 1870s. Others interpret it as real but essentially limited to the clarification of elite goals, i.e., to be non-goal threatening. Thus Harootunian's premise is that 'the existence of a strong strategic elite in the initial stages of modernization is more important than a comprehensive design for change of a variety of politically articulate groups competing with each other'. Although in the 1870s there appeared to be a chaos of approaches and fundamental perspectives, in fact, according to Harootunian, this was simply the reflection of 'renovation' processes, with the truly strategic elite (the 'goal attainment' elite) articulating policy and separating out from the 'adaptive elite' which was composed of a variety of diplomatic, military and political sub-elites whose activity served to remove disruptive elements, and whose sphere was in turn distinct from that of the 'integrative elite' operating in such areas as education, religion and so on, or from that of the 'system maintenance elite' in literature, entertainment and other fields, the latter reflecting rather than creating a general moral-political tone. So, in this Parsonsian world all seeming disruption was structural-functional to the adaptability and continuity of the total system. The hierarchy and integration of these levels was further conditioned by the existence of 'external' pressure i.e., a supposedly disruptive elite activity (e.g., political opposition?) at the level of the internal processes of the political system may in fact have clarified ideology or aims and reduced threats emanating from the outside.[1] Such an approach, unfairly schematized above, not only requires some acceptance of a portfolio of structural-functional relations, it is also in danger of being too selective of material and ahistorical in its mode of analysis – working backwards from a conclusion to assert that all the forces under consideration contributed to the maintenance and adaptability characteristics of the 'system'. Generally, there is a problem involved in evaluating the strength of such a procedure when considering national comparisons. This becomes clear when we introduce the popular contrast between China and Japan. Several historians of late nineteenth-century China argue that intellectual and political conflict was a sufficient cause of retardation in national leadership and that this was crucial in determining the overall process of underdevelopment.[2] At precisely what point does debate over fundamental issues have the effect of political 'retardation'? How do we begin to measure this? In what manner might it be argued that political debates in Japan were less than profound and did not indeed represent a challenge to elite goals? In fact, the debates of the elite intellectuals – from amongst which group the historians and biographers most often select their subjects – were reflections of debates at a far more diffused level and took place throughout a period of violent civil unrest. We must, therefore, ask why the intellectual elite failed to act as effective leaders of a more *widespread* reform movement, why they did not broaden their basis of support once the more spontaneous agitation was attacked from above, and why the physical violence of the time failed to generate a sustained radical political opposition? Such questions may not be adequately answered, but our point is that they ought to be proposed.

A number of the *Hōchi Shinbun* for November 1878 attempted to rationalize the physical uprisings of the decade as non-political and an offshoot of the personal discontents of ex-government officers. There was more than a grain of truth in this. Former employees who led physical agitation included Mutsu Munemitsu, Ōye Taku, Iwagami Kō, Okamoto Kenzaburō, Furusho Kamon and Major Okamoto. But there was undoubtedly a large political and social element in the famous Satsuma rebellion of 1877, and we may claim something of the sort for the less well-known civil uprisings and conspiracies of Kumamoto, Tosa, Take-bashi (Tokyo 1878), Kagoshima and the several tentative rumblings which followed the influential memorial of January 1874 on representa-tive government. Rebel activists were by no means limited to Tokyo or other large administrative areas: a survey of 1,137 southern rebels who were sentenced by the Kyūshū Special Court from May to December 1877 shows that some 208 (18 percent) were from the Tokyo *Fu*, 151 (13 percent) from the Miyagi *ken*, 88 (7 percent) were from Niigata, but that the remaining 690 (61 percent) were divided fairly evenly amongst the 12 *ken* of Yamanashi, Gunma, Saitama, Tochigi, Chiba, Ibaragi, Ishikawa, Akita, Iwate, Yamagata, Tokushima and Awomori (Aomori), each of these prefectures providing between 4.4 percent arfd 6.8 percent of the total.[3] The political element in the Satsuma rising is well known, but in a more general sense much of the physical unrest was political and social (e.g., involved many ex-samurai) and gave rise to longer-term local political traditions in crucial areas. Tosa witnessed *unrest* in 1877 and 1878 but by 1879 appears to have become a centre of *political* ferment. The ex-samurai (*shizoku*) gathered in large numbers as members of such societies as the *Risshisha* (Self-Help) and the *Aikokusha* (Patriotic), the former boasting 2,000 *shizoku* members, and composed those audiences numbering 500 daily who attended at discussions on political rights, freedom of the individual and representative government.[4]

These were precisely the issues which concerned the elite intellectual opposition. In his treatment of political thought in early Meiji Japan, Pittau confines his detail to the *Meirokusha* group, 'the most articulate elite of the day'. Some members, such as Fukuzawa relied upon their intellectual reputations as a basis for their political and cultural influence, others such as Katō Hiroyuki, 'the prototype of the bureaucratic scholar' were seldom outside of an official position, and this at once suggests some limitation to the range of debate possible within such a format. Katō's early democratic radicalism of 1868–70 was soon diluted in a solution which served to separate *Kokutai* (the national polity) from *Seitai* (the politico-administrative system) in the formula that Japan's backwardness prevented democracy at the *Seitai* level even though the *Kokutai* might include it as a political goal. Education was needed in order to increase the efficiency of the *Seitai*, a position not far from that arrived at through other means by Fukuzawa, for whom *Kokutai* was, essentially, the fact and spirit of national 'independence', a treasure lost if a prematurely democratic state had to withstand the Western onslaught.

If we narrow down to the role of the 'professors' in politics, as does Marshall, then the institutionalization of opposition is even more finely drawn. His 'establishment' intellectuals were the group which emerged from *Tōdai*, in turn trained or influenced later intellectuals and bureaucrats of the 1880s and provided intended and 'highly important integrative functions'. If this group was involved in 'stormy political controversy', this was mostly formulated within the concerns of government and arose from their activity as advisers in matters of education, economic development and social legislation. Marshall's article is not required reading for the student of Meiji political opposition which, mainly by definition, took place outside of central institutions and networks, integrative, system maintaining or otherwise.[5] What of Passin's 'progressive, politically oriented intelligentia' who debated such matters as currency reform, extension of the treaty ports, the separation of church and state, the social and intellectual role of education, the introduction of a popular assembly, appropriate industrial investment and local government, through to more abstract problems associated with Confucian thought and theories of nature, history and progress? Government action was in most cases sufficient to alter their stance or strength as political groups, and relatively mild legislation could effect their closure altogether. Thus the 1875 Press Law which increased the accountability of editors was fatal to the *Meiroku Zasshi* and eventually emasculated the *Meirokusha*. We might also agree with Braisted that this sort of effect coincided with the absorption of the elite opposition into the institutions of the bureaucracy – the authorities were defining political opposition in a legal manner, shifting the centre of the national polity, and in the process absorbing and transforming the elite intellectual opposition.[6] If physical violence did not adequately express the political opposition, then neither did the somewhat precocious debates of the elite societies.

Historians have often stressed the 'integrative' functions of the Meiji press as disseminators of government opinion and advertisers of government notifications.[7] However, during the 1870s the political press presented a fairly cogent set of alternatives to government goals, and were repressed accordingly. There was a substantial political public in early Meiji Japan. According to the sixth annual report for 1877 of the Japan Postmaster General, over 38 million newspapers, letters and books were annually transmitted, an increase of 8 million over 1876 and of 13.5 million over 1875.[8] By 1878 there existed 236 newspapers in Japan with a total annual circulation of 33.5 million. If we dismiss the a-political city 'dreadfuls' such as the *Yomiuri shinbun* with an annual sale of 6.5 million, and focus on the leading political press there would appear to have been an enormous growth in the years 1874 to 1878. Whereas the 'elite' *Meiroku zasshi* might have been sold to only 3,000 persons and the Tokyo *Nichi Nichi Shinbun* to 8,000 in 1874, by 1878 the *Hōchi Shinbun* and *Chōya Shinbun* were reaching over 2 million copies each, slightly trailing the 3 million of the *Nichi Nichi*, and were closely followed by the *Akebono Shinbun* and the Ōsaka *Nippō Shinbun*. The fact that wild fluctuations in circulation followed political events suggests a political role: According to the official statistics of the Department of the Interior the *Chōya Shinbun's* opposition during the Satsuma rebellion gained it a circulation for the year of 5 million, whilst the increased pressure brought to bear on the press during the Kumamoto outbreak of later 1876 did not prevent the stategically placed *Kumamoto Shinbun* from increasing its daily sales to between 5,000 and 6,000.[9] The repressive Press Laws of 1874 allowed the government to interpose directly in controlling opinion e.g., by April 1876 there were at least thirty leading newspaper editors and writers in prison for minor offences 'prosecuted with arbitary ignorance'.[10] Despite an increase in the intensity of pressure as embodied in the new legislation of June 1875 and again in 1876 the papers continued to form; during 1878 although eighty papers ceased publication, many because of government action, sixty-six new papers were started, including twenty-five in Tokyo, two in Kyoto, ten in Osaka, five in Kyūshū and three in Shikoku in the face of legislation embracing fines, imprisonment and the confiscation of the type and machinery of the establishments. Radical publishing continued well into the early 1880s. For his two articles of May–June 1880, a staff writer of the *Hokushū zasshi* and his publisher were imprisoned for one year. The articles, entitled, 'Powerless at Home and Powerless Abroad' attacked the government violently and generally. Later in the same year Kawasaki Saburō, editor of the *Akebono Shinbun* received one year's imprisonment for his publication in number 304 of that paper of an article on 'Popular Rights', which coincided with the arrest and conviction of Nakajima Katsuyoshi under the fifth article in the penal code of the Press Laws and the fourteenth article of the Newspaper Regulations for his popular pamphlet *Kokkai no kumitate*, (Constructing a National Assembly).[11]

By 1878 the press could not be referred to as performing a goal specifying function, at least not without a liberal stretching of terminology. Even the mild *Nichi Nichi Shinbun* of April considered the 'Evils of Government Interference', which severely questioned the administration's activity in industry and education, 'forced upon the [people] at the expense of personal convenience ... the public should remember that there is a certain limit to the exercise of benevolent intentions which if carried too far becomes a kind of tyranny', whilst the *Hōchi Shinbun* of the same month advocated 'Freedom of Speech' and the right of public criticism of government programmes.[12] In December the *Sei Dan* continued the agitation for a foreign loan in place of a reduction of government enterprise as an aid to deflation, suggesting an inflow of 10 million yen. Perhaps it is a matter of judgement as to how far the press challenged the goals of the bureaucracy, but legislative suppression, clamping down on particular papers and themes, and the very flavour of the political press all suggest that the government recognized effective pressure. On the promotion of industry – a tenet of bureaucratic policy – the *Chōya Shinbun* of October 1880 managed to make a link with wider political representation by declaring that:

If we have any industries [which will promote exports and reduce imports] they should most certainly be encouraged, but we do not consider this to be the duty of the government. Its duty is the protection of the community. Considering the matter practically, it is evident that the encouragement of industries by the Government would entail unnecessary officials and expenses, which would all prove prejudicial to any advance being actually made. It is not in conformity with the principles of reason that the Government should interfere with the undertakings of the people. The only course is to let the people understand that Japan is their own, and the only way to accomplish this is to establish a National Assembly. Those who talk about encouraging industry, instead of obtaining representative institutions first of all, are decidedly wide of the mark.[13]

An extreme challenge from a different direction came from Kubota Kan'ichi in his article on 'Barbarian Expelling' and the rejuvenation of a *Jōi* mentality in the *Nichi Nichi Shinbun* of September 1878. At a time when the authorities were doing everything in their power to introduce foreign technique, education and social institutions, Kubota Kanichi was advising that:

To hope to maintain national interests intact and to preserve our independence by having recourse to ethical principles [with foreigners] in this Age of Force, and by making morals the basis of our intercourse with a Society of Artifice [the foreigners] would be to hope for the impossible.[14]

Perhaps of even greater import was the direct link between the press and the popular political opposition through societies and lecture meetings. Well-known leaders of political societies were at times appointed to the editorship of newspapers, e.g., in 1879 Numana's acceptance of the *Mainichi Shinbun*. On 25th August 1881 the directors of the *Nichi Nichi* and the *Mainichi* joined with others to lecture publicly on the much-debated question of the selling up of property belonging to the *Kaitakushi* (the government organization for the development of Hokkaidō) to audiences of some 5,000 people at the Shintomiza Theatre, and appear to have thrown the whole plan into disruption.[15]

In early 1876 the government hardened its attack on popular political societies through new legislation specifically directed at 'public lectures', and from that time many popular societies were operating beyond the political pale. Until the coalescence of opposition into the 'popular assembly' movement around 1880, which served to focus yet limit much of the popular radical energy, the existence of a widespread opposition is somewhat shadowy and thus difficult to judge. For instance, the *Risshisha* Society at Kōchi, 'finding that they were prevented from delivering their seditious lectures, joined the *Kurosumi* sect as priests and gave their offensive lectures under the guise of sermons', an activity which resulted in the *ken* authorities prohibiting the priests from preaching at vulnerable venues. By 1880 the societies seem to have been very well organized. In April of the year the *Mainichi Shinbun* claimed of the Ōsaka *Aikokusha* that it,

has despatched agents throughout the length and breadth of the land to dissemi-nate the views of the society, thus making Ōsaka the centre of the national movement for the accomplishment of the great object – the establishment of representative institutions ... Here in Japan, the political societies number thousands and tens of thousands of adherents, and are to be found in every part of the Empire ... [if repression is increased] all concord and harmony between the Government and people will be destroyed, and the latter will come to look upon the former with hatred and, in fact, as if they were suffering under the evils of foreign domination.[16]

So the editor of the fairly mild-tempered *Mainichi* judged the spirit of the times. In fact new regulations respecting political meetings did come into force in April 1880. The securing of a government licence was no guarantee of immunity, for (clause 6) 'if the lecture or debate is considered prejudicial to the public interest, or liable to incite the hearers to commit offences against the laws, or if persons are in attendance who ought not to have been admitted ... then it shall be lawful for the police officers to close the meeting'. By clause 7 all professors and students were 'forbidden to attend or take part in political lectures or meetings, or to become members of political societies' i.e., with or without licences, radical or revolutiona-ry.[17] The societies were further suppressed by being forbidden to advertise in any way, to combine, to hold any open-air lectures or debates, or to in any way encourage new members, the penalty for the first of such offences being imprisonment for one year and closure of the society. Such legislation was clearly designed to halt the activity of such organizations as the *Risshisha* at Shikoku, who in December 1878 were reportedly 'giving open-air lectures in the streets at night; and that they have versified passages from the works of Rousseau, and from the history of the American Revolution, set to well-known Japanese airs; that they distribute printed copies of them . . .'[18] In some areas, such as Kanagawa, the authorities seem to have greatly curtailed popular activity, in others, such as Kyoto and Yokohama, activity seems to have been stimulated. The police effectively stopped the lectures at Tokyo, and the Ōsaka *Aikokusha*, which took a particularly constitutionalist stand, was closed by *Fu* authorities.[19] In contrast, in Shimane, Shikoku, a political group issued over 50,000 circulars arranging a public meeting to petition the government on the subject of a popular assembly.

At Kōchi, where their meetings were frequently closed by authority, over thirty lecturers organized a series which at times attracted audiences at any one lecture of over 2,000 persons, 'many of whom had flocked in from the adjacent villages'. Other societies were more out of hand:

The directors of the Aikokusha appear to be already experiencing the difficulty of guiding forces which they have once let loose. Osaka journals aver that the debates of the body are characterised by great clamour and disorder, and that the President has much difficulty in governing the speakers.

In Okayama public lectures became more vocal and violent, speakers drawing their swords and proclaiming that 'This is what is called the prop of the popular rights'. When in December 1880 the *Senyūsha* of Nagasaki organized a lecture before some 2,000 to 3,000 people, the police intervened half-way through and were thereupon attacked and thrown out by the audience at large. Nor did the legislation prohibit the formation of new societies, e.g., next door to the capital itself the newly formed *Kenyūsha* met in the Sagamiya Restaurant at Yokohama.[20] Prior to 1880 the societies were not solely concerned with a wider representation. The *Hokushinsha* of Yokohama organized lectures on diverse political topics at the Concert Hall in Bashamichi and their general range was duplicated in several others e.g., the *Seigisha* (Kanazawa), the *Kaimei* and the *Doshū* (Kagoshima).[21]

CONFLICT, CONFLICT MANAGEMENT AND CONTROL PROCESSES

Conflict and the context of control

1   From 1873 to 1877 the number of notifications issued by government departments, excluding public works, war and navy, amounted to

1,011 in the first year, 850 in the second, 1,116 in the third, 1,061 in the fourth and 517 in 1877, giving a grand total of 4,555.[22] The very weight of administrative formality circumscribed political behaviour. In addition, a tentative suggestion is that the wide ranging agitation of the mid-1870s gave way to a channelling of energies into the demand for a popular assembly. When this latter aim appeared to be realised, the regional or urban activists were unable to seek further objectives. As early as May 1878 at least nominal representation in local matters was increased through the introduction of a general assembly of provincial governors. When in early 1880 the administrative power of the local assemblies was furthered, many of the popular societies who continued to press for democratic institutions were, by legislative implication, pushed further and further into an outside position, one from which they were vulnerable to direct government intervention – arrest or closure or both. For reasons such as these the amount of popular opposition appears to have decreased greatly during the early 1880s, and was possibly absorbed in the 'party' movement of 1882.

2   The provincial bureaucrats and the non-elite popular opposition possessed 'mental maps' which at best were national in scope, whose metropolitan centre was not the West, but Tokyo, or the Emperor or the Meiji bureaucracy itself, and this served to delimit their activity as an effective, coherent intellectual opposition. A possible alliance was never forged. On the one hand there was the bureaucracy and the intellectual elite, composed of 'persons whose "maps" had been widened by education, experience, and a more open and active curiosity ... much more disposed to see their centre as a metropolitan space of which they might be a part and towards which, in any case, their attention and respect are drawn'.[23] Japan was to become part of the 'metropolitan space' through technological borrowing, Western institutions, and industrialization, which brought opportunity to any skills and perspectives which were themselves fundamentally facets of the worldwide centre. On the other hand, the non-elite political opposition of the 1870s was provincial in spatial and social terms, and the problems it promoted were principally (not wholly) related to a local-national framework, rather than one which was national-international in essence. With such widely differing mental maps it was hardly surprising that the elite groups failed to join the popular agitation, even though there existed at times a coincidence of political statements and issues between them.

3   The failure of the popular opposition was also a reflection of other forces. Silberman's argument that characteristics of the local bureau-

crats meant that they worked effectively with the central administration undermined the non-metropolitan opposition, as did the directing of that opposition into the narrow confines of the popular assembly agitation.[24] The removal of the means of diffusion and sustenance of radicalism (the press, public meetings) and the selling-up of government enterprises and introduction of subsidies during the later 1870s, which provided occupational avenues for the *shizoku* and others, finally shattered the opposition. The success of the latter policy, which was in operation well before 1882, further reinforces the idea that the opposing groups were not all populist or conservatist but included highly 'progressive' individuals.

4   In all cases of serious physical threat the government was very quick to act, to mobilize resources efficiently and to remove the power of leaders without creating martyrs. For reasons such as these the amount of popular opposition appears to have decreased greatly during the early 1880s, and was possibly absorbed in the 'party' movement of 1882. But what explains the contemporaneous decline of the elite intellectual opposition, which should have been less vulnerable to direct legislation, and was more subtle in its political statements and more continuously organized around a comprehensive range of issues? One key to Meiji political history lies in the failure of the elite intellectuals to either forge an alliance with the popular movement in its prime (1875–8), to counteract the channelling of popular political energy into the popular assembly issue, or to manufacture a new basis for radical popular politics in the 1880s.

Conflict management – the undermining of the intellectuals

We may draw up a shopping list of discrete factors. For serious conflict to have emerged the intellectuals would have had to have had much less in common with the bureaucratic elite. *Tōdai* or other institutions and the 'feedback' mechanism they provided converted antagonism of the 1870s into consensus in the 1880s. Similarly, we might drop into our bag the real change in the industrial and commercial base, which cut the ground away from any critique which centred on alternative modes of development, as full employment of intellectuals generally, in education, government and trade, prevented class alienation. The influence of the foreign communities, with whom leading political intellectuals interacted quite frequently, served further to pull intellectuals, who were otherwise critical of the regime or particular cabinets, towards a commitment to foreign knowledge, science and technique, which was identical to that of the bureau-

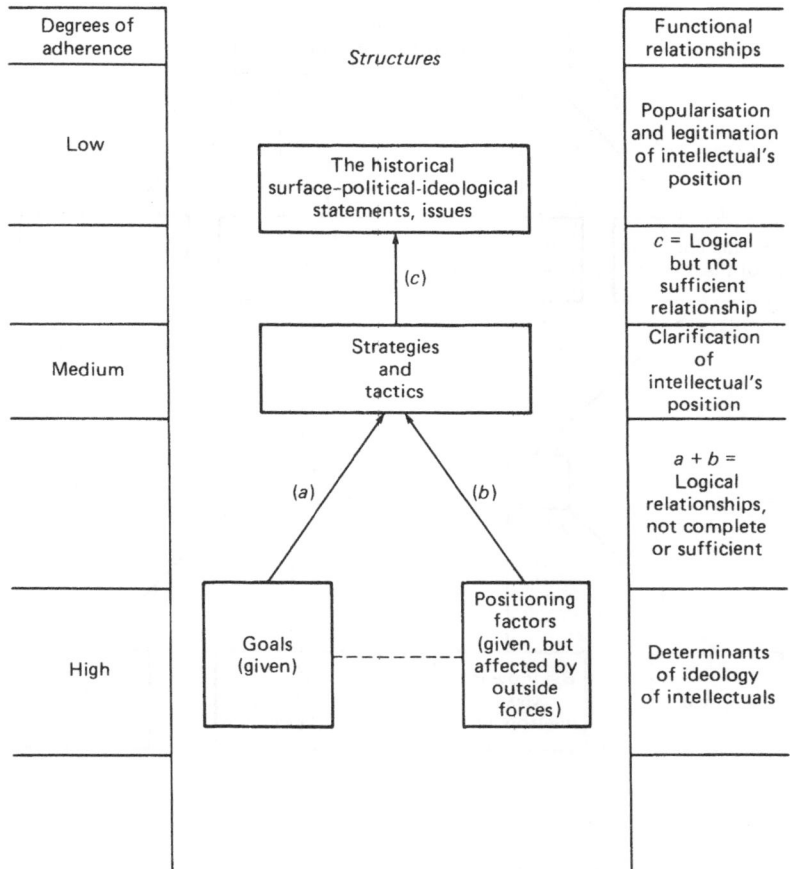

Figure 1 The place of politics

cratic strategic elite. Lastly, with relative ease, a radicalism which stemmed from the thought of Spencer and Darwin could, through the notion of societal evolution and a belief in the moral functions of science, be found compatible with involvement in technocracy, social engineering and bureaucracy.

From even a cursory survey of groups and individuals we could identify all of these 'factors' operating on the intellectual elite in early Meiji Japan. But at this point we are interested in change – from relative intellectual choice and conflict to relative consensus, and in how far the Meiji experience may be compared with the historical or contemporary experience of other nations. In addition, we wish to arrive at some understanding of differences between intellectuals as well as identify their position or

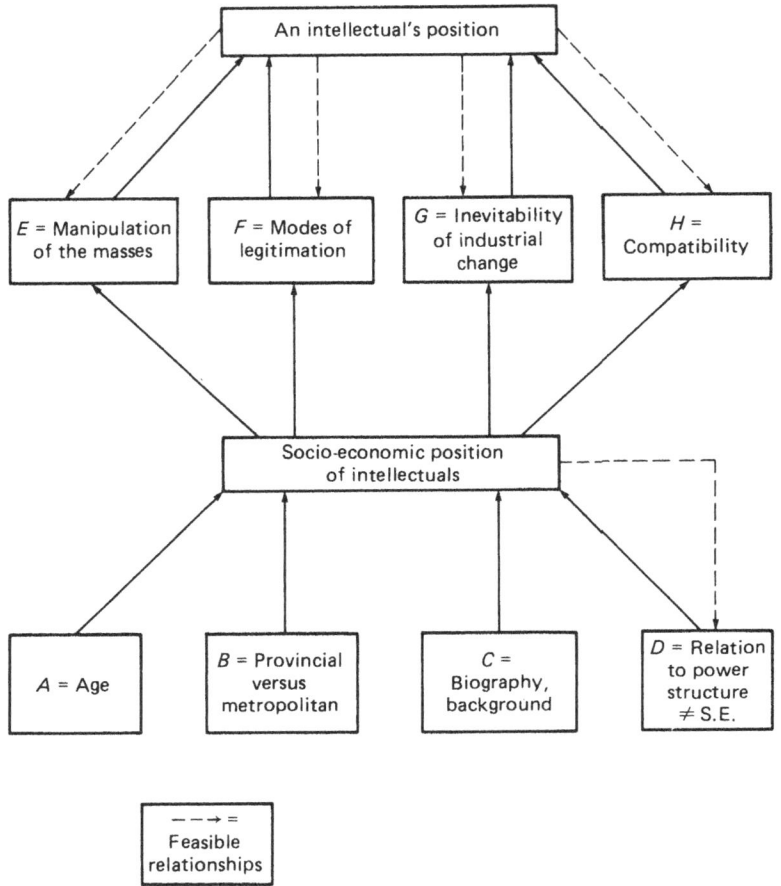

Figure 2 The position of the intellectual

the forces working upon them *viz-à-viz* those associated with the strategic elite. One argument here is that the political intellectual elite was 'integrated' into the nationalism of mid-Meiji not through a sudden 'change of mind' or conversion of ideological conviction, but as a result of alterations in the *structural conditions* which defined their overall identities as individuals in society. So-called ideological clashes between intellectuals as expressed in political statements may not be used as definitions of their difference of underlying position, but should be thought of as reflections of those positions, further refracted by varying perceptions on optimum strategies and tactics.

Figure 1 suggests that political statements and issues may be formulated by intellectuals who feel only a fairly low degree of attachment to them. This is not because intellectuals are particularly frivolous – though they too often find themselves compromised – but because particular state-ments are rarely a sufficient representation of underlying positions and goals, and are likely to be more or less logical expressions of social-politi-cal strategies, which themselves serve to clarify the intellectual's position yet are not held with resolute conviction. Figure 2 elaborates on the elements which feed into the overall configuration of an intellectual, are considered fairly universal during periods of extreme social change, and are far less easily abandoned than tactics or statements of the moment. Briefly, we might argue that factors *A* to *D* served to pull together the political intellectuals of Meiji Japan, however, factors *E* to *H* produced variety and dissent amongst them and between them and the administra-tion. Crudely, *E* to *H* may be given positive or negative values on a scale, and may be considered as independent elements; an intellectual either believes that elites are able to manipulate the masses or that they are not; an intellectual either thinks that the industrialization process is incompat-ible with existing social structures or he believes in some degree of compatibility. Figure 1 suggests that even given a more or less common and well-defined set of ultimate goals, the combinations possible at *E* to *H* together with the variety of possible strategies or modes of action may lead to a veritable host of debates, statements and 'attitudes'. As we have pointed out, amongst the non-elite political opposition goals might not have been so determined, and by the early 1880s this type of activity was therefore selected out by government for special repressive treatment. Why did the elite intellectuals not develop alternative goals? The answer lies not only in their communality of socio-economic characteristics with the bureaucrats (*A* to *D*) but also in the other determinants of their general positions (*E* to *H*), which altered in nature during the period, leading to changes of tactics which ultimately reduced cleavage amongst them and brought them further into line with the administration. When forces *E* to *H* joined ranks with those labelled *A* to *D* the removal of the conditions for effective opposition had been effected.

It is possible to grasp the range of such processes by focussing on just two elements or 'positioning factors', as in Figure 3. Here, intellectuals falling into area 2 would be the least visible of the four possible groupings. Secondly, it is possible to visualize accommodation between groups 2, 4 and 1 if there existed other forces tending towards internal cohesion e.g., appropriate changes anywhere in the range *A* to *F* of Figure 2. Thirdly, groups 3 and 4 would be least likely to arrive at mutual compromise on such matters. Lastly, a factor explaining 'integration' – given that we are only dealing with *G* and *H* – could have been the almost complete absence

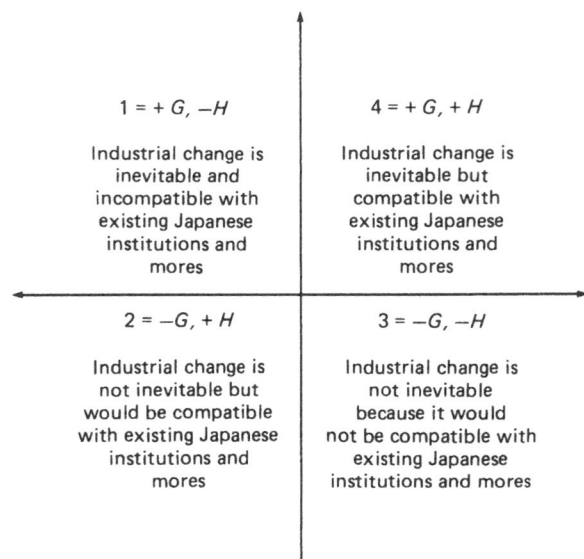

Figure 3 Hypothetical levels of conflict between positioning factors G and H

of group 3 in Meiji Japan. Although there was a revival of a modified *Jōi* mentality around 1878 this did not seem to have involved a position over the non-inevitability of industrial change, and indeed the movement as it appeared seems closer to group 1 than 3. Contrariwise, in late nineteenth-century China group 3 represented a strong anti-Western intellectual movement. The contrast points not to the argument that Japanese intellectuals were pro-Western when Chinese were anti-Western, but rather suggests that these political *results* were caused by the differing positions which were taken over *more* fundamental factors not *directly* related to the issue of Western encroachment. The matrix might be related to specific issues or 'statements'. Groups with a strong $+H$ element in their intellectual configuration would specifically address themselves to the removal of pre-industrial 'vestiges' which would be visualized as barriers to modernization e.g., the traditional family, political institutions, rural social structures etc. As such, belief in the inevitability of industrial development – groups falling in areas 1 and 4 – would lead to arguments that such 'barriers' did not in fact exist, were a figment or a fetish or would wither away e.g., the evolutionary element in Meiji thought. Thus members of group 4 would be most likely to emerge as social reformers rather than political revolutionaries, although those of group 1 would adopt a more radical stance on social institutions. As industrialization progressed – by 1881–4 and the Matsukata measures – the strategic positions of intellectuals falling in space 2 would become seriously eroded, just as that of the Populists or Narodniks of Tsarist Russia during the 1890s. On the other hand that of space 1 individuals might be strengthened in so far as members struggled to accommodate themselves to the new conditions through the creation of new strategies and statements. As such adaptation was attempted some groups would suffer greatly from the growth in the gap between their declared positions and their real positions, and in terms of Figure 2 would be caught between the needs of inner-clarity and the needs of outer-popularity and legitimization. The underlying dynamic of such intellectuals would be weakened. For instance, those intellectuals falling naturally into space 2 would be utilizing a political rhetoric by the 1880s which was either quaint or incoherent in terms of both their fundamental beliefs and their earlier political statements i.e., seeming positions.

An investigation of intellectual groups and their changing 'political positions' suggests that initial divergences amongst the intellectuals and between them and the bureaucracy were a result of forces operating at E to H, with factors A to D and basic goals, (the improvement of the 'people', self-strengthening against the West) roughly in common. As the 1880s unfolded the factors operating from E to H ceased to generate divergence, debate, dissent and conflict.

The intellectual emergence of Fukuzawa Yukuchi is a case in point, for his altering rhetoric is characterizable in terms of its relation to the postulates of the *Jōi* writers, the *Kaikoku* philosophy and the *Keimō* formula developed during the 1870s, and also with heuristic reference to the 'positioning factors' G and H in Figures 2 and 3. The simple 'expell the barbarians' cry of the *Jōi* group of writers was in fact the reflection of a strategy which followed their underlying belief that the state was in process of dissolution, and that such aspects of industrialization as the profit motive or Western science were no substitute for and were incompatible with Japanese values and Confucian ethics. When it became clear that the industrial process could not be halted *Jōi* was reformulated to embrace limited Westernization. Similarly, *Kaikoku* (ultimately expell the West through the immediate use of their techniques) also had at its base the notion of the incompatibility of traditional modes with new (Western) industrial ones, the possibility of protecting the former and limiting the latter by excluding the pervasive influence of Western morals and concepts of virtue. Western technique could be used to cope with the external threat as Eastern ethics were incorporated to address internal dissolution. No doubt, Fukuzawa was caught between these positions during the period of his early intellectual development, that is, from 1866 with the publication of *Conditions in the West* to circa 1872–3 and the formation of the Meirokusha group. But both *Jōi* and *Kaikoku* represented *tactical* systems which could be found inappropriate when other conditions changed. When Western science was seen as but part of a more general 'spirit', which included diverse elements such as Christian ethics, democracy, freedom of the individual and social mobility, then Fukuzawa and others were forced to realise the incompatiblity of 'science' with many traditional modes and institutions and to press for the wholesale reform of these, especially focussing on education and the family in their 'political statements'. The fact that the *Jōi* notion of *kyōri* (science) involved its supposed underlying metaphysic, meant that its rhetoric was somewhat hollow at a very early date as it became obvious that specific engineering-science skills and training could be adopted without cultural disorder.[25] In the meantime, Fukuzawa's political framework was supported by Western evolutionary theory (Buckle, Spencer) which by introducing the idea of stages of evolution solved most of the very real 'compatibility' and 'inevitability' problems: because of the direction of universal progress Japan was bound to industrialize, and because of the evolutionary adaptation of social systems, institutions and modes would through time accommodate themselves to economic changes. So, this is not to say that by reading Spencer, Fukuzawa underwent a revolution of thought regarding Japanese institutions. It is to suggest that, given the real problem of evolving a strategy which could satisfy his belief regarding inevitability and compatibility in the long run yet at the same time serve to popularize and legitimize his statements, Fukuzawa chose certain theoretical constructs as most appropriate. Although this puts the case too boldly or crudely, it is a step towards a perception of early Meiji intellectual life which is not full of unexplained reverses, renovations and rejuvenations. Through the conclusion that civilization was Western, but fortuitously and temporarily so, Fukuzawa evolved a series of propositions which fell between those who saw Western technique as detachable from a broader commitment and those who uncritically adopted everything Western. Not surprisingly, this was precisely the position in real practice of the bureaucracy. There is a certain amount of bureaucratic dissimulation and intolerance in Fukuzawa's famous phrase: 'How futile to allow wild guesses about moral relations to distort the laws of physics.'[26] Science was neither a technique nor a philosophy, but more a mode of learning. Rather than select those facts of Japanese history which seemed to directly justify or exemplify traditional moral precepts, the task of the intellectual concerned with the theme of 'compatibility' was to discover within Japan those innate characteristics which were fundamental to the notion of 'civilization'. Such a compromise was increasingly relevant from the early 1870s to the 1880s, and the dissolution of the *Keimō* movement left little in the path of the strategic elite's reassertion of Confucianism and nationalism, culminating in the Imperial Rescript on Education of 1890.

By the 1880s industrial progress meant that the 'incompatibility' and the 'inevitability' themes could best be handled by rejection of Japanese culture, art and institutions as inappropriate to the now materially successful Western-World view – a position adopted by Tokutomi Sohō and his followers.[27] The alternative was a search for a 'usuable past', as formulated by the *Seikyōsha*, founded in 1888. By the 1890s the strengths of the two formulas had been tested, and the more nationalist position of one group meant their better ability to resolve and popularize their given political statements in terms of both inevitability and compatibility themes. That is, the position of the nationalist intellectuals meant that their political positions as publicly stated increasingly conformed to their underlying beliefs. First, institutional substitutions which by-passed personal, individualistic communication systems served to enhance nationalism – the railways, telegraphs, postal system and newspapers. Between 1874 and 1892 the number of magazines and newspapers handled annually by the post office grew from 3 million to 50 million.[28] Secondly, it became possible to emphasize the elements of 'traditional Japan' which did seem to be compatible with the industrial process – the continuation of the traditional sector of the dual industrial economy meant that the village, the family and associated values of obedience and solidarity were neither decimated by industrial change nor barriers to that change. Fukuzawa's or Tokutomi's Spencerism could not adequately address this development. Thirdly, Pyle rightly discerns a turning point in 1889–90 as a sense of 'national mission' enfolded the new constitution (February 1889) and the meeting of the first Diet (November 1890). This was sustained by the feeling that unanimity of position (i.e. strategic positions and political statements) was required if matters which involved external issues were to be properly addressed – the unequal treaties, international political relations. Retaining their actual positions on inevitability and adaptability, the *Seikyōsha* group were in a strong position to evolve appropriate political strategies and legitimizing and popularizing statements. *Sekai no bunmei* (world civilization) became a composite of differing yet progressive national cultures which grew through competition between nations. Therefore, the obligation of the Japanese was 'to preserve and develop their distinctive talents and values in order to supplement the contribution of Western culture [to world civilization]'.[29] Cultural nationalism was not only natural and national, it was honourable and international.

Social control and appropriate technology

From Figure 2 we might argue that the opposition to the bureaucracy which arose from intellectuals was dampened by a communality of socio-economic positioning characteristics. The political opposition at the elite level possessed a social profile very similar to that of the bureaucrats. Factor B in Figure 2 requires some elucidation. We may postulate that both the central bureaucracy and the political intellectuals possessed 'worldwide maps', frames of mind which helped them locate themselves spatially and socially. The 'crisis of identity' of the intellectuals represented the process whereby they estimated their distance from the Western metropolis, and this crisis was eventually resolved by their participation in the development process, after taking up intermediate positions which involved real opposition. Others, e.g. the provincial bureaucrats and the non-elite popular opposition possessed 'mental maps' which at best were national in scope. Thus Fukuzawa and others 'distanced' themselves from 'the people' and considered them something apart from their own universe of concerns. Given that the strategic elite was generally characterized in terms of training and skills rather than wealth or initial power, the contrast was that much more pronounced, for as Shils puts it, 'the culture of a foreign metropolis is the only modern culture [the metropolitan intellectual in a provincial setting] possesses. If he denies that culture, he denies himself and negates his own aspiration to transform his society into a modern society'.[30]

Similar consideration of the positioning factors *A*, *C* and *D* of Figure 2 leads one to the conclusion that socio-economic factors served to coalesce the political intellectuals, to relate them rather finely to the bureaucratic elite, and to truncate them structurally from the non-elite opposition. On the other hand, the forces listed *E* to *H* tended to create a variety of political statements and issues amongst the intellectuals in the 1870s and promoted some overlap between intellectual elements and popular politics and agitation over specific issues. But as Meiji progressed such forces were insufficient to prevent the intellectual elite from converging with the various groupings of the strategic elite.

The *detachment* of the elite political opposition was one important cause of the failure of popular politics to retain its radicalism into the 1880s. But, in addition, the 'technological imperative' was of such a kind as to dampen or remove altogether the politico-economic tensions which arose in such countries as Tsarist Russia during the Witte industrial programme of the 1890s. Politically, this was somewhat fortuitous. *Economically*, there were excellent reasons for the creation of a 'technological imperative' with a relatively low political profile. At least to an extent, the technology available from the West represented a continuum of technological choice, i.e., this was not a case of a colonial relationship forcing *a* technology onto a powerless partner. On the problem of how a backward nation open to the international economy secures the best available technique to close the technology gap, Mansfield has written that, 'An ... important factor is the social system and attitudes towards technical change in the backward country, the relevant attitudes being those of the people as a whole as well as government planners.'[31] Given that the gap facing late nineteenth-century Japan was *technologically* less than those facing today's underdeveloped nations (particularly if we exclude non-transferable techniques utilized in enclaves), then we might argue that the pressure on the political system resulting from transference was less in Japan than in nations otherwise analogously placed at the present time. In addition, Murphy has argued that late nineteenth-century 'machinofacture' *decreased* the amount of *specific* skills required of operators, reducing the likelihood of political opposition arising from either the unemployed or the alienated. Kranzberg has stressed that the transfer of technology 'involves a disruption of previous work habits and thought patterns', that is, demands of the system an increase in non-specific skills, (time keeping etc.).[32] Yet in even the leading-sector cotton industry of the 1880s there was surprisingly little required in the way of either technical or management skills, and the use of female labour removed the potential for truly proletarian unrest.[33] In addition, we may argue that the growth of military technology during the 1870s served to indigenize the capital goods industries in Japan. Such industry has historically served to institutionalize the very processes of technological change, learning and diffusion.[34] The importing of capital goods of a far greater sophistication and far larger scale into today's underdeveloped nations, often without a related policy of import substitution, separates initial transfer from diffusion, causing socio-political tension, inequality and regional retardation, as well as continuing economic backwardness. Amongst other possible final points which concern the technological determination of Meiji political development, we must mention that the 'technological imperative' did not inhibit the ability of potential dissidents to take the industrial opportunities offered by the regime during the late 1870s and the 1880s.

Technological borrowing and the process of its application in the economy affected political development in a multitude of ways. The business elite was incorporated into the strategic elite relatively quickly because the comparatively simple level of even 'best technique' allowed them to move from their traditional 'commercial' functions to a more modern 'management' function. At the same time, bureaucrats returned from Europe or North America and utilizing the skills of hired foreigners could comprehend and control technological-industrial developments very readily. At the level of opposition or potential opposition to the regime, the nature of technology was such as to minimize 'displacement' effects on employment, skills and regional balance, (indeed, technology planning in Hokkaidō was seen as a means of relieving employment problems as well as creating growth), and to permit new skills to be developed under less-than-traumatic conditions. The subsequent emergence of dualism between and within particular industries during the 1880s further permitted a continuity of economic activity and organization in older industries and newer leading sectors, and a retention of non-urban commercial and social institutions.

Developments in Meiji Japan suggest that it is not sufficient to rely on a strong bureaucracy as a means of political control during industrialization. If technology and capital is to be deployed from without then so will external ideas and ideologies become an invisible import, to be balanced by internal measures. In Meiji Japan during the crucial 1870s imported ideology was by its nature not so disruptive of internal patterns as is the radical political thought of today. If the relation between decision making at the centre and implementation in the regions is not close, a large amount of effort and expenditure will be involved in controlling the divisive forces unleashed by development through industrialization, especially if such forces are linked with non-metropolitan political opposition. Pre-industrial social groupings must be acknowledged and integrated into the new political economy unless the armed force of the new state is sufficient to expel them altogether, and this further adds to the non-productive expense of the industrial drive. Accepting the importance of technology, as argued here, an alternative might be to consider 'appropriate technology' as more than an economic affair. Technology as imported or created may be appropriate not only to factor endowments, the extent and nature of markets, existing technique, and structural and spatial balance, but also to the political requirements of the nation concerned. In Japan, non-political economic factors forced a technology which created relatively few political strains. In today's underdeveloped nations economic criteria combine with international political factors to encourage a technology which possesses a very high political profile. If the capital cost of a technology is great, and if the capital cost of controlling the social unrest created by its introduction is also great, then the real and opportunity cost of transfer is huge and the process of industrialization will not bear a family resemblance to that of Meiji Japan.

CONCLUSION

Today's under-developed economies may despair at the Japanese example – with a non-colonial past, an efficient amelioration of the conditions of pre-industrial social groups, an intelligent bureaucracy, and faced with a much lower-level external 'impact', Japan suffered significant problems of control and legitimacy, most of which were satisfied on one hand by autocracy and on the other by fortuitous and non-reproducible circumstances. 'Appropriate' technology may be today's political solution, but itself will upset strategic internal interests (often previously fostered by formal colonialism), be difficult to justify in the international political arena, and difficult to evaluate economically. Finally, such a technology may not in fact exist.

Without revolution, even democratization has to be decided upon. Surely it is not a coincidence that in the relatively economically backward nations of the late nineteenth century profound disturbance, (e.g. China), profound crisis, (e.g. Russia), and disturbance followed by eventual renovation, (e.g. Japan), amongst intellectuals and politicians were outstanding features of their 'incorporation' into that network of relations which economic historians depict as the International Economy, a construct composed of flows of goods, capital, technology, knowledge and persons. Furthermore, if we consider the 'tension' between relative backwardness and relative forwardness in this period as acting in the role of an 'enabling agent' for reaction and response,[35] then we may argue that tension between indigenous socio-economic structures and *enclave* modern sectors within today's under-developed nations acts as a severe 'disabling' agent, and that this contrast is one worth examining in some historical detail.

NOTES

* The present paper is a much reduced version of a report prepared under the auspices of the *United Nations University Project on Technology Transfer – The Japanese Experience*, undertaken jointly by the UNU and the Institute of Developing Economies in Tokyo. I would like to acknowledge the helpful comments of several colleagues who attended seminars on or relating to this paper at Hitotsubashi University (June 1980), the Department of Economic History, University of Melbourne (September 1982), Legal Studies, La Trobe University (March 1983) and the Nissan Institute of Japanese Studies, University of Oxford (December 1983). I would also like to thank Drs. James Goode (Vancouver) and Jane Devitt (Sydney) for their helpful comments upon the original script.

1 Harootunian, in Introduction to B.S. Silberman and Harry D. Harootunian (eds.), *Modern Japanese Leadership*, Tucson, 1966, pp. 16–20.

2 Marion Levy, 'Contrasting factors in the modernization of China and Japan', in S. Kuznets, W.E. Moore and J.J. Spengler (eds.), *Economic Growth: Brazil, India, Japan*, Durham, 1955, pp. 496–537, and W.W. Lockwood, 'Japan's response to the West: The contrast with China', *World Politics*, vol. 9, 1956, pp. 37–54.

3 Calculated from information in the *Mainichi Shinbun* and the *Japan Weekly Mail*, 12th January 1878 (henceforth *JWM*).

4 *JWM*, 22nd March 1879, p. 354, and further references detailed in the present writer's forthcoming more detailed account of popular politics, entitled, 'The Politics of Technology Transfer – Protest, Agitation and Intellectual Debate during Meiji Industrialisation'.

5 Byron K. Marshall, 'Professors and politics: The Meiji academic elite', *Journal of Japanese Studies*, vol. 3 no. 1, 1977, pp. 71–97.

6 William R. Braisted, *Meiroku Zasshi, Journal of the Japanese Englightenment*, xli-xliv, Tokyo, 1976.

7 James L. Huffman, 'The Meiji roots and contemporary practices of the Japanese press', *Japan Interpreter*, vol. 11, no. 4, 1977, pp. 448–66, Albert A. Altman, '*Shinbunshi*: The early Meiji adaptation of the Western-style newspaper', in W. G. Beasley (ed.), *Modern Japan, Aspects of History, Literature and Society*, Tokyo, 1976, pp. 52–67.

8 *JWM*, 5th January 1878, pp. 2–3.

9 *Osaka Nippō*, 9th and 10th January 1878, article entitled 'A short history of the Japanese press', *JWM*, 19th April 1879, p. 474.

10 *JWM*, 22nd April 1876, p. 351.

11 *JWM,*, 8th May, p. 586; 5th June, p. 731; 23rd October, p. 1,384; 20th November, p. 1,506, 1880.

12 *JWM*, 6th April, pp. 313–14; 13th April, p. 335, 1878.

13 *Chōya Shinbun*, 28th October 1880, translated as 'On the encouragement of industries', *JWM*, 30th October 1880, p. 1,412.

14 Kubota Kan'ichi, 'Barbarian expelling', 2nd September, 'Explanation of the discourse of barbarian expelling', 24th September of *Nichi Nichi Shinbun*; *Japan Herald*, 21st September, p. 2; *JWM*, 28th September, p. 1,011 etc.

15 *JWM*, 29th November 1879, p. 1,596, 3rd September 1881, p. 1,029.

16 *JWM*, 1st November 1879, p. 1,462.

17 'The new regulations respecting political meetings or societies' reproduced in *JWM*, 10th April 1880, p. 471.

18 *JWM*, 21st December 1878, p. 1,392.

19 *JWM*, 28th February, p. 278, 24th April, p. 536, 1880.

20 *Mainichi Shinbun*, 16th December 1880, p. 3; *JWM*, 15th May, p. 633; 10th July, p. 891; 18th December, p. 1,633, 1880.

21 *JWM*, 29th November 1879, p. 1,597; 7th August, 1880, p. 1,020.

22 *Kinji Hyōron*, 3rd January 1878, p. 2 (figs, calculated).

23 E. Shils, *The Intellectuals and the Powers*, Chicago, 1972, p. 356.

24 B. S. Silberman, *Ministers of Modernization*, Tuscon, 1964.

25 *Kyōri* is now more properly rendered as doctrine, creed or teachings.

26 Carmen Blacker, *The Japanese Enlightenment, A Study of the Writings of Fukuzawa Yukichi*, Cambridge, 1969, pp. 1–11. See also Albert M. Craig, 'Fukuzawa Yukichi: The philosophical foundations of Meiji nationalism', in Robert E. Ward (ed.), *Political development in Modern Japan*, Princeton, 1968, pp. 99–148; Koizumi Shinzō 'Fukuzawa Yuckichi', *Japan Quarterly*, vol. 11, no. 4, 1964, pp. 486–93.

27 Kenneth B. Pyle, *The New Generation of Meiji Japan*, Stanford, 1969.

28 *Ibid.*, p. 82.

29 *Ibid.*, p. 151.

30 Edward Shils, *The Intellectuals and the Powers*, p. 23.

31 Edwin Mansfield, 'Comments on Kmenta' in D. L. Spencer and A. Woroniak (eds.), *The Transfer of Technology to Developing Countries*, New York, 1968, pp. 57–60.

32 J. J. Murphy, 'Retrospect and prospect' in *Ibid.*, pp. 6–30; Melvin Kranzberg, 'What constitutes an industrial revolution?', *Foreign Policy Bulletin*, vol. 39, 1960.

33 K. Seki, *The Cotton Industry of Japan*, Tokyo, 1956.

34 Nathan Rosenberg, 'Capital goods, technology and economic growth', *Oxford Economic Papers*, vol. 15, 1963, pp. 217–27; Yamamura Kōzō, 'Success illgotten?: The role of Meiji militarism in Japan's technological progress', *Journal of Economic History*, vol. 37, 1977, pp. 113–33.

35 See the discussions in Ian Inkster, *Japan as a Development Model?*, Bochum, 1980, Ch. 2–4.

7 Nuclear power and the labour movement

YUKI TANAKA

ECONOMIC BACKGROUND

The most severe over-production crisis in Japan hit the economy between 1974 and 1975 and this was accompanied by sudden price rises caused by the 'oil shock' of autumn 1973. Technically, such a phenomenon is known as a 'structural depression'. As a result of this economic crisis, a 'rationalization' policy, also called an 'adjustment of employment' policy was adopted in almost all sectors of industry. Irrespective of the name, this was nothing but the retrenchment of large numbers of employees. It included a significant reduction in new recruitments, resort to temporary lay-offs and voluntary retirement, reduction of over-time work, the retrenchment of seasonal and part-time workers and the cutting down of large numbers of day labourers. Industries also took various measures to reduce investment in production plants and machinery and to economize by amalgamating or dissociating affiliated companies.

The people hit most severely by this 'rationalization' were temporary workers and day labourers at the bottom of the Japanese industrial structure, who subsequently found difficulty in finding work. In 1960 temporary workers and day labourers made up 12.5 percent of the total Japanese workforce. In 1973 this figure rose to 17 percent; reflecting the rapid economic expansion. However, in 1975, when the economic recession following the first oil crisis hit its bottom, this figure fell to 14.4 percent.[1] It should be noted also that the largest group of unemployed temporary workers and day labourers was and still is those people of middle age or more. Two major reasons for this phenomenon are the recent increase in Japanese life-expectancy and the sharp decline in birth rates. Japan is rapidly becoming a society of aged people at a speed hitherto unparalleled and as yet not experienced by any other 'advanced country'.[2]

Where did – and do – these unemployed, unskilled workers of middle age or more seek new jobs in this grim, economic atmosphere of 'structural depression'? Since the first oil crisis, it has been the nuclear power industry – one of very few sectors which has enjoyed rapid growth – that has been able to provide these people with great opportunities for new jobs.

INDUSTRIAL STRUCTURE

In 1970 there were only three nuclear reactors throughout Japan. By 1975 at the worst time of the economic depression the number of reactors increased to ten. Subsequently, this figure increased to nineteen by 1980 and as of mid-1985 there were thirty reactors operating. In addition to these, ten are already under construction, four are planned and another gigantic complex known as the 'Mutsu Ogawara Development', in the far north of Aomori Prefecture, is on the drawing boards. This last complex comprises a 'Spent-Fuel Reprocessing Factory' as well as two more nuclear power plants, one of which is a plutonium power plant. Construction of this project is expected to be completed in time to commence operation in the early 1990's.

The nuclear power industry which seems rational and scientific, characterized as it is by new high technology, is based upon the unique 'dual-structure' of Japanese socio-economic organization. That is, it has at its base countless sub-contractors who work under the monopoly of a few big electric power companies. In Japan, there are nine of these, namely Hokkaidō, Tōhoku, Tokyo, Chūbu, Hokuriku, Kansai, Chūgoku, Shikoku and Kyūshū, as well as the Japan Atomic Power Company. Working in close connection with these utility companies are five large nuclear power industry groups which are composed of Mitsubishi, Hitachi, Sumitomo, Tōshiba, Kawasaki Heavy Industry, Mitsui, Itō-chū and many other big electric and trade companies.[3]

Through the complex network of the dual-structure these electric power companies, in cooperation with the nuclear power industry enterprises, hunt for workers needed to maintain and repair the various parts of the nuclear power plants. These plants rely heavily upon the so-called 'Jinkai Senjutsu' ('strategy of throwing large numbers of men into action') in order to keep operating.

An electric power company enters into a contract with several prime contractors ('moto-uke'), such as Mitsubishi, Tōshiba, Hitachi, Sumitomo or any other of a number of major companies, in order to maintain its nuclear power plants. These prime contractors secure the necessary workforce for actual maintenance and repair of the plants by requiring numerous sub-contracting companies ('shita-uke') to recruit temporary

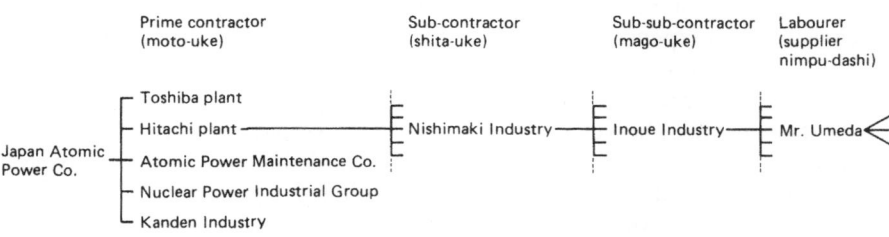

Figure 1

workers and day labourers. The sub-contractors again hand over this recruitment job to sub sub-contracting companies ('mago-uke'). Thus, when a temporary worker or a day labourer is actually offered a job, there are many sub-contracting companies involved in the recruitment. In extreme cases, six or seven sub-contractors are under a prime contractor. For example, Mr. Takashi Umeda from Kita-Kyūshū city, who was exposed to a large dosage of radiation while working at the Tsuruga Plant in 1979, was recruited through Inoue Industry in Shimonoseki City. This company was under Nishimaki Industry in Tokyo, which was in turn sub-contracted to the Hitachi Plant, one of the prime contractors for the Japan Atomic Power Company at Tsuruga.[4] In the case of the Tsuruga plant of the Japan Atomic Power Company there are five prime contractors and numerous sub-contractors under them, composing 'an intricate network of nodular roots'[5] extending from the nuclear power company through the mass of temporary workers and day labourers. At the bottom of this pyramidal structure are many people like Mr. Umeda who are called 'nimpu dashi' (labourer suppliers), although they themselves are often labourers. (See Figure 1.)

An electric power company usually pays a daily wage of approximately 40,000 Yen for each worker. However, the real wage each labourer receives is one-quarter or less of this amount, because each sub-contracting company takes its percentage. For instance, it is said that Kanden Kōgyō – a company capitalized at 300 million Yen, mostly coming from Kansai E.P.C., and which is a prime contractor for J.A.P.C. – takes between 15,000 Yen and 20,000 Yen daily per worker from the wages paid by J.A.P.C. It is widely believed that sub-contracting companies also take percentages amounting to about 10,000 Yen per worker each day.[6] Yet it is not only the sub-contracted workers who are exploited through salary deductions, but also the full-time workers of prime contracting companies. One example is the following.

The Atomic Power Maintenance Company, a company responsible for decontamination and control of radiation at nuclear power plants, kept receiving large sums of money from Chūba E.P.C. as a special allowance to cover various expenses, such as overtime – regardless of whether or not the workers did overtime. In addition, it received a daily allowance of 7,500 Yen per worker for cost of accommodation. Until an employee discovered this fact in 1976, however, the company did not pay the employees any special allowance for overtime work and even deducted 2,000 Yen from the daily accommodation allowance.[7] Another situation where workers are exploited through sub-contracting companies occurs when big businesses are faced with the possibility of bankruptcy and thus 'dispatch' their own employees to nuclear power plants. For example, since 1976 the Tsuruga branch of Tōyōbō, one of the largest textile companies in the Kansai area has been sending workers in their fifties to a sub-contracting company of J.A.P.C. until these workers retire at the age of fifty-five. Seventy per cent of their wages are paid by the sub-contracting company, while Tōyōbō contributes only 30 percent. Tōyōbō has thus adopted an extraordinary 'rationalization policy' by reducing its expenditure on salary payments. Although this company is no longer threatened with bankruptcy, it continues to employ this policy.[8] Incidentally, there is no age limit for sub-contracted workers at nuclear power stations. They therefore also provide retired local workers with the chance of re-employment.

A further problem regarding the recruitment of temporary workers and day labourers for nuclear power plants is the involvement of Yakuza (Japanese gangster) syndicates which operate some of the sub-contracting companies. These organizations are financed by the percentages they take from workers' wages, drug sales and loan-sharking to workers. In Tsuruga there are many branch offices of prominent Yakuza groups, such as Masaki-gumi and Kanamaru-gumi of the Yamaguchi syndicate, and they are responsible for sending a large proportion of the workforce to nuclear power plants in this area.[9]

It is stipulated by law that each nuclear reactor must be stopped once a year and the safety of over seventy items examined. This is called 'Tei-ken' (Regular Inspection) and it takes approximately one hundred days. During such periods nuclear power plants need a few thousand extra temporary workers. It is the Yakuza syndicates which have an important role in finding these workers, who are recruited by various syndicates throughout Japan, from places including the day labourers' quarters in Kamagasaki, Osaka and Sanya, Tokyo.

In the worst cases, Yakuza members force workers by intimidation to go to nuclear power plants in order to make up the numbers required during Regular Inspections. For example, in 1980, two young juniors working for a Yakuza group in Tsuruga were arrested because they attacked some workers in an attempt to force them to work at J.A.P.C.'s Tsuruga nuclear power plant.[10]

Table 1. *Number of nuclear power plant workers (man)*

Year	A. Company employees	B. Sub-contracted workers	C. Total	B/C (%)
1970	823	1,675	2,498	67.1
1971	904	4,339	5,243	82.8
1972	1,056	4,753	5,809	81.8
1973	1,512	6,960	8,472	82.2
1974	2,076	10,282	12,348	83.2
1975	2,282	13,798	16,080	86.8
1976	2,555	17,241	19,796	87.1
1977	3,233	22,129	25,362	87.3
1978	3,578	30,577	34,155	89.5
1979	3,759	30,495	34,254	89.0
1980	3,976	31,978	35,954	88.9
1981	4,374	36,158	40,532	89.2
1982	4,688	35,941	40,629	88.5
1983	5,367	41,072	46,439	88.4
1984	5,784	45,726	51,510	88.8

During the Regular Inspection the initial daily wage paid by an Electric Power Company is slightly higher than at other times. Therefore, the amount of profit for the Yakuza syndicates also increases, although this margin is not passed on to the sub-contracted workers.

Over the years the number of sub-contracted workers from all walks of life recruited to work at nuclear power plants has increased. There are ex-miners who lost their jobs in coal mines because of the drastic change in the Government energy policy, day labourers from Kamagasaki and Sanya, so-called discriminated *buraku* people, farmers away from their homes during the slack season and local retired workers. They are all unskilled and relatively older workers.

According to official figures published by the Central Registration Centre for People Engaged in Radiation Work – a foundation controlled by the Agency of Natural Resources and Energy – there were only 1,675 sub-contracted workers in 1970. However, in 1974, soon after the first oil shock, the number increased to over 10,000 and in the following decade it quadrupled. It is said that more than 50,000 sub-contracted workers are now working at thirty nuclear power plants in Japan. Not only has the total number of sub-contracted workers increased, but the ratio of sub-contracted workers to electric power company workers has been extremely high, right from the beginning. Since 1971 sub-contracted workers have occupied well over 80 percent of the workforce at nuclear power stations in Japan. In 1984 the number of sub-contracted workers was eight times that of the electric power company employees. (See Table 1.) As a result of exposure to high doses of radiation, a number of diseases – particularly cancer, leukemia and brain tumors – have repeatedly appeared amongst sub-contracted workers at nuclear power plants.

GENPATSU BUNKAI

On 1 July 1981 a trade union organization called 'Genpatsu Bunkai' was set up in order to protect sub-contracted workers exploited under frightful working conditions at nuclear power plants. Composed solely of such workers, however, it is both economically and politically weak. Officially, it is affiliated to the Tsuruga Branch of the All Japan General Transport Workers' Union, one of the most powerful organizations in Tsuruga City, although its political activities are independent and it is regarded as an autonomous trade union. It has a policy that allows any contracted worker to become a member, irrespective of his political inclinations.[11]

Genpatsu Bunkai began with 183 members, the majority of whom were workers from nuclear power plants in the Wakasa area of Fukui Prefecture. Generally, the names of members are kept secret, to avoid possible threats or sacking, although those of the chairman, Mr. Seiji Saitō, and the Chief Secretary, Mr. Hajime Nawa, are publicly known.[12] Today, Saitō works full-time in his role as chairman, although previously he spent many years as a sub-contracted worker at Mihama and other nuclear power plants. Born in Okayama Prefecture, he moved around Japan doing temporary work at construction sites for several years prior to going to Mihama in 1966. After marrying a local he decided to settle in the area and work at the nuclear power plants. This lasted until April 1984 when, while sick in hospital, he was secretly sacked by his employer, a sub-contracted company under Kanden Industry, a prime Contractor for Japan Atomic Power Co.[13]

Immediately following the establishment of this trade union for sub-contracted workers, Genpatsu Bunkai demanded that the Tsuruga offices of Japan Atomic Power Company and Kanden Industry accept collective bargaining and meet the following twenty conditions.[14]

A Guarantee a life of work without fear
1   Abolish dismissal without notice; compensate for suspension from work; guarantee re-employment.
2   Stop illegal kick-backs taken by sub-contracting companies; establish a standardized, fair wage that is uniform throughout the country.
3   Stop making workers do dangerous work in highly irradiated areas; if it is really unavoidable, pay a special danger allowance.
4   Pay the usual summer and winter bonuses.
5   Adhere to the law, including the Labour Standards Act.
6   Offer comprehensive compensation to commercial and industrial traders.

B Establish Safe Working Conditions

1 Make it compulsory to record irradiation dosages shown on pocket counters at checkpoints in pen, not pencil.

2 Do not falsify records of irradiation dosages in the various work areas.

3 Do not force workers to engage in illegal activities such as concealing accidents.

4 Turn off the nuclear reactor when work is required in highly irradiated areas such as inside the supply water heater and the condenser or when checking the core for defects.

5 Adhere to the law (Regulation to prevent Ionizing Radiation Sickness) by making the irradiation data taken from pocket radiation counters and film badges with TLDs available to workers at any time.

6 Allow workers to have scientific, appropriate medical checks, not only at hospitals designated by the companies, but also at public hospitals. The companies should meet all the necessary costs for this.

7 In the case of a worker becoming sick because of irradiation, provide accurate information so that he may receive benefits under the 'Workers Compensation for Ionizing Radiation Scheme' and under the 'Compensation for Injury at Nuclear Power Plants Law'.

8 Make it compulsory for the companies to contribute to general accident compensation insurance, industrial insurance and health insurance for subcontracted workers.

9 Reduce the Japanese 'permissible dosage of irradiation' to one tenth of the current level.

10 Based upon the principle of independent, democratic, public openness, abandon all secretivism and, in the case of regular inspections and accidents, stop the companies' practices of preventing workers from talking to outsiders, issuing confinement orders, keeping workers under observation and forcing workers to lie to government inspectors.

11 Allow workers working in highly irradiated areas to stop work for the day immediately their radiation exposure reaches the maximum daily limit, even if this happens within a very short time.

12 Stop discriminating between permanent company workers and prime contractor workers on the one hand and we subcontracted workers on the other.

13 Enforce attendance at their posts of both company and prime contractor technical officers and radiation control officers.

14 Allow subcontracted workers to keep their own 'Workers Irradiation Record Book' and let them check this when data is being recorded.

Both Japan Atomic Power Company and Kanden Industry refused to accept these demands on the grounds that 'there was no employment relationship' between the sub-contracted workers and the companies. Thus, they argued, the companies 'are not in a position to become involved in collective bargaining'.[15] At the same time, these companies ordered sub-contractors and sub-sub-contractors to eradicate Genpatsu Bunkai and to force worker-members to dissociate themselves from the organization. In addition, secretarial staff of the organization were threatened by Yakuza. For example, on one occasion, the front door of Mr. Saitō's house was smashed in during the middle of the night. On another, Yakuza members intruded on Mr. Nawa's house one night, threatening that his family would not be able to stay in Tsuruga unless he resigned his post.[16]

After overcoming these difficulties, however, Genpatsu Bunkai repeated their demands four times during the following two and a half months. As there was no response at all, the organization proceeded against Japan Atomic Power Company and Kanden Industry by presenting a 'Statement Demanding the Cessation of Unfair Labour Practices' to the Fukui Local Labour Relations Commission on 19th September 1981. The statement pointed out that as sub-contracted workers' tools, uniforms, helmets and all necessary equipment are provided by the companies, as their work is also supervised by employees of these companies and as the doses of irradiation to which the workers are exposed are checked and controlled by these companies, the sub-contracted workers must indeed be in an 'employment relationship' with the companies. It further noted that each month the wages paid to sub-contracted workers are categorized as 'Labour Wages' in J.A.P.C.'s accounts. Moreover, it concluded, as the above-mentioned twenty demands can be fulfilled by no companies other than J.A.P.C. and Kanden Industry, these companies must admit to being the true employers of these sub-contracted workers and therefore must accept promptly the collective bargaining demanded by Genpatsu Bunkai.[17]

Since the first hearing on 25th February 1981, at the Fukui Local Labour Relations Commission, thirty inquiries have been held to examine the real employment relationship between sub-contracted workers and the companies, but as yet no conclusion has been reached. Undoubtedly, part of the problem is that J.A.P.C. and Kanden Industry still do not really take Genpatsu Bunkai seriously. Not only is it the first trade union organization of sub-contracted workers in the history of the Japanese labour movement, it is unprecedented that so small and weak an organization appeal to a Local Labour Relations Commission against one of the largest companies in the country. That Genpatsu Bunkai has had the courage to undertake such an appeal is indicative of the serious nature of their complaints.

Although the inquiry at the Local Labour Relations Commission is progressing very slowly – due to delaying tactics by the two companies – Genpatsu Bunkai has already achieved some limited improvements in the working conditions of sub-contracted workers. For example, not long after the establishment of the organization, J.A.P.C. voluntarily began using pens instead of pencils to record radiation dosages to which workers are exposed. 'Workers' Irradiation Record Books', which the company had previously kept, were also suddenly made available to individual workers.[18] In addition, some of the sub-contracting companies began to introduce health and industrial insurance schemes for their workers.[19] Of course, the main reason that the companies improved certain working conditions is that, because there is no list available of Genpatsu Bunkai members, they have no way of knowing which workers belong to the organization. Consequently, they fear reports to the union by worker-members forced to accept unreasonable working conditions or to engage in highly dangerous operations.

Genpatsu Bunkai has also received the support of ordinary citizens and in August 1983 The Association to Protect the Rights of Nuclear Power Plant Sub-contracted Workers was set up in Fukui Prefecture. The Association is composed of people from a wide variety of backgrounds, including academics, religious groups, school teachers, representatives of various trade unions and people involved in the anti-nuclear movement.[20] The immediate aim of this diverse organization is to support Genpatsu Bunkai's case at the Local Labour Relations Commission by exerting pressure upon the electric power companies. As a consequence, the companies can no longer simply ignore Genpatsu Bunkai and the influence it is having on various sections of society.

The political activities of this organization are not limited to the Fukui area, but extended to other parts of Japan too. In August 1981 Mr. Saitō and Mr. Nawa attended the World Peace Conference in Hiroshima and presented a report on their activities in a panel entitled 'Irradiation of Workers and Dumping of Radio-Active Nuclear Waste in the Ocean'. Later that same month they conducted a 'Caravan Campaign' around nuclear power plant sites in Fukushima, Ibaragi and Shizuoka prefectures. They travelled to these areas in a mini-bus and at each place organized meetings with sub-contracted workers, where they promoted their ideas and exchanged opinions with various people. During this trip they distributed 12,200 leaflets explaining their movement.[21] In another project in 1982, they produced a one-hour-long film, '*Genpatsu wa Ima*' (Nuclear Power Plants Today), with the help of the *Gekidan Mingei* theatrical troupe. The film depicts the working conditions at nuclear power plants, the health and living conditions of sub-contracted workers exposed to high dosages of radiation and the political activities of Genpatsu Bunkai. So far, this film has been shown more than one hundred times in various parts of Japan, although mainly in areas where power plants already exist or where they are planned.[22]

There are two fundamental weaknesses in the Genpatsu Bunkai organization. One lies with the workers, the other with the union itself. As most sub-contracted workers are employed on a temporary basis and thus move from one power plant to the next in search of work, they tend to lose contact with their trade union and break away from it when they move away from the Fukui area. This is because the labour movement of this organization is at present centred in the Tsuruga and Wakasa areas of Fukui prefecture. Although there is a potential for workers who move away from the area to spread information about Genpatsu Bunkai's activities throughout Japan, this does not seem to happen, for the very reason that the workers lose contact with the organization.

The other problem associated with this trade union is the basic acceptance of nuclear power plants, both by the sub-contracted workers – who, despite knowing the dangers involved, earn their livings there – and by the union itself which, although critical of conditions in the power plants, accepts the existence of the nuclear power industry. This crucial point makes the Genpatsu Bunkai movement fundamentally incompatible with anti-nuclear civilian organizations. Thus, although the union does maintain contact with such groups, it tends to be isolated because of this basic policy. Indeed, this was abundantly clear at the World Peace Conference in Hiroshima in August 1981. There, the views presented by Genpatsu Bunkai directly conflicted with the theme of the Conference, which aimed at the immediate abolition of nuclear power plants and the so-called peaceful use of atomic energy. Following the Conference, the union members had to admit that their participation 'ended by deepening the discrepancy'[23] between the basic Conference philosophy and Genpatsu Bunkai's policies.

DENRYOKU RŌREN

So far, we have looked at Genpatsu Bunkai in detail, but how does it relate to the official trade union for employees of electric power companies? Let us first examine the history and nature of *Denryoku Rōren* (Confederation of Japanese Electric Power Workers), a gigantic organization comprised of unions from the nine major electric power companies. Ninety-seven percent of the 130,000 employees of these companies belong to unions under the control of Denryoku Rōren.[24] In January 1966 this organization announced the 'First Proposal for Atomic Power Development', reflecting its initial ideas on the nuclear energy policy designed by the Government and the economic circle. This proposal was made because the first commercial nuclear reactor – of about 160,000 kw and operated by J.A.P.C. – was about to start in July of that year at Tōkai Village in Ibaragi Prefecture. In addition, it was planned to commence construction of plants at Tsuruga and Mihama in April and December of the same year.

In this proposal, Denryoku Rōren stated that it would 'spare no effort to cooperate in developing electricity generated by atomic power'.[25] Although this has been its basically consistent policy ever since, there have been some changes in Denryoku Rōren's attitude towards 'cooperation'. Up until 1975 it had been critical of the Government and the electric power companies regarding the way in which the nuclear power industry was being developed. This criticism was augmented in the Fifth Proposal of 1975 when it announced that 'it is doubtful whether atomic energy can be so easily seen as an alternative to oil' and that the existing nuclear reactors must be improved technically, as they cannot be regarded as perfect commercial reactors in their present state.[26] Of course, the reason for such critical views was simply the reality that faced the workers: on the one hand, they were suddenly exposed to greatly increased dosages of radiation, and on the other, they could no longer endure the work inside the concrete shield building at the plants. Despite the organization's conservative nature and its tendency to cooperate with the companies in the development of the nuclear power industry, it could not ignore the danger that faced its members. In fact, in the 1975 Proposal, it took a hard line and threatened not to cooperate with increasing the number of nuclear power plants, unless the government and the power companies adopted greater safety measures to counter the irradiation problem.

But as a result of this line of argument, considerable internal conflict followed Denryoku Rōren's Fifth Proposal. The three leaders, including President Inagaki were criticized by workers from various unions and eventually they were forced to resign. Mr. Kōichirō Hashimoto, Chairman of the Chūbu Electric Power Company Workers' Union, took over the position of president in August 1975. Under the new leadership, Denryoku Rōren adopted a different stance which openly supported the national policy of developing the nuclear power industry. The reason for this dramatic change of attitude can be understood from a glance at official statistical data on workers' exposure to radiation. Prior to this, the amount of radiation to which company employees (that is, permanent staff) were exposed was significantly greater than that to which sub-

Table 2. *Total irradiation dosage (man-rem)*

Year	A. Company employees	B. Sub-contracted workers	C. Total	B/C (%)
1970	236	326	562	58.0
1971	370	896	1,266	70.7
1972	464	1,433	1,897	75.5
1973	596	2,098	2,694	77.9
1974	701	2,427	3,128	77.6
1975	716	4,283	4,999	85.7
1976	769	5,473	6,242	87.7
1977	726	7,399	8,125	91.1
1978	782	12,418	13,200	94.0
1979	858	10,872	11,730	92.7
1980	828	11,105	12,933	93.6
1981	785	11,933	12,718	93.8
1982	733	11,767	12,500	94.1
1983	661	11,206	11,867	94.4
1984	621	11,534	12,156	94.9

contracted workers were exposed. After 1975, however, this situation was reversed and the steady decrease in the exposure to radiation of company workers coincides with an increase to such exposure of sub-contracted workers (see Table 3).

As a result of the re-election of Denryoku Rōren secretarial staff in 1975, Mr. Ken'ichi Aoki, Chairman of the J.A.P.C. Workers' union was elected to the position of Secretary General. He is a keen promoter of the nuclear power industry and his ideas led to the creation, in 1976, of a special section in the Annual Policy Report, entitled 'The Development of Nuclear Power (Securing a Source of Energy)'. Since then, promotion of the nuclear power industry has escalated each year, such that Denryoku Rōren did not dream of changing its pro-nuclear policy following the world's (till then) worst nuclear power plant accident at Three Mile Island, Pennsylvania in March 1979. Indeed, at the 26th Annual General Meeting the same year, President Hashimoto even proposed 'self-control of strike action at nuclear power plants', and advocated that the organization firmly maintain its policy to actively promote the development of this energy form, on the grounds that an alternative to oil was urgently needed.[27] His perception of this urgency led him to criticize grass-root anti-nuclear movements and to say: 'it is unforgivable that some people are spreading false information throughout the nation.'[28] Referring to the Three Mile Island accident, he added, 'As the details of the Three Mile Island accident are now clear, I am convinced of the safety of the nuclear power plants in our country. I think their safety will be improved even further in the future.'[29]

Table 3. *Average irradiation dosage (rem)*

Year	A. Company employees	B. Sub-contracted workers	B/A (times)
1970	0.27	0.19	0.66
1971	0.41	0.21	0.51
1972	0.44	0.30	0.68
1973	0.39	0.30	0.77
1974	0.34	0.24	0.71
1975	0.31	0.31	1.00
1976	0.30	0.32	1.06
1977	0.22	0.33	1.50
1978	0.22	0.41	1.86
1979	0.23	0.36	1.57
1980	0.21	0.38	1.81
1981	0.18	0.33	1.83
1982	0.17	0.33	1.94
1983	0.12	0.27	2.25
1984	0.11	0.25	2.27

However, less than eighteen months after this speech, which overflowed with confidence about the safety of Japanese plants, a serious accident occurred at J.A.P.C.'s Tsuruga plant. In this case, a large quantity of highly irradiated water flowed through a crack in the fourth supply water heater, and many sub-contracted workers were exposed to high dosages of radiation in the 'clean up' that followed.[30] But this accident had no effect on Denryoku Rōren's policy either. In fact, it is still usual for its members to visit areas where nuclear power plants are planned in order to promote the notion of 'safe nuclear technology', by handing out leaflets and talking to the local people.

As is clear from this brief historical analysis of Denryoku Rōren, its fundamental attitude is one of discriminatory control, in which company employees are engaged exclusively in clean and safe work and all dangerous operations are left to sub-contracted workers. This is clear from the fact that since 1975 and the decrease of radiation dosages to which company employees are exposed, Denryoku Rōren has become more and more conservative. Similarly, it has never shown any concern for the working conditions of sub-contracted workers. Nor has this basic discriminatory attitude changed since the establishment of Genpatsu Bunkai. Indeed, it appears that a few days after Genpatsu Bunkai was formed, the Chairman of the Kanden Industry Workers' Union – which is affiliated to the Kansai Electric Power Company Workers' Union – visited the Tsuruga plant and made a speech to sub-contracted workers who were assembled by the company for the occasion. In his speech he is reported to have said that:

The General Transport Workers' Union, to which Genpatsu Bunkai belongs, is a Commie's union under the control of Sōhyō [General Council of Trade Unions of Japan]. This organization is incompatible with nuclear power plants. The nuclear power industry secures your work and lives. If you have joined Genpatsu Bunkai without careful consideration, please withdraw from it as soon as possible.[31]

Genpatsu Bunkai claimed that this action by Kanden Industry's union leader was clearly illegal and not in accordance with Article 7, Item 3 of the Trade Union Law. Hence, this matter was included in the 'Statement Demanding the Cessation of Unfair Labour Practices', presented to the Fukui Local Labour Relations Commission.

Soon after the establishment of Genpatsu Bunkai, Denryoku Rōren adopted a new plan to 'organize sub-contracted workers' and to 'modernize the employment relationship between the companies and sub-contracted workers'. At its 28th Annual General Meeting in September 1981, President Hashimoto clarified the meaning of 'modernization of the employment relationship between the companies and sub-contracted workers' and explained that Denryoku Rōren would make an effort to 'secure long-term jobs for local workers [i.e. sub-contracted workers], so that they can settle down'.[32] But he made no reference whatsoever to the 'modernization' and improvement of working conditions for sub-contracted workers. Clearly, for Denryoku Rōren 'modernization' and 'organization' meant nothing but the guarantee of a constant supply of contracted workers, in order to lessen the danger of irradiation to company employees. Incidentally, it is interesting to note that a few years ago, Denryoku Rōren sent delegates to the Australian Council of Trade Unions to ask its leaders to change their anti-uranium export policy.[33] Without doubt, this already gigantic, strongly pro-nuclear organization is seeking to gain yet more power and to further promote the nuclear industry, irrespective of its effects on workers and the environment.

CHŪGOKU DENSAN

While the vast majority of electric power company workers belong to the unions controlled by the pro-nuclear Denryoku Rōren, only about 3,000 company workers belong to the four different unions which form *Densan* (abbreviation of Nippon Denryoku Sangyō Kumiai, Confederation of Japanese Electrical Industrial Workers' Unions). Unlike Denryoku Rōren, Densan adopted a very strong anti-nuclear policy right from the beginning. Generally speaking, Densan is known as a very militant labour organization, not only because of its stance on the nuclear issue, but also with regard to other aspects of the labour movement too. It was set up as early as 1947, not long after World War Two, not as a confederation of company unions, but as the single national electric power industrial union organization.[34]

Following the beginning of the Korean war in 1950, however, this union was virtually destroyed by the 'Red Purge' carried out by the U.S. Occupation Forces. By 1957, with the exception of a small number of workers from Chūgoku Electric Power Company who remained in Densan, nearly all the Japanese workers engaged in the electric power industry joined the unions controlled by the new organization, Denryoku Rōren.[35] At the end of the 1960s, some workers from Hokkaidō and Kyūshū responded to the appeal by a minority of workers at Chūgoku E.P.C. and joined the Densan Organization. Today, Densan belongs to Sōhyō (General Council of Trade Unions of Japan), which is under the influence of the Japan Socialist Party.

The core of the Densan organization is still the Chūgoku E.P.C. Densan Union (hereafter Chūgoku Densan), which has only 650 members. In 1973, when Chūgoku E.P.C. started experimental operations at its first nuclear reactor in Shimane Prefecture, Chūgoku Densan issued a statement saying that 'a nuclear power plant will cause pollution by irradiation and destroy the lives of farmers and fishermen'.[36] The following year when the Shimane Plant commenced full-scale operation, worker members of Chūgoku Densan blockaded and picketed the gate of the plant, stopping the plant's operations for half a day.[37] This may be the first case in the world in which workers from an electric power company have actually stopped the operation of a nuclear power plant. Another example of this organization's strength and determination occurred between 1977 and 1978 when Chūgoku Densan cooperated with the fishermen of a small town, Hōhoku in Yamaguchi Prefecture, and eventually succeeded in preventing Chūgoku E.P.C.'s plan to build a new nuclear power plant there. During this struggle it had to contend with many problems. Not only did it have to develop strategies to prevent construction of a new power plant, but it also had to fight the interference of fellow company workers belonging to Denryoku Rōren.[38]

This organization does not only take an actively anti-nuclear stance within its own company, but it participates in anti-nuclear movements in other areas too, predominantly by cooperating with regional anti-nuclear civil groups.[39] Its social conscience is also strong and it is concerned about the plight of various minority groups. Among these are sub-contracted workers – both those at nuclear power plants and other power plants – and workers of *buraku* background who are typically discriminated against.[40] Densan's accurate awareness of social problems in general and its political activities are beginning to attract the interest of workers at other electric companies, however its numbers are increasing only slowly.

CONCLUSION

The trade union organization for nuclear power plant workers is thus divided into three: Denryoku Rōren; Chūgoku Densan; and Genpatsu Bunkai. Each has a very different emphasis and is aimed at a particular clientele. Denryoku Rōren is the gigantic, conservative organization, which takes a positively pro-nuclear stance and caters for permanent company workers. Chūgoku Densan, on the other hand, is a very small, yet remarkably powerful, left-wing union, actively opposed to nuclear power. It is supportive of both permanent and sub-contracted workers, not only regarding nuclear issues, but other matters too, particularly those related to discrimination. The third trade union, Genpatsu Bunkai, shares aspects of both Denryoku Rōren's and Chūgoku Densan's policies, and yet it remains separate and isolated.

Like Denryoku Rōren, it adopts a pro-nuclear policy, although its specific *raison d'être* is concern for the welfare of sub-contracted workers, an aim it shared with Chūgoku Densan. Despite these common elements of policy, however, Genpatsu Bunkai is also in conflict with certain aspects of each of the other union's policies. Whereas sub-contracted workers are not the concern of Denryoku Rōren, which sees them more as a means of diverting the danger of exposure to high doses of radiation away from its own members (permanent electric power company employees), the conditions and circumstances in which these labourers work is the paramount concern of Genpatsu Bunkai. Thus, despite their common pro-nuclear stance, the two unions are in conflict.

The situation with Chūgoku Densan is similar, as although there is a shared concern for the welfare of sub-contracted workers, this organization is categorically opposed to nuclear power itself. Genpatsu Bunkai, on the other hand, does not and cannot take this stance, for although well aware of the dangers of working in nuclear power stations – and as a consequence, fighting for higher safety standards and better working conditions for workers therein – it recognizes and accepts these plants as a viable source of livelihood for its members, and thus adopts a pro-nuclear policy. Clearly, this is the ultimate contradiction and problem for this organization. Its pro-nuclear policy isolates it from Chūgoku Densan and humanitarian and conservation-oriented civil movements which are otherwise sympathetic to its worker support and activities, and hence makes it difficult for Genpatsu Bunkai to strengthen and expand.

NOTES

1 Nishikawa S., 'Sangyō kōzō no henbō' ('Changes in industrial structure'), in *Kigyō to rōdō* (Industry and labour) Juristo sōgō tokushū, no. 14, Yūhikaku, 1979, p. 66. 'Temporary workers' are people under contract of less than one year and 'day labourers' are people employed on a daily basis for the duration of several months. 'Seasonal workers' are categorized, therefore, as 'temporary workers'.
2 *Ibid.*, p. 69. The average age of the Japanese workforce in 1985 is estimated at 41.3 years for males and 40.6 years for females.
3 Senda H., *Enerugii sangyōkai* (The energy industry world) (Kyōikusha), 1984, pp. 202–5; Higuchi K., *Yami ni kesareru genpatsu hibakusha* (Irradiated nuclear power plant workers who face into oblivion), San'ichi shobo, 1984, pp. 210–4.
4 Shibano T., *Genpatsu no aru fūkei* (A view of nuclear power plants), vol. 1, Miraisha, 1983, p. 66.
5 B. Nee, 'Sanya: Japan's internal colony' in *Bulletin of Concerned Asian Scholars*, vol. 6, no. 3, 1974, p. 14.
6 Genpatsu Bunkai, *Genpatsu nyūsu* (Nuclear power plant news), no. 1, 1981, and mentioned by Mr. S. Saitō, Chairman of Genpatsu Bunkai, during an interview in Tsuruga in Japan, 1985.
7 Morie S., *Genpatsu hibaku nikki* (The diary of an irradiated nuclear power plant worker), *Gijutsu to Ningen*, 1982, pp. 158–60.
8 Horie K., *Genpatsu jipushii* (Nuclear power plant gypsies) Gendai shokan, 1984, p. 107; *Shūkan Posto*, 17th November 1978.
9 Shibano, 1983, p. 160.
10 *Fukui Shinbun*, 1st February 1980.
11 *Genpatsu nyūsu*, no. 2, 1982.
12 Mentioned by Mr. Saitō, S. during an interview in Tsuruga in January 1985.
13 Shibano, 1983, vol. 2, pp. 182–3.
14 A leaflet issued by Genpatsu Bunkai (no date).
15 *Genpatsu nyūsu*, no. 4, 1982.
16 Shibano, 1983, vol. 2, p. 187.
17 A leaflet issued by Genpatsu Bunkai (no date).
18 Mentioned by Mr. Saitō, S. during an interview in Tsuruga in January 1985.
19 A leaflet issued by Genpatsu Bunkai (no date). It is reported, however, that since these schemes were introduced, workers' daily wages have been cut.
20 *Mamorukai nyūsu* (The protection society newsletter), no. 1, August 1983.
21 A leaflet issued by Genpatsu Bunkai (no date).
22 A leaflet issued by Genpatsu Bunkai (no date).
23 *Genpatsu Bunkai nyūsu* (Genpatsu Bunkai news), no. 2, September 1981.
24 Shimizu H., 'Denryoku rōdosha no han-genpatsu tōsō' ('The anti-nuclear power plant struggle by electric Power workers') in Nishio B. (ed.), *Han-genpatsu mappu* (The anti-nuclear power map), p. 178.
25 Horie K., 'Genpatsu no uchi to soto', ('Inside and outside nuclear power plants') in Watanabe E. (ed.), *Rōdōsha no Sabaku* (Labourers in the desert), Tsuge shobō, 1982, p. 150.
26 *Ibid.*, pp. 150–1 and *Yomiuri Shinbun*, 13th January 1975.
27 *Denryoku rōren*, January 1981.
28 *Ibid.*
29 *Ibid.*
30 *Fukui Shinbun*, 3rd April 1981.
31 Shibano, 1983, vol. 2, p. 197. Sōhyō (General Council of Trade Unions of Japan) is not under the influence of the Japanese Communist Party, but of the Japan Socialist Party.
32 Horie, 1982, p. 115.
33 Han-genshiryoku hatsuden jiten henshū kai (ed.), *Han-genpatsu jiten*, vol. 2, Gendai shoten, 1979, p. 251.

34 Shimizu H., 'Genpatsu to denryoku rōdōsha' ('Nuclear power plants and electric power workers') in *Han-genpatsu jiten*, vol. 2, p. 185.
35 *Ibid.*
36 *Ibid.*, p. 190.
37 *Ibid.*, p. 190.
38 *Han Genpatsu rōdō undō: Densan Chūgoku no tatakai* (Anti-nuclear power plant labour movement: The struggle of Densan Chūgoku) edited and published by Gogatsusha, 1982, pp. 202–13.
39 *Ibid.*, pp. 27–30 and 166.
40 *Ibid.*, pp. 54–7, 151 and 155–7.

8 Street labour markets, day labourers and the structure of oppression

MATSUZAWA TESSEI

THE NATURE OF *YOSEBA*

The *yoseba* is a peculiarly Japanese kind of slum. In the *yoseba* the contradictions of Japanese imperialism are exposed with the utmost intensity; it is, one could say, the epitome of Japanese society. The majority of its dwellers are day labourers who, as Marx put it, form 'a disposable industrial reserve army ... for the changing needs of the self-expansion of capital, a mass of human material always ready for exploitation'.[1] *Yoseba* labourers suffer from the 'uncertainty and irregularity of employment, the constant return and long duration of gluts of labour [known as 'abure' in *yoseba* jargon] all these symptoms of a relative surplus population ...'[2] The *yoseba* has the following distinct features:

(i) concentration of flophouses in one quarter of the metropolis (called *Doya-gai*). Their tenants are day labourers employed in manual or physical labour, who pay rent by the day and stay on for an indefinite length of time. Inhabitants of '*Doya*' range from paupers unable to work, through those able to work, and the poverty-stricken of the active working class, to skilled '*tobi*' workers.

(ii) The majority of these inhabitants are middle-aged or aging men who are cut off from family life in one way or another.

(iii) The 'average citizen' and public officials of all ranks and sectors discriminate against and try to segregate the *yoseba* and day labourers from 'civilian life'.

Yoseba were established in the process of the formation of Japanese capitalism in the last half of the nineteenth century. *Yoseba* in Sanya (Tokyo), Kamagasaki (Osaka), Kotobuki (Yokohama) and Sasajima (Nagoya) are well known. Others existed in Sapporo, Aomori, Sendai, Kanazawa, Kōbe, Hiroshima, Kita Kyūshū, Fukuoka (Hakata) and

147

Okinawa, expanding with the recession and the depression of the 1920s and the 1930s. The *yoseba* and the day labourers nationwide were the first to bear the weight of Japanese fascism on their shoulders, being exploited through forced labour to establish a repressive domestic system and to carry out external aggression.

In the late 1950s *yoseba* were created anew to meet the growing demands of the 'high growth economy' brought forth by the dismantling of the rural community and reorganization of the industrial structure. The 'rioting' sixties were followed by the tightening of the domestic security control system. This, along with the depression triggered off by the 'energy crisis' of 1974, has considerably affected the *yoseba*; they are now reduced in size and qualitative change has been forced through. The *yoseba* are now gasping for breath under strict control, high-level exploitation and increased hostility and violence such as could be seen in the Kotobuki case where between December 1982 and February 1983 a group of adolescents assaulted and murdered day labourers. Another recent trend is the steady increase in the number of aging day labourers. In Sanya, 66 percent of the labourers are in their forties or fifties whereas only 16 percent are in their thirties; 99.7 percent are single males. This is what distinguishes the *yoseba* from other large slum areas in Asia such as Klong Toi (Thailand) and Tondo (Philippines) where the inhabitants lead a family life, their numbers evenly distributed into both sexes and different age groups. Beneath their chronic poverty a world of self-subsistence may still be detected, which some evaluate as 'an intricate community in itself'. Natural environment, economic conditions and historical circumstance within the Japanese imperialist system make survival harder than ever. However, these same pressures work to strengthen the bonds and mutual cooperation among the day labourers. Coming from all walks of life, the labourers are nevertheless united as individuals ostracized by the political, economic and social system of Japanese imperialism. This common factor, of shared oppression, has been transformed by the day labourers into a source of energy for the anti-imperialist struggle, bursting out in the form of 'riots' and 'taking the law into their own hands'. For example, *yoseba* labourers are among the few workers who continue to take a firm stand against 'collaborationist [company aligned] trade unions' introduced by Japanese imperialism to undermine the working class.

The historical formation of the *yoseba* is intrinsically linked with that of the Burakumin (outcaste) and the Korean-in-Japan communities, as well as with that of the Ainu and Ryūkyū Islanders (*Uchinanchū*). The imperialist oppression of the Japanese minorities becomes visible at the *yoseba* where the contradictions of imperialism are projected. It also indicates that the minorities in Japan, including most of the Burakumin and Koreans (estimated to be between 600,000 and 800,000) form the bottom strata of the Japanese system, along with the day labourers of the *yoseba*. The core population of *yoseba* is made up of the following:

(i)   Displaced peasants: when the Organic Law of Agriculture came into effect in 1961 the ratio of the agricultural population was drastically reduced to make way for industrialized-capitalist agriculture. Many peasants were affected by this structural change.

(ii) Discharged colliers: many coal miners were discharged as a result of changes in energy policy.

(iii) Workers discharged in the process of rationalization of primary and other industries.

(iv) Junior high school graduates of rural areas who were employed en masse.

(v)   The unemployed in general and the so-called social misfits.

The minorities of Japan may be added to these categories. All together they form the bottom layer of Japanese society.

Here we must point out the fact that the bottom layer of society is connected both to the prisons and to the mental institutions. The triangle of *yoseba*-prison-mental hospital is a familiar course to many day labourers whose overwork supported Japan's high economic growth. Now they are faced with long-term unemployment *'abure jigoku'* (the hell of unemployment) while many suffer from internal diseases caused by hard physical labour and alcohol. So harsh is the discrimination that it is surprisingly easy for a *yoseba* labourer to end up in prison or mental hospital: a drop of alcohol may become the proof of alcoholism, scavenging is enough to justify a charge of theft. The brutal treatment of *yoseba* labourers was given extensive coverage by the mass media in relation to the Utsunomiya Hospital patients murder case of Spring 1984.[3]

Yoseba or *doya-gai* basically function as a transit port for day labourers moving from one labour camp and/or construction site to the next, and as a home ground to return to. At present there are about 11,000 labourers in Sanya, 30,000 to 40,000 in Kamagasaki, 6,000 to 7,000 in Kotobuki, 500 in Sasajima and 1,000 in Chikko (Hakata), who go out daily to work in national and local projects or private enterprise. From calculations based on the 150,000 registered day labourers in the public employment offices the total number of *yoseba* labourers in Japan is estimated to be over 400,000. As well as the ordering body (usually state, local or private enterprise) called *'motouke'* or prime contractor, there are various intermediary sub-contractors who milk the labourer's wage. At the end of this multi-layered sub-contract system – structure of multiple exploitation of labour – lies the *'tehai shi'* or the unlicenced recruiting agent who is almost always linked, either directly or indirectly, with a gangster organization (*yakuza*). He serves as a valve to control the flow of labour power supply for the sake of big business. In order to secure a job, the day labourer must rely on the following contacts:

(i)   *Tehaishi* or *oyaji* (boss). At a fixed point of the *doya-gai* the labourer negotiates directly with the recruiters.
(ii)   Public Employment Security Office.
(iii) Semi-official job agencies.
(iv)   Fellow day-labourers.

Others hold jobs that are more or less stable.or regular. In recent years, workers attracted by newspaper advertisements and the honeyed words of recruiters at railway stations tend to be exploited under extremely poor working conditions. Common jobs are those which only require unskilled manual labour, e.g. construction, public works, wharf labour and transport, where the labourer remains disposable anytime. There are also some *tobi* labourers living in the *yoseba* who hold specific construction skills, e.g. scaffolding, steel erecting, bridge building, etc. The *yoseba* serves as a safety valve (or the holder of relatively surplus population) essential to the capitalist system in order to control the supply of labour power according to the fluctuations of the market. But this also permits the day labourer to 'escape' at any time knowing he can eventually find another similar job elsewhere. Hence *yoseba* labourers often change their field of labour (*genba*); some go back and forth from Sanya to Kamagasaki; others may work in a distant construction camp (*hanba*) living in a framehouse for long periods; still others return to the *yoseba* after moving around from one camp to another: they subsist in the floating form. Basically they work for a day's wage (*dezura*), but at the *hanba* payment is effectuated every ten days and sometimes further delayed for another ten-day period. *Hanba* wage is relatively low and expenses (*shoshiki*), e.g. food, laundry, bath, etc., are deducted. Where shops are not accessible alcohol is more expensive than usual. 'Fundamentally, [they] are only hired by the day, and therefore under the most precarious form of wage.'[4]

Without house or decent living, outside the bonds of the existing system, this floating existence becomes a source of resistance against the establishment and authority. Alternating tides of overwork and enforced idleness confront the day labourer. This continuous contraction and expansion accumulates in him in the form of strong resentment against the system and a deep comradeship with his fellow sufferers; and in a burst of anger he cries out: '*I have had enough!*'. The day labourers of the *yoseba* are the incarnation of what Marx described a hundred years ago as 'the proletariat in revolt'. More efficient exploitation of labour power forces more day labourers into the sphere of pauperism while capitalist wealth accumulates. The law of accumulation of capital functions in its pristine form only in the *yoseba* 'as if it were a law of nature'. It takes on a political facade and is therefore twisted when applied to organized workers in a large enterprise. Capitalism has bought off the upper strata of workers (those who think of themselves as 'middle'-class) and is now dumping the expenses on the shoulders of the bottom strata, including part timers, piece workers and temporary workers, seasonal workers, and particularly the inhabitants of the *yoseba*.

The old tactic of 'divide and rule' is present here. Those in power have openly discriminated against the day labourers of the *yoseba* by isolating them from the 'average citizen', and now the friction between the two is being raised to the heights of hostility. Assaults against day labourers by ordinary citizens are condoned and even incited by those in power.

The Ministry of Justice is currently working on a revision of the Penal Code which would include a Preventive Detention (*hoan shobun*) system, under which persons seen as endangering public order could be compulsorily consigned to prison or mental hospital. These revisions have serious implications for human rights, yet they would not constitute anything new for residents of the *yoseba* for whom the oppressive circuit of *yoseba*-prison-mental hospital is already a grim reality.

These measures taken by the state are backed by the discrimination, contempt and prejudice of the 'average citizen' against the *yoseba* and its day labourer inhabitants. The whole of the mass media functions to further amplify and exaggerate this popular prejudice. The 1982 assault committed by a group of junior high school students on day labourers in Kotobuki is an extreme case but not the only one. Similar cases have occurred in Sanya and Kamagasaki: *chōnaikai* or town associations in Sanya submitted a petition to the ward council verbally attacking the *yoseba* and the day labourers; a film director shooting a documentary film on Sanya was stabbed to death by an ultra-rightist who was affiliated with the group in Sanya calling for the strengthening of the Tennō system. Another leading activist of the militant leftist Day Labourers' Union movement in Sanyo was shot to death by members of the same gangster group. The situation in the *yoseba* today could be described as pre-fascist, since it is symptomatic of fascism that ordinary people become politicized so that, on their own initiative and without orders from above, they resort to direct and forceful measures to suppress those struggling against the establishment. However, the struggle against this is growing, and the militancy carried forward from the 1960s is finding support among radical students, workers, and citizens of the new-leftist trend.

THE DAY LABOURERS' STRUGGLE

Historical background – Edo period

The word '*yoseba*' derives from '*ninsoku yoseba*' (navvy's camp) established in Ishikawajima, at the mouth of Sumida River in Edo City, in 1790 by Matsudaira Sadanobu, a senior member of the Shogunate Council of Elders. The context was one of external crises and a decline in the patrimonial-feudal system as a commodity economy grew. The peasants who were the bedrock of the patrimonial-feudal system rebelled against relentless exploitation. Many fled to the cities to break away from severe structural controls such as *goningumi* (five-family neighbourhood unit which served as the lowest unit in a system of mutual surveillance, responsibility and control) and the census registration. At the end of the eighteenth century measures were taken to detain the growing pauper population in Edo in one place, as part of an attempt to reorganize the feudal system which was on the verge of falling apart. Totally innocent vagrants and ex-convicts were 'hunted down' by the policeforce and relocated to the *ninsoku yoseba* as 'subversive elements'. The aim was that of preventive detention and intimidation of the rest of the population. The internees were forced to work in jobs like oil pressing or construction. More *yoseba* were created in Ibaraki Prefecture for reclamation work, followed by Nagasaki, and towards the end of the Shogunate in Hakodate and Yokosuka. These were the prototype of the present *yoseba*.

Oyakata (boss) – *Hanba* system and the *Yoseba*, from the late nineteenth century to the end of the 1920s

The essential features of *ninsoku yoseba* were passed on to the Meiji Era as Japan experienced the genesis of capitalism. The first capitalists regarded the *yoseba* dwellers and convicts as a handy source of low-wage labour power. Convict labour was exploited to the full in the shipyards and steel mills of Yokosuka, the collieries of Hokkaidō and Kyushu, and in reclamation and public works in Hokkaidō. Merciless over-working led many to attempt to escape. With the development of capitalism more and more wage labour power was needed, so the *oyakata* (boss-supervisor of labourers) system was created. *Oyakata* used recruiters all over the country to gather the bond labourers making them 'tenants' and dumping them into 'coolies' quarters'. This was known as the '*hanba*' or '*naya*' system. Daughters of poor peasants and the fallen samurai classes, bound by long-term contracts, were forced to work in confinement in textile mills and silk reeling. Their hard labour greatly contributed to the prosperity of the Japanese textile industry and Japanese capitalism itself.[5] Even Yahata Steel Mill, equipped with the best imported machinery, kept about 3,500 bond labourers at the mills and 2,000 more under *oyakata* supervision. In both cases the lodgings were 'authentic slave pens'.[6]

Between 1885 and 1890, *yoseba* were established for the first time in back slums and flophouse quarters to keep up with the growing *hanba* demands for day labour power. Peasants who were affected by the first Japanese depression and had settled in the city slums, were absorbed into the *hanba* system through *yoseba* recruiting agents. In this way the day labour power supply structure was established, though full-scale development of *yoseba* came between the late 1890s and the early twentieth century, coinciding with the development of Japanese capitalism and Japanese imperialism. Yokoyama Gennosuke[7] identifies 'the areas of the lowest stratum of Tokyo' or 'back slums'[8] and 'flophouse quarters' 'where paupers of all kinds flock together'.[9] Here 'pauper' included the poor active day labourer. They worked as navvies, carters, ricksha men, scavengers and buskers. The Tokyo City 'Survey on Freeworkers' (in 1923)[10] confirmed the existence of these *yoseba* as well as public employment services, contractors camps for labourers, in places nearby to labour sites. According to this survey conducted in 1922, Tomikawachō had the largest *yoseba* in Tokyo with approximately 2,500 free labourers based in the area.[11] (Yokoyama points out the difference between the day labourer without a fixed boss or job and the navvy who works in large construction projects under a single boss with whom he develops a close relationship.)

What was the state of the day labourers' struggle then (late nineteenth century to the 1920s)? Already in 1882 a *Shakai-tō* (or Ricksha Sector Party) was founded by Okumiya Takeyuki and others 'to defend the riksha workers' right to live'. But it failed to attract many workers and was eventually dissolved. The reformist *Rōdō kumiai kiseikai kiseikai* (Society for the Formation of Labour Unions)[12] succeeded in organizing the iron workers, among whom there was a high percentage of day labourers, but otherwise organizing was slow and the trade union movement static. In the era which followed the Russo-Japanese War (1904–5) labour disputes involving many day labourers in collieries, shipyards, metal mines, textile mills, silk industry and railways, were frequent, some of them culminating in uprisings. However, with the passing of the notorious Public Peace Police Law in 1900, the orthodox labour movements were suppressed. Riots at the Ashio mines in Tochigi Prefecture, Horonai colliery in Hokkaidō, and Besshi Copper Mines in Shikoku were met with military intervention. In the 1910s, the *Yūaikai*[13] became the core of a new tide of the labour movement. Around the time of

World War One, the number of strikes and workers on strike increased considerably. Stevedores, miners and metal workers were affected.[14] In 1919, when labour strikes were at their height, 497 strikes were reported, mainly in the printing, electric, shipbuilding, artillery, railway and mining (Ashio, Kamaishi, Hitachi, etc.) industries.

INFLUX OF IMMIGRANTS FROM THE COLONIES INTO THE *YOSEBA* AND THE BUILD-UP OF THE WARTIME SYSTEM

One consequence of Japan's imperialistic aggression and colonization was the influx of Korean and Taiwanese immigrants, who found work as free labourers in construction, transport, gravelyards, factory services, etc.[15] A high percentage of these immigrants went to work in the *hanba* system throughout the country, in mines, collieries, ports and reclamation. In 1923, 17,000 Koreans were employed in 'stable jobs' – for example, mines, textile and chemical factories, agriculture and other industries – while the remaining 62,000 became irregular labourers. In 1931 this grew into 105,000 Koreans holding stable jobs and 525,000 irregular labourers.[16] In the 1920s, Korean immigrants began to establish their own communities on the fringe of the cities, and to constitute a major force within the bottom layers of Japanese society.

The Ryūkyū Islands were assimilated by Japan as a prefecture in 1879 as a result of armed threat, and 'land registration' continued there until 1903. Emigration to foreign lands and to mainland Japan began at this stage and the latter increased rapidly after World War One, reaching a peak of 50,000 immigrants in 1920. Eighty percent of young Ryūkyū females worked in the textile mills while males were generally hired as day labourers in construction, etc.[17] They also belonged to the bottom layer of Japanese society, oppressed along with the Koreans; thus the common notice hung out by landlords said 'No Ryūkyūites, No Koreans'.

The Ainu, weakened after a long history of armed struggle against the Yamato race or 'Shamo' (the Ainu word for 'Japanese') and brutal exploitation under the feudal Shogunate-clan system, were no longer considered a menace to the Japanese. The entire 'Ainu-moshiri' region was taken over by the Shamo and the name changed to Hokkaidō after the Meiji Restoration. Ainu were subjected to total colonial rule: they were forced to convert from hunting and fishing to agriculture as means of subsistence; Ainu language was banned; the use of Japanese was forced on them and the official appellation changed to '*kyūdojin*' – or primitives. The Hokkaidō *Kyūdojin* Protection Act was passed in 1899 with the aim of confining the Ainu in 'reservations', as in the case of native Americans and the Australian Aborigines.[18] Those Ainu who survived the brutal repression lived a submerged life within the bottom strata of society, expressing their ethnicity whenever they could (e.g. editing ethnic newspapers and fighting lawsuits over land ownership).

The Korean colonies, flophouse quarters and the back slums were, in some areas, contiguous to each other, and burakumin villages and prostitution quarters located nearby. Sanya fits this description[19] and Kamagasaki had a similar arrangement. 'Each morning, Korean, Ainu and Ryūkyūite day labourers gathered at a fixed place [in the *yoseba*] mingling with the wandering Japanese labourers, finding their jobs for the day through an agent [*ukeoisha*] or *oyabun*.'[20]

Meanwhile, Japanese labour and peasant movements were entering their most dynamic era. But the Japanese socialists (including the anarchists and the bolsheviks of later years) inflexibly followed the doctrinaire Marxist theory: 'With the defeat of Japanese capitalism/imperialism, the colonies will be liberated, and in conjunction minorities such as the Koreans in Japan will be liberated too.' Their failure to see 'national liberation' as a distinct issue stunted the budding ethnic solidarity between Japanese and Korean labourers in some sectors of industry. The 'Kanto Free Labourers Trade Union' made some effort to organize day labourers but failed to produce any powerful impact (in 1929 it had only 170 members).[21]

It was the Korean labourers who took the initiative. The 'Korean Labour Alliance' was founded in Tokyo and Osaka for 'the establishment of the right to livelihood' and 'stable employment'. Its objective was to carry out a national liberation and class struggle firmly rooted in the actual living conditions and consciousness of the day labourers.

'The General Alliance of Korean Labourers in Japan' (*Chōsen Rōsō*) was founded in 1925, starting with 800 members and growing to 24,000 at the time of its dissolution. It supported the New Trunk Society (*Shin Gan Hoe*) in the Korean Peninsula and aspired to the foundation of a single all-Korean national party to defend the well-being of the Korean people as a whole. They viewed colonized Korea as one gigantic factory, and all the Korean people, including those in Japan, as common victims of Japanese imperialist exploitation and oppression. In other words the whole Korean people was seen as proletarian.

The national liberation struggle was therefore identified with the class struggle against Japanese imperialism; supported by many active Korean labourers in Japan, the General Alliance demanded better working conditions, the right to work, an end to social discrimination: guarantees of political freedom and access to class-conscious education. It also maintained links with the labour movement on the Korean Peninsula and the socialist movement in general, while at the same time collaborating with the *Rōnō* (Worker-Peasant) Party and *Nihon Rōdōkumiai Hyōgikai* (National Council of Trade Unions of Japan).

Unfortunately, this direction taken by *Chōsen Rōso* was cut short by the erroneous 'One Nation – One Party' line set by the Comintern and the Profintern in 1928. According to this, foreign and colonial labourers in capitalist nations should be affiliated to the trade unions and the Party of the resident nation. In Japan, after a heated debate, *Chōsen Rōsō* was disbanded in response to this decision, and from early 1930 the Korean workers were reabsorbed by the (Japan Communist Party-affiliated) National Conference of Trade Unions of Japan (*Zenkyō*). Korean communists of that time were forced to join the Communist Party of the country they lived in, whether in Japan, China or the Soviet Union.

In spite of the dissolution of their own trade union and the enforced reorganization, Korean labourers continued to play a major role in the day labour movement in Japan. At the end of 1932, approximately 8,000 Koreans, were organized in *Zenkyō*-affiliated unions, making up half the total membership. During 1933 many Korean trade union activists became leading figures in branches of *Zenkyō* such as construction and public works, chemicals, printing and metal industries.[22]

The individual struggles carried out by *Zenkyō* and its branches at this stage are instructive and worth thorough investigation and reassessment. Their activities included giving aid and guidance in the establishment of independent organizations by and for the unemployed, such as the Fukagawa Employment Association for the Betterment of Registered Labourers, and joint mass actions with Koreans to put a stop by all means to rent increases or evictions.[23]

Japan invaded China in 1931 and a systematic build-up towards war developed in the mid 1930s: trade unions and the labour movement in general were forced to convert to support the war policy of the state and industry; then they were organized into '*Sangyō hōkoku kai*' (Patriotic Industrial Society) which was to be the main Japanese labour organ to help carry out the Fifteen Year War. Day labourers, Korean and Japanese alike, were organized by the state, registered in '*Rōmu hōkoku kai*' (Patriotic Labour Society) and violently forced to work.

The demand for a labourforce grew extraordinarily as the war developed, and Koreans and other colonial people were brought into Japan entirely by force (for example, taken from the street or from the fields at random, dragged to the port and then confined in the hold and transported to the different parts of the nation), to work under confinement particularly in collieries, mines and factories of the military industry.

At this stage, the function of the *yoseba* as a buffer was no longer necessary. Direct state coercion, and often violence, was used through the *hanba* system to mobilize labour as required. Here the true nature of the war-time system bared its teeth upon day labourers: the unmistakable character of Japanese fascism may be seen in the violent lynching of Korean labourers by *oyakata* (and often by ordinary Japanese people), along with the direct intervention of state power in the labour exploitation process.

THE FORMATION AND REORGANISATION OF THE POST-WAR *YOSEBA*

In the context of trends in the post-war period as a whole, the post-war *yoseba* history may be divided into three phases. First, the collapse of Japanese imperialism in 1945 initiated a period in which the continuance of the status quo, based on the emperor system and capitalism, was called into question. This period, which lasted till the revolutionary possibilities arising from the defeat were finally crushed and bourgeois control reestablished in the late 1950s, amounts to a prelude to post-war *yoseba* history.

When this period began, day labourers were still organized in the Patriotic Labour Society, which continued from wartime. They were forced to live in *takobeya* (concentration shacks for forced labour) and to do heavy labour – those from Sanya being employed to do wharf labouring work in the Shibaura warehouses, portering hoarded goods, and in various works connected with the dissolution of the old military or for the U.S. Army. At the beginning of 1946 moves were made towards dissolving the *takobeya* and formally the branches of the Patriotic Labour Society were renamed labour unions (in Sanya the 'Tamahime Free Labour Union') and subsequently came under the umbrella of the 'Self-Governing Federation of Tokyo Labour Unions'.[24] From October 1947 till February 1948 the organization continued under another new name 'Labour Society' (*Rōmu kyōkai*) and the administration of the employment advisory service was transferred to the Employment stabilization (*shokuan*) administration. (The Labour Ministry was established in September 1947.) However, the gangster-labour boss system was retained and strengthened at the same time, and as the sub- and sub-sub-contracting system was adopted to supplement the Employment stabilization administration, there were early signs of the revival of the *yoseba*. There were limits to the volume of labour power that could be gathered through the Employment stabilization offices, and greatly increased labour supply was made necessary as a result of the 'success' of the very uneven economic recovery policies of increased coal production and favoured treatment for specially designated enterprises. In short, for the labourers in the *yoseba* the candle gleam of post-war revolution shone only faintly and for the briefest of moments. This was most clearly shown by the way that labour unions were called into existence from above, and were completely lacking in autonomy.[25]

This period, especially between 1945 and 1947, saw a burgeoning struggle for food and work, and 1949 saw the development of a movement to protest against the inadequacy of government unemployment measures and to demand security of livelihood. The day labourers participated in these movements and constituted an important part of the post-war revolutionary stream under Japanese Communist Party leadership, but because the Communist strategy of aiming at peaceful revolution under U.S. Army occupation was fatally flawed, these movements were destined to fail. Furthermore, while the J.C.P.-affiliated Daitō Free Labour Union (which later merged with the All-Japan Free Union of *Zennichi jirō*) deserves some credit for its involvement in the struggle to secure work and to effect improvements in the *doya*, this was always within the framework of the struggle for specific demands; it was marked by a narrowness in that it lacked any orientation beyond the improvement of conditions for registered day labourers.

In January 1950, the Japanese Communist Party's basic line of peaceful revolution under the U.S. Occupation was criticized by the Cominforn. The Party was torn by divisions over how to respond to this criticism, and so serious were the divisions that for a time it virtually collapsed. For a short time the Party Central Committee adopted the line of revolution by armed struggle. 'Bloody May Day', with its struggle to liberate the square in front of the Imperial Palace, followed. Also, many Korean workers from Japan's *yoseba* and *hanba* took part on the front lines in the Korean War, seeing it as the struggle to liberate South Korea. This was all part of the J.C.P.'s armed struggle line. At its sixth National Conference in July 1955 the J.C.P. switched to parliamentary activity as an opposition party within the system, but without resolving the issues of the previous line. By this time, many had left the party. From this time on there has been no active left-wing revolutionary party in Japan.

Also in 1955, the bourgeois parties (Liberal Democratic Party or LDP and Japan Socialist Party or JSP) were reorganized and they worked out a demarcation of spheres of influence, and the hegemony of Sōhyō Democratization Faction (*Sōhyō mindō*) was established in the labour movement. The *yoseba* and its day labourers were frozen out of this 1955 system which was established on three fronts. The *yoseba* itself was stimulated by the special procurements arising from the Korean War beginning in 1950, but the day labourers were left under the dominance and exploitation of gangsters and the labour bosses, and they were ignored on all sides.

The second phase is from the late 1950s to the early 1970s. From around the late 1950s the Japanese economy entered its so-called 'high growth' period, overseas economic aggression was resumed, and the *yoseba* developed its full form. The day labourers expressed themselves much more fiercely, and often in violent ways, for being frozen out of the system. From the end of the 1960s through the early 1970s groups began to appear in the *yoseba* which were firmly based on the ranks of the day labourers and which aimed at transforming the status quo. Japanese capitalism by this time had revived and was gaining strength as imperialist.

The first characteristic of this period was the dissolution of agriculture and the villages. In 1960 the Basic Law on Agriculture was adopted, and mechanization and large-scale agriculture was promoted. Large numbers of farmers abandoned agriculture. While this led to an abundance of ex-farmer workers and of ex-agricultural land, under the 'Reconstruction of the Japanese Archipelago' scheme huge amounts of capital investment were used for the construction of factories and the laying of expressways. Japan's economic structure was totally transformed as a result. This is the second characteristic of this period. In other words, the weight of the primary sector was reduced, secondary industry – construction, manufacture (especially heavy and chemical industries) – greatly expanded, and service industries also expanded. As heavy and chemical industrial goods accumulated – from electrical and transport machines to automobiles and electronics – a change to a favourable balance of international trade resulted. One factor which greatly influenced both industry and trade was the so-called energy revolution. Reliance on domestic coal was set aside and Arabian and other oil adopted as a source of power. The *yoseba* took its final shape as a result of these two factors of dissolution of the villages and the change in economic structure.

The people uprooted from agriculture or the coal industry flowed into the big cities like Tokyo and Osaka. From the *yoseba* they went out on a daily or temporary basis to labour in the construction or engineering industries, on the wharves or in transport, or the steel or shipping industries. Because of the sudden need for large quantities of capital and labour, the importance of the labour boss was enhanced, the control over *yoseba*, *hanba*, and work sites by the labour boss-gangster was reinforced, and the level of exploitation was intensified through the establishment and strengthening of a multiple sub-contract system. Furthermore, during this period, young (ordinary) workers in the automobile, electronics, and petrochemical industries in particular were forced to work long hours and at high intensity. This is also the time when the system of temporary employment of 'outworkers', or seasonal workers, and the system of multiple sub-contracting, became standard practice. The combination of these factors constituted the secret for achieving 'high growth'.

Various enterprise groups, linking important industries, also became established at this same time. The market monopoly share of these groups was extremely high. While between five and ten companies held a controlling position in each sector, competition between these groups was extremely intense. This has been called an oligopoly; in any event, it represents a new kind of monopoly capitalism.

After the Japan-Republic of Korea Treaty of 1965, Korean residents in Japan who became citizens of the Republic of Korea gained various legal rights hitherto denied them. One outcome was that many Korean capitalists entered upon cooperative relations or became affiliated with Japanese companies or banks and, under the facade of helping their compatriot Korean workers, exacted longer hours of work or more intense conditions of labour from the Korean workers who stood at the bottom of the work force. The merging Japan-South Korean relationship saw a deepening of chauvinism and discrimination among workers and people in general, and introduced divisions into the Korean community in Japan.[26]

Faced with this situation, the *yoseba* labourers, including among them various nationalities, rose up in riots, resolutely refusing to be passively assimilated within such a system. Beginning with the first Sanya riots of August 1959, there were four riots in Sanya in 1960, involving a total of 10,000 people, followed in 1961 by a violent uprising which lasted for three days at Kamagasaki, the largest *yoseba* in Japan, and which involved a total of 15,000 workers. Attacks were launched on police stations and sub-stations and on gangster-labour boss offices. Since that time, dozens of violent uprisings have been fought out in these two locations. Funamoto Shūji, later an activist in *Kama Kyōtō* (on which see below), has described these riots in these terms:

What is common to the Kamagasaki and Sanya riots is that they began as explosions of workers' anger against discriminatory and inhuman treatment of their comrades by the police. The reality of the riots was that of mass retaliation against a background of daily humiliation, resentment and anger experienced by the workers as individuals against the victimization of their comrades, and their arming themselves was the collective expression of the class hatred felt by the lowest strata of workers.

The riots were the passionate expression in struggle of the resentment of the 'weak', who had been made to swallow their daily humiliation, against their 'strong' enemy, the authorities.[27]

In the summer of 1967 a demonstration by the Sanya Independent Amalgamated Labour Union (*Sanjirō*) triggered off a riot which heralded a phase of activity by left-wing activists using riots as pressure to gain their demands. Apart from the *Sanjirō*, the Tokyo Day Labourers' Union (*Tonichirō*) was also founded in 1969 and struggled especially for reforms in municipal administration. Eventually, the *Sanjirō* secretariat faction (the so-called Hiroshima University group), a left-wing offshoot of *Sanjirō*, proposed a rough outline plan for the building of a workers' 'Combat Commune' which would turn the *doyāgai* into a combat base, they could launch attacks from it, capture districts or streets and make ready for a general strike and occupation of factories. In other words, they planned a fundamental change in the system. In 1969, they formed an 'All-Metropolis United Labour Union' (*Zentorō*) and planned to link up the reformist struggle in Sanya and the strike by lower level workers in the electric companies. However, the sudden leap beyond the limits of radical reformist struggle proved impossible to implement at this time.

Kamakyō (Kamagasaki Joint Struggle Committee against Corrupt Labour Bosses) and *Gentō* (Site Struggle Committee against Corrupt Labour Bosses) were both established in the summer of 1972. Both inherited directly the achievements and the limitations of *Zentorō*, *Tonichirō*, and the Wharf Labourers Union Sanya Branch, and both were created out of the mass struggle against labour bosses such as the *tehaishi* and the *bōshin* (site work control) system (both of which are frequently connected with gangsters). *Kamakyō* and *Gentō* opposed by principled means and by force the hitherto completely untouched system of control of day labourers by gangsters, and were able, by victories in this struggle, to win the overwhelming support and sympathy of their fellow workers. The day labourers came to realize their own strength, and to put into actual practice the slogan, 'We will decide by ourselves our own labour conditions.' Under the slogan 'Turn the circumstances of our oppression into a weapon', *Kamakyō* and *Gentō* succeeded in uniting Korean, Chinese and Asian residents in Japan, Ainu people, people from Okinawa and from the discriminated *buraku*, workers in small and tiny enterprises and temporary and out-work workers, seasonal workers and small farmers, with the drifting workers of Sanya and Kamagasaki, and having them rise up in riots. The struggle developed in the direction of an urban people's war, and was a move in the direction of universal liberation.

The *Kamakyō* and *Gentō* struggles possessed an epochal strength, and even now are a repository of rich ideological significance. However, by about the mid-1970s, they both collapsed because of the repressive and containment tactics of the authorities headed by security bureaucrats under the public peace strategy of the 1970s, and because of the various weaknesses of political ideology and organizational theory on the part of *Kamakyō* and *Gentō* themselves.

THE PRESENT JUNCTURE

The present is a continuation of the third stage, characterized by the restructuring of the *yoseba* that accompanied the oil crisis, recession, and slow growth. The 1973 oil shock brought a renewed realization that high growth was a castle built on sand. The economy of Japanese imperialism faced a turning point. Steps taken by capital to deal with the crisis included diversification of sources of supply of essential materials, intensification and heightening of rationalization, retrenchment, and exploitation as part of administrative reform, export of pollution, and greater economic aggression. From the beginning of the 1980s, government and business have been hammering out the idea of comprehensive security. While on the one hand aiming at the establishment of a system in which the individual people can be controlled and mobilized at will down to the level of their livelihood, on the other hand, externally it relied on negation, hostility, and discrimination. At *yoseba*, labour sites, and *hanba*, the result was that there were frequent occurrences of violence, labour coercion, greater work pressure, extremely low wages or non-payment of wages. Monopoly capital planned the reorganization of labour sites as a whole, including *yoseba*, in order to put them under the direct control or strong influence of gangsters.

The response on the part of those struggling in the *yoseba* was very belated. From about the mid-1970s (in Kamagasaki), or the end of the 1970s (in Sanya), the struggle began by mass bargaining for an improvement of *hanba* or sites where labour conditions were particularly bad, virtually '*takobeya*', and against middle men who would only negotiate work under such conditions. From 1980 there were constant discussions between various *yoseba* movement groups in an effort to achieve unanimity of direction, or unification of line, and eventually in 1982 the *Hiyatoi zenkyō* (National Council of Day Labourers' Unions) was set up. *Zenkyō* was heir to the traditions of *Kamakyō* and *Gentō*.

The struggle by *Hiyatoi zenkyō* has now entered an unprecedented phase. In just over a year from the end of 1984 till the beginning of 1986, two from the *Zenkyō* side were attacked and killed.[28] The superheated discrimination and chauvinism of the social order is clearly demonstrated by incidents such as these, as by the 1982 beating to death of a day labourer by middle school students at Yokohama. *Hiyatoi zenkyō* sees the corruption of the present Japanese imperialist system as the root of such actions. It aims at prevailing ideologically against this, while keeping up a many-fronted struggle – forcibly resisting the gangsters who constitute the shock troops of discrimination and chauvinism, resisting repression by the authorities, and developing a programme of intelligence, propaganda, and organization towards the reactionary residents of certain districts. At present the struggle in Sanya is directly confronting Prime Minister Nakasone's 'Liquidation of Post-War' politics, in conjunction with the struggling farmers of Sanrizuka and Kitafuji, the students struggling against the new right organizations like the 'Student League against the Constitution' at Nihon University and other universities, the liberation movements of handicapped persons and of women, and the people of the discriminated *buraku* struggling to get rid of discrimination. They constitute the front line of the struggle in Japan to overturn and transcend imperialism and fascism.

NOTES

1 Marx, *Shihonron* (Capital), Tokyo, Aoki bunko edition, translated by Hasebe Fumio, Aoki shoten, vol. 4, p. 981.
2 *Ibid.*, pp. 1,084–5.
3 The Japanese psychiatric hospital system, and this hospital in particular, were the subject of representations to the Japanese government by international organizations, including the Human Rights Sub-Committee of the United Nations (in August 1984) and of some sensational court actions in 1984–85 arising out of allegations of forcible detention of patients, various ill-treatments including beatings, indiscriminate drug dosages, and lynching. A number of Sanya day labourers, picked up in the streets for drunkenness or vagrancy, had been consigned to Utsunomiya Hospital. See Utsunomiya byōin mondai o kangaeru kōza jikkō iinkai (ed.), *Utsunomiya byōin, sono shuju to saihen no kōzō o tou* (Questions about the Reorganization of Utsunomiya Hospital), Utsunomiya, 20th April 1985; and 'Nihon shūyōjo retto' (Japan detention centre archipelago), *Inpakushon*, special issue, no. 31, September 1984. [Ed. note.]
4 *Capital*, Aoki bunko edition, vol. 4, p. 1084.
5 See, for example, Hosoi Wakizō, *Jokō aishi* (The tragic history of female factory workers), 1925, Iwanami bunko.
6 Yamada Moritarō, *Nihon shihonshugi bunseki* (Analysis of Japanese capitalism), 1977, Iwanami bunko edition, pp. 76–7.
7 Yokoyama Gennosuke, *Nihon no kasō shakai* (The bottom levels of Japanese society), 1899, Iwanami bunko edition, pp. 22–7.
8 Yotsuya Samekawabashi, Shitaya Mannenchō, and Shiba Shin-ami were three places designated by name, along with Nagomachi in Osaka.
9 Hanamachi and Koume-narihirachō in Honjo-ku, Asakusachō in Asakusa-ku, Tomika-wachō in Fukagawa-ku, etc. Yokoyama, pp. 54–9.
10 In Tokyo city and suburbs the following organs controlling labour supply and demand or having similar functions, such as control of the day labour market in general, may be found:
    i   the street labour market around the flophouse quarters,
    ii   the public employment service and labour camp,
    iii the contractor of day labourers,
    iv   the labour market adjacent to work sites.
11 Yokoyama, pp. 69–70. (The map here illustrates the layout of the *yoseba* in Tomi-kawachō.)
12 Leaders of the *Dōmeikai* included Takano Fusatarō and Katayama Sen (ibid.).
13 On the *Yuaikai*, see Stephen S. Large, *The Rise of Labour in Japan – the Yūaikai 1912–19*, Tokyo, Sophia University, 1972. [Ed. note.]
14 Kyōchōkai (ed.), *Saikin no shakai undō* (The recent state of social movements), Kaizōsha, 1929, pp. 323–4.

15 Tōkyōto shakaika, *Zaikyō chōsenjin rōdōsha no genjō* (The conditions of Korean labourers in Tokyo), April 1929, pp. 63–6.

16 Kin Jyu-sei *Zai-Nippon Chōsen rōdōsha no genjō* (The conditions of Korean workers in Japan), 1931, in *Kindai minshū no kiroku* (Records of the people in modern times), Shin-jinbutsu ōraisha, 1978, vol. 10, p. 148.

17 Aniya Masaaki, 'Imin to dekasegi' (Immigrants and seasonal labour), *Kindai Okinawa no rekishi to minshū*, (History of modern Okinawa and the people), expanded and revised edition, Heibonsha, 1977, pp. 144–65.

18 Cf., for example, Arai Genjirō, *Ainu no sakebi* (The Ainu call), Hokkaidō shuppan kikaku sentā, 1984.

19 So far as Sanya was concerned, there were discriminated *Burakumin* villages, a *yoseba* and Korean villages in the area around Tamahime, Senju-Ōhashi and Minowa, and the Shin-Yoshiwara brothel quarter just over Nihonzutsumi.

20 Higuchi Yūichi, 'Zainichi Chōsenjin shakaishi kenkyū shiron' (Preliminary sketch of the social history of Koreans in Japan), *Kaikyō*, no. 12, Shakai hyōronsha, 1984, p. 34.

21 Kyōchōkai, 1929.

22 Mun Kung-ju, *Chōsen shakaishi jiten* (Dictionary of the history of Korean society), Shakai hyōronsha, 1981, pp. 404–11.

23 See, for example, Watanabe Tōru, *Nihon rōdō kumiai undōshi-Nihon rōdō kumiai zenkoku kyōgikai o chūshin ni* (History of the Japanese labour union movement, with special reference to zenkyō), Aoki shoten, 1954.

24 Furukawa Yoshimori, 'Henshūsha maegaki' (Editorial preface), *Jitsuryoku-dokō Tamakichi* (Chronicle – Navvy Tamakichi), Taihei shuppan, 1974, p. 17.

25 Tokyo Jōhoku fukushi sentā (Tokyo Jōhoku Welfare Centre), *Sanya – genkyō to rekishi* (Sanya: present situation and history), 1971, pp. 19–20. (This section is by Takeuchi Yoshinori.)

26 Ibid. Miwa Ryōichi, *Nihon keizaishi* (History of the Japanese economy), Nihon hōsō shuppan kyōkai, 1985. Also Santama, yama no kai, *Yoseba no rekishi kara mirai o mitōsu* (Looking at the future from the past of the *yoseba*), 1984.

27 Hiyatoi zenkyō (ed.), *Damatte notarejinu na – Funamoto Shūji ikōshū* (Don't go like a lamb to the slaughter – Posthumous essay of Funamoto Shūji), Renga shobō shinsha, 1985, p. 144.

28 On 22 December 1984, an independent film maker, Satō Mitsuo, who was also a well-known supporter of the workers' struggle, was killed while working on a documentary film project in Sanya, and on 13th January 1986 a Sanya labour leader, Yamaoka Kyōichi, was shot dead on a Tokyo street. A week later, the leader of a rightist gang group was arrested as suspect for the murder. [Ed. note.]

IV Sex, politics and 'modernity'

9   The Japanese women's movement: the counter-values to industrialism

UENO CHIZUKO

THE RISE OF THE JAPANESE WOMEN'S MOVEMENT

The 1960s, the decade of Japan's economic 'miracle', was also marked by a new wave of the Japanese women's movement, which was deeply rooted in the structural changes of society at the time. It can be said that the Japanese women's movement arose soon after Japan had completed its modernization, as its inevitable side effect.

Although Japanese society had been going through the process of gradual modernization since the Meiji Restoration (1868), this process accelerated rapidly during the 1960s: for example, the urbanization of the population passed the halfway mark, the number of nuclear families increased to 60 percent of all households, and the average number of family members suddenly dropped from five to three (see figure 1).[1] The year 1960 is generally regarded as the first year of 'rapid economic growth', when the number of workers employed in industry exceeded that of workers in family businesses (see figure 2). These processes combined to alter the family lifestyle: the urban nuclear family with the sex-role division of working husbands and housewives was established at a mass level.

In terms of family issues, the typical life-cycle pattern of the Japanese people has greatly changed. The Japanese have experienced the so-called 'biological revolution' of lower birthrate and longer lifespan (see figure 3).[2] These demographic changes have been universally experienced by most developed industrial societies but what is notable with the Japanese case is that this change was truly a 'revolution', because it took place during a very short while. Japan is now a rapidly aging society.

These changes affected the reality of women's lives earlier than that of men's lives. Compared to the old Japanese women who had from six to seven births during fifteen years of their lifetime, women in the 1960s stopped giving birth after the second child. Because of this, most children

167

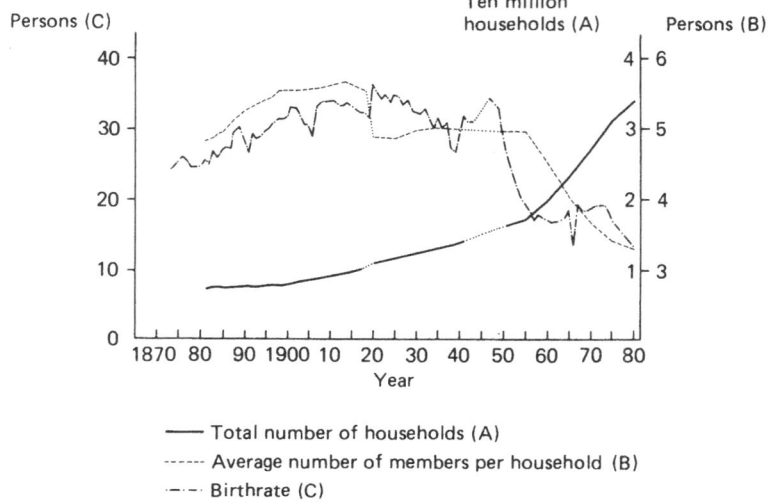

Figure 1 Trends of number of households, average number per household, and birthrate. *Source:* Sōrifu, 1983 (see note 1)

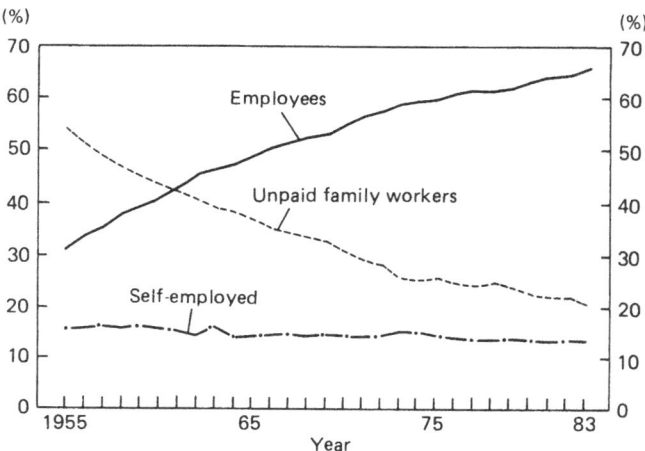

Figure 2 Trend of proportion of female labourforce by class of workers. *Source:* Sōrifu, 1983

were in school by the time the mother turned thirty-five, which meant women had to start the post-nurturing stage much earlier than their mothers. In other words, women experienced, as it were, 'retirement' from motherhood, or at least shrinkage of their mother-role, at the age of thirty-five, which everyone would agree is too early for retirement.

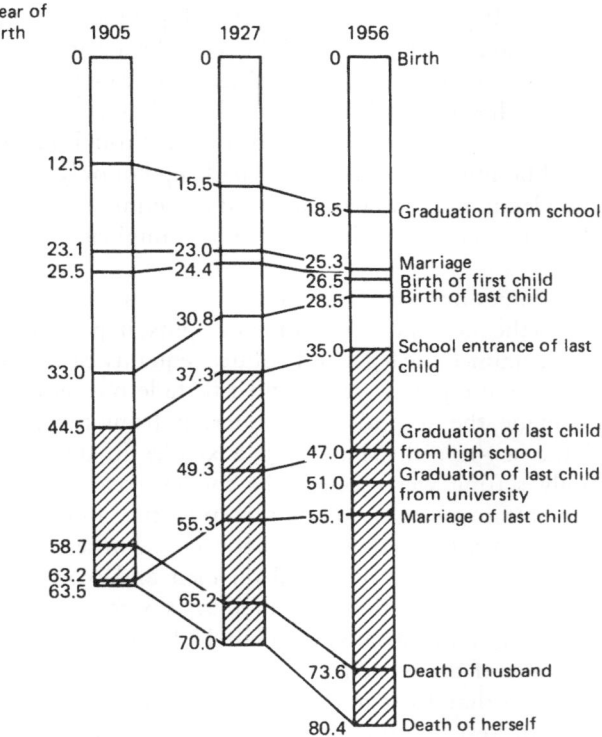

Figure 3 Average life cycles of Japanese females. *Source:* Ministry of Labour, *Dai 35-kai fujin shūkan kōhō shiryō* (Public relations data for the thirty fifth women's week)

The completion of the modernization of the family was the completion of women's confinement to the home. Their isolation from society and from one another, and their subjugation to modern style patriarchy, was disguised as the 'happiness of the home'. The Japanese housewives started to resent this situation, and to share the same complaints and resentments as those described in 1963 by Betty Friedan in *The Feminine Mystique*.[3]

It is not surprising that Japan had a radical women's movement in the early 1970s. Contrary to the views of conservatives who were antagonistic to this movement, it was not just an import from Western feminism, but rather had its innate *raison d'etre*. By the end of the 1960s, Japan had reached the stage of highly developed industrialization and as the product of this society, Japanese feminism had much in common with its Western counterpart.

Japanese feminism has a long tradition, dating back more than half a century. Fujieda Mioko divides the history of Japanese feminism into two stages: the first wave in the 1900s and the second wave in the 1970s.[4] The former was based on and limited to the highly educated women in urban middle-class families in the Taisho period, while the latter reached the grass-roots level after urbanization had spread throughout the country. The women's liberation movement was made up not only of the so-called 'radicals' who led it and who were inclined sometimes to be extreme, but also the 'silent majority' who gave their implicit support to the movement.

As Fujieda points out, the first wave of Japanese feminism tended to confine itself to the movement for women's rights, in particular, women's right to vote. It claimed the legal and political equality of women and, like most Western counterparts, was inclined to fade out as it reached this goal. What made the second wave different from the first is that it questioned the sex-role division itself. Sex-roles in the modern urban nuclear family setting were seen to be the very cause that prevented women from attaining real equality despite the attainment of legal equality. This sex-role division between worker-husbands and homemaker-wives is in fact imposed on men and women by modern industrial society. To question this modern sex-role assignment is to question the modern industrial system in which we live.

This new wave of the women's movement was rooted in the counter culture movement that arose in the late 1960s and the early 1970s as an expression of protest against and resentment against the outcome of industrialization. The women's liberation movement logically paralleled the anti-pollution movement: both raised questions about the nature of industrialism that put priority on production rather than on reproduction.

FAMILY: THE 'SHADOW' OF THE MARKET SYSTEM

Let me briefly follow the analysis of this logically parallel structure between women's issues and industrial pollution issues. According to the Marxist feminist problematic, the industrial system as a whole may be seen as a composite of patriarchy and capitalism.[5] But what has been neglected by analysts of the industrial system, even by Marx, is the fact that the industrial system cannot be a closed system. Although modern industrial society looks autonomous, and it is assumed that it is controlled by a self-regulative market mechanism, in reality it depends on the external 'environment' made up of 'nature' in terms of production and 'family' in terms of reproduction.

When industrial society confronted industrial pollution the limits of the market system in relation to its external environments were revealed. From the very beginning, industrialization has suffered from its undesirable by-product – pollution.

Interestingly enough, no sooner had Japanese industrialization been accomplished, during the 1960s, than serious pollution issues such as Minamata burst in view as conspicuous social problems.

The market system assumes that the natural environment surrounding this system is some sort of a 'black box' without any limit, that is, that natural resources are infinite and nature somehow has a self-cleaning function. The system takes in resources from the outside and puts out industrial waste instead. However, today people have come to realize that there are limits in our eco-system; industrial societies are now collapsing under the weight of the resource and energy crisis on the one hand and industrial pollution on the other.

Along with nature, the invisible domain outside the market system, there is another invisible domain which is the provider of human resources and a dumping ground of human 'industrial waste' or those in retirement – the domain of the family. Industrialization organized male members of the society as wage labourers but it left women in the private sphere with children, the elderly, the diseased and the handicapped. Ivan Illich calls it the 'shadow' domain.[6] Invisible as it is, the family sector is there in the 'shadow' of the market system. As Marx said, capitalists left reproduction to the 'instinct' of labourers; the market system is literally a *'laissez-faire'* – let it be – mechanism, both in production and in reproduction.

However, under the market system, the domain of the family did not automatically work as a provider of workers for the next generation and as a shelter for the old and the handicapped.[7] Maintaining this family system is only possible at great cost; in particular it demands the sacrifice of women. The sex-role division of the modern family is based on women's unpaid labour in the domestic sector. Women began to express their objections to and frustration with this unfairness soon after it had been established.

THE BIRTH OF HOUSEWIVES

When women's studies was first introduced as *Joseigaku*[8] to Japan with the impact of feminism in the 1970s, *shufu*, or housewives, became for the first time the focus of study.

The study on *shufu* had a great significance in women's studies. Discrimination against women in the workplace and sexual slavery is visible; on the other hand, women's subjugation and unpaid domestic labour are taken for granted under the guise of complementary sex-role division, and therefore the oppression of women as *shufu* is rather invisible. Besides, *shufu* represents the majority of women. Invisible as they are, *shufu* are there like the 'black continent'.

As Itō Masako writes:

The problem of *shufu* is essential when one considers women's issues. Women are more or less obsessed by the idea of *shufu* in so far as they assume that women are happy as housewives. This is true with almost all women – not only those who are now housewives, but also those who have not yet or will not or can not become housewives, and those who used to be housewives. At least, it seems that most women tend to define themselves by measuring their distance from the ideal *shufu*.[9]

The first book that included the word *joseigaku* in the title was *Joseigaku kotohajime* (Introduction to women's studies) in 1979,[10] in which one of the editors (Hara Hiroko) wrote an article entitled 'Invitation to the study of *shufu*'. The main article in this book was 'A hundred years of the Japanese women: considering the name of *shufu*' by Segawa Kiyoko, a woman successor of the renowned founder of Japanese folklore studies, Yanagida Kunio.

Segawa points out that the term *shufu* first appeared in the late Meiji (1868–1912) period and spread among urban middle-class families through the Taishō (1912–26) period. By the year 1907, when the first woman's magazine for housewives, *Shufu no tomo* (Housewives' companion) was published, the number of people called *shufu* had grown sufficiently to support this sort of journalism for women.

In support of Segawa's argument, Umesao, an anthropologist, describes his memories of his mother, a merchant wife who was born in the Meiji period. She worked for their family cosmetics business. Umesao observed that his mother was very careful to call only a certain kind of woman customer *okusan* (a term for middle-class housewives) – wives of the new rising class of officers, businessmen and teachers, or in other words, the wives of employed workers without any family business.

My mother was not called *okusan*. Who were considered *okusan* then? *Okusan* were those who lived in cheap official residences, often bargained with the merchants arrogantly, and spent all day long talking together with other *okusan* in the neighbourhood. In fact, wives of such employees as policemen, teachers and businessmen tended to call each other *okusan*. (. . .) What made them different from most women like my mother was that they didn't work.[11]

To women, modernization meant becoming wives of this new middle-class of employed workers. Being a housewife meant upward mobility: marriage was seen as the one and only way for women to move up the social ladder. Men by contrast had the means of education which women lacked. Girls dreamed of becoming businessmen's housewives, a dream which later turned out only to be a trap.

FROM FULL-TIME HOUSEWIVES TO PART-TIME HOUSEWIVES

The modernization of Japanese society by the 1960s resulted in the completion of the 'housewife-ization' of women, and a status reserved for a 'lucky' minority in the past was now extended to an overwhelming majority of women. However, once this housewife-ization was achieved, it began collapsing, and this too occurred in the 1960s. By the early 1970s, the word *sengyō-shufu* or full-time housewife appeared. Where *shufu* usually had till then meant a woman who stayed at home all day long without a job, now a *shufu* would often introduce herself as 'just a housewife' as though ashamed of herself. This showed that by then *shufu* did not mean a woman without a job. In the early 1970s, there was already a significant number of working housewives.

In 1960 nearly 60 percent of working women were unmarried, and 12 percent were widowed or divorced.[12] This means that the majority of working women stayed outside marriage. It used to be generally true that women were either married or working; the life of the working woman was incompatible with homemaking. However, in 1975 this was reversed. By 1981, 58 percent of working women were married and 51 percent of married women were working, which means that the assumption that working women were single was no longer true [see figure 4].

This working housewife is called *kengyō-shufu*, or part-time housewife, as opposed to *sengyō-shufu*, borrowing the term from one used to describe the social change in peasant families.[13] What is happening with women is that they are shifting from *sengyō-shufu* to primary *kengyō-shufu*, or part-time housewives with part-time jobs. This change might precede a complete shift to secondary *kengyō-shufu*, or working housewives with a full-time job as is the case in such countries like Sweden and some socialist countries, but it seems unlikely in Japan for various reasons beyond the scope of this paper.

Let me go into more detail on the matter of primary *kengyō-shufu*. Primary *kengyō-shufu* are literally 'part-time housewives and part-time workers'. At first glance, this new arrangement appeared liberating for women, for it seemed to solve the incompatibility traditionally thought to exist between being a homemaker and working outside the home by providing women with the 'optimal mixture' of work and home. It enabled women to work outside without damaging their household

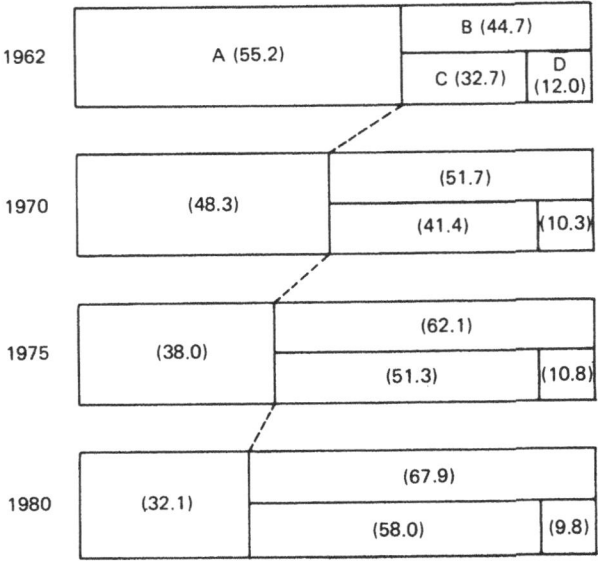

A: Single
B: Married at present or in past
C: Currently married
D: Divorced or widowed

Figure 4 Trend of marital status composition of female employees (percent).
Source: Sōrifu, 1983

responsibilities so that they could preserve their 'mother-wife' identity.

However, it should be noted that this part-time working arrangement was invented by entrepreneurs for their profit, not by women for themselves. It was really a great innovation by Japanese capitalists so as to invite married women to the workplace. As the Japanese economy grew they needed to pull women out of the home into the labour market.

The part-time working style was a 'Columbus egg' in developing the employment system. If workers had stuck to the idea of nine-to-five full-time work only, they would have kept forcing women to choose either (full-time) work or home. So, instead of adjusting women to the traditional working arrangement, they tried to adjust working hours to women's life-style.

The increase of working women did not take place until this 'pull' factor in the labour market was created — no matter how many 'push' factors women have on their side to 'push' them to the workplace from the home, such as higher education, consciousness for equality, time and

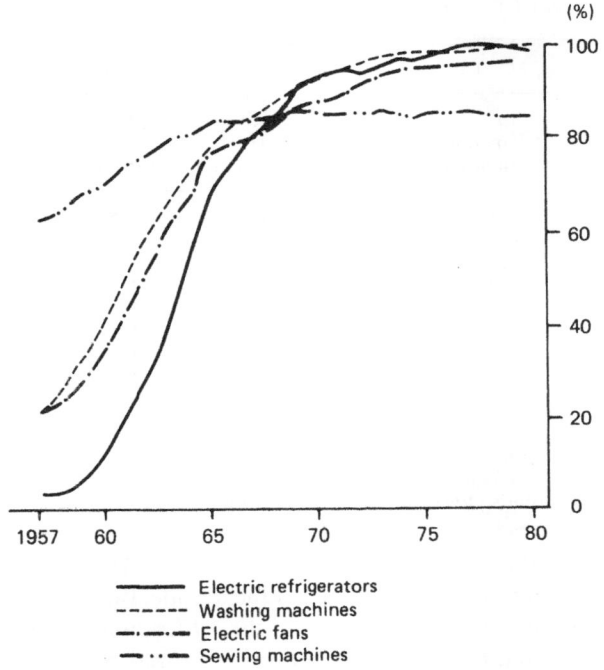

(%)

100

80

60

40

20

0

1957   60   65   70   75   80

———— Electric refrigerators
- - - - Washing machines
—·—·— Electric fans
—··— Sewing machines

Figure 5 Diffusion rate of consumer durables. *Sources: Gekkan NIRA* (monthly NIRA), February 1985, Nippon sōgō kenkyūsho

energy saving appliances for housekeeping, shrinkage of the mothering burden with the decrease of the number of children, and so forth (see figure 5). Though women tried to knock at the doors of business and industry, the doors would not open so long as they were not needed. But throughout the phase of 'rapid economic growth', Japanese industry continuously suffered from a labour shortage, especially at the bottom of the labour hierarchy. One reason lay in the trend of mass education, which resulted in a shortage of young, single, low-educated, low-paid manual workers such as had supported the bases of the Japanese economic prosperity. This shortage was partially solved by automation in factories, but there still remained many manual or service positions to fill with low-waged workers.

Japanese capitalism had three alternatives with which to fill these slots: women, the aged, or immigrant workers.[14] What they actually decided was to choose both women and the aged, that is, middle-aged housewives, thus achieving a double exploitation. In this way, the part-time working arrangement was first introduced to some workplaces when women

Table 1. *Trend of part-time employees*

	Total			Female		
Year	(A) Total employees	(B) Part-time employees	B/A (%)	(C) Total employees	(D) Part-time employees	D/C (%)
1975	3,556	353	9.9	1,137	198	17.4
1976	3,623	314	8.7	1,174	192	16.4
1977	3,682	321	8.7	1,221	203	16.6
1978	3,715	330	8.9	1,251	215	17.2
1979	3,793	366	9.6	1,280	236	18.4
1980	3,886	390	10.0	1,323	256	19.3
1981	3,951	395	10.0	1,359	266	19.6
1982	4,013	416	10.4	1,386	284	20.5
1983	4,119	433	10.5	1,451	306	21.1

Notes: Figures in columns (A), (B), (C) and (D) are in 10,000.
        Figures do not include agriculture and forestry.
Source: Prime Minister's Office, *Labour Force Survey*.

workers were taken on at some time in the early 1960s, and it rapidly spread to most enterprises by the end of the decade. But part-time workers did not appear in labour statistics until the 1970s, when the government came to realize that they could no longer be ignored. Sometimes they had not even been counted as employees, lacking a job contract, security, health insurance, etc., and had been paid lower than the minimum wage guaranteed by law. By the end of the 1960s, the number of part-time workers had reached 10 percent of all the workers and it has been continuously growing since. By 1983, the proportion of part-timers of all women workers reached 20% (see table 1).[15]

Now, the majority of women have both family and work outside the home. But is that what feminists wanted to reach as a goal of liberation? Natalie Sokoloff, an American Marxist feminist, calls being a worker and a housewife a woman's 'dual role'.[16] As is already well known, what Japanese husbands say when their wives ask if it is all right to go out to work is, 'you can do anything you like, as long as you fulfil your household responsibilities'. So women are still in full charge of homemaking, and their husbands do not share housekeeping and child-rearing duties. Any statistics on the use of time in daily life prove that wives work longer than husbands and sleep less. Being a housewife is by no means a good job with a lot of free time and three meals a day.

Sokoloff points out one more interesting fact: women go out to work in order to fulfil their mother-wife role. It was long believed that there was a role conflict in women's dual role as worker and housewife; but being a part-time worker and a part-time housewife is no longer incompatible. Women start working after their children go to school at the age of seven. Their child-rearing responsibilities have shrunk in terms of daily physical care, but have increased in terms of educational expenditures instead. Their income is spent mainly on education and housing. The majority of Japanese families have already become double-income families: wives' supplementary incomes are now necessary to maintain a middle-class standard of life which will allow them to send their children to universities and colleges, and to obtain a so-called 'my home' with a life-long loan. So part-time work can be considered as a part of woman's fulfilment of the mother-wife role.

Just ten years ago, a working housewife was unusual and exceptional, and women had to justify themselves when they wanted to work outside the home; today women must justify themselves when they want to stay at home after their children start going to school. This change has taken place only in the last decade.

Higuchi Keiko, a leading Japanese feminist, calls this the 'new sex-role division' among families in the 1980s, comparing it to the former one between working husbands and full-time housewives with a single income. She writes:

I wonder if I am right in seeing the emergence of a new sexual division of labour. Although the idea of men as workers remains stable, the idea of women as homemakers has gone through some changes: on the one hand, married women are needed in the labour force, and on the other hand, as the society rapidly ages, they are also relied on to supplement insufficient government welfare by caring for their husband's parents as well as their own. Japanese society will not work any longer if women stay at home.... Since the UN treaty for women was ratified without any real concern for equality, the new sexual division of labour will maintain the male-dominant family.[17]

This sex-role division is new in various ways: women can earn money for themselves, though ironically this is spent in fulfilling their mother-wife role, and their income assures them more decision-making power than before *vis-à-vis* their husbands. Their husbands, in turn, are able to enjoy the middle-class standard of living that the women's supplementary income provides, without paying any of the cost of sharing household chores. At the workplace, men, as bosses, can also exploit the cheap labour of women who are in the same situation as their wives.

This 'new sex-role division' is double-edged for women: it has affected the traditional power relationship between husbands and wives, but women have ended up being doubly exploited by men, both at home and at the workplace. As both part-time workers and housewives, women must be content to be 'second-class citizens', while feeling guilty for being neither sufficient workers nor perfect mothers-wives. In short, 'dual-role' has not solved the 'women problem'. It is simply a new compromise between capitalism and patriarchy in the developed stage of capitalism.

THE DECEPTIVE TRADE: EQUALITY VERSUS PROTECTION?

Women's issues have now reached the complicated stage in which discrimination against women is more invisible. Today more than 35 percent of women agree that working outside the home is not necessarily liberating for women. After the majority of women became part-time housewives, they came to realize that working outside as part-time workers was not liberating after all. Before starting to work, it seemed as if work might solve all their problems, but this turned out to be just a fantasy. Women became both workers and housewives only to be exploited both inside and outside the home. They sell their labour at a cheap price to gain a little money at the cost of their free time. Today, some women are deciding to stay in the home as full-time housewives to participate in community activities and women's groups, etc. This is ironic in that working housewives do not have time to participate in such activities, whereas full-time housewives are provided with the resource of time which their husbands can afford them. This tendency is reinforced by conservatism among the younger generation who want to justify their desire to become full-time housewives by claiming that homemaking should be a respected occupation in society. There is of course a hidden desire to get married to wealthy husbands who can provide them with sufficient money and time. To them, being housewives looks more 'liberating', as it guarantees their 'freedom'.

Japanese feminism has come to face a contradiction: if working outside the home is not liberating for women, and women's 'dual role' oppresses women more than before, should women stay in the home as housewives just as before? Feminists for long believed that having a job liberated women, but now this strategy is being discarded. Within feminism there is actually an argument that the home is a liberating shelter from capitalist oppression, and that homemaking is a valuable vocation in that it provides family members with this sheltered environment. A liberal socialist thinker, Matsuda Michio, encourages women to stay in the home.[18] Ironically, this is no different from what the right-wing conservative, Fukuta Tsuneari, insisted some thirty years ago.[19] During the UN Decade for Women, we have reached the national consensus that equality between the sexes is desirable, though till then most people, including women, believed that women were inferior to men by nature. But this notion of equality tends to take the form of 'separate but equal', which reinforces the sex-role division at home and in the workplace.

In order to be on time for the UN conference on women in Nairobi at the end of the UN decade for women, the Japanese Diet rushed to ratify the international treaty for women in May, 1985, and this was followed by the readjustment of domestic laws that discriminated against women. Following these changes was the enactment of the Equal Employment Opportunity Law, that will be effective from April, 1986. If one reads this law carefully, one will understand that it does not aim at women's equality in the workplace; its effect will be the polarization of women workers between full-time elite women workers and the majority of part-time women workers, by forcing them to choose either working as hard as men, or being contented with low-paid part-time work if they want to keep their mother-wife role. This law justifies discrimination against women who do not adjust to work conditions regulated to male standards, including overtime, frequent business trips, and the possibility of transfer. Because of this, women will receive lower status, not because they are women, but because of their own choice.[20] 'Either equality or protection' is the 'false question' that has been long imposed only upon women, while the question of men's responsibility for homemaking is ignored.

Not only is this law a great retreat from the original in terms of the 'Equal Employment Law';[21] worse, it is accompanied by modifications to the existing Labour Law under which the protection of women workers such as the prohibition of late night labour, limitation of extra work, and menstrual leave are abolished. What the Japanese male-dominated government aims at is to force women to choose either 'equality' or 'protection'. In response to women activists, who insist on equality between the sexes in the workplace, the conservatives point to examples in Western industrial societies to illustrate that working women have no protections like menstrual leave. They force women to sacrifice protection for equality. Even worse, this is in fact a deceptive trade, because the Equal Employment Opportunity Law does not assure the effective enforcement of equality because it lacks a viable penalty clause.

Moreover, Japanese women still do not agree to giving up protection: they are demanding 'both equality and protection' instead of 'either equality or protection'. This attitude may sound contradictory and may be hard for Western feminists to understand. But Japanese women have sufficient reason to insist on this, if one considers the working conditions of both male and female workers in Japan. The annual labour hours of Japanese workers in 1983 reached 2,136 hours, 450 hours longer than German workers, who work for 1,882 hours per annum.[22] Even part-time workers work just one or two hours less a day than full-time workers. Though they are often asked to work nearly as long as full-time employees, they are paid less only because they are categorized as part-time employees. Thus, part-time employment is now used as an excuse to utilize women's labour at low cost. Under such labour conditions, equality at the cost of protection will certainly end up by reinforcing the exploitation of women's labour.

The forcible alternatives of 'either equality or protection' will compel women to work like male full-time workers, or otherwise drive women workers out of the workplace. Without protection, working housewives have no other way to continue working other than as part-time workers, content with low pay, and little job security. Despite its pretence, the Equal Employment Opportunity Law will create greater inequality in the workplace; it will reinforce the sex-role division between full-time working husbands and part-time working housewives which enables men to take advantage of women's labour outside the home as an employer and inside as a husband.

Japanese feminists are opposing the Equal Employment Opportunity Law. They insist that instead of abolishing protection, which tends to serve as an excuse to keep women's wages low, this protection has to be extended to male workers. Considering the Western standard of labour hours, Japanese men still work too long. The higher payment of male workers mainly consists of compensation for their extra labour hours. Without this additional income, their basic salary would not be so high. Japanese women are asking that men come back to the family sector instead of devoting themselves to the economic imperialists. Women have already gone out into the work force as *kengyō-shufu* or part-time workers and housekeepers. Now it is man's turn to come back home to be a 'working househusband'.[23] Men have to reduce their labour hours to be part-time workers and to share housekeeping and child-rearing as part-time househusbands.

The labour union of the civil servants of Tanashi in suburban Tokyo demanded equally guaranteed child-rearing leave for men (thirty minutes each morning and afternoon). Their request was approved by the city government last year. The working fathers who first proposed this request to the labour union had to face a lot of discrimination and misunderstandings especially among their male colleagues just because it was not supposed to be manly to share in child-rearing. When their proposal was passed by the City Council, the Ministry of Domestic Affairs was shocked, and started to worry that this precedent might influence other cities.

As in the case of Tanashi, sharing household responsibilities is no longer seen as detrimental to manhood. In fact there is no alternative to men's participating in the home in the modern urban nuclear family setting, because household responsibility is too heavy a load for women to carry. Although the domestic labour in general has been greatly lightened, women are loaded with the entire responsibility of child-rearing, which was not the case in the traditional community where communal parenthood was alive. Women had never been isolated from one another until the modern industrial society destroyed this 'women's world'. However, the feminist strategy of inviting men to accept child-rearing responsibilities is rather exceptional and extraordinary in human history. Many feminist anthropologists have tried to find a model in pre-industrial societies where men and women equally shared household responsibilities, but their attempts have so far been in vain. We had better recognize that we live in an unusual society without precedent among our ancestors. Since the society has already been broken up into nuclear family fragments, there is no adult in the family other than the husband upon whom a woman can call for help. This will necessarily lead to the abolition of the sex-role division.

THE JAPANESE FEMINIST STRATEGY

Michelle Rosaldo, an experienced and deliberate feminist anthropologist, suggested that there are two kinds of feminist strategies to attain equality between the sexes. The first one is: '... two sorts of structural arrangements elevate women's status: women may enter the public world, or men may enter the home'.[24] She also suggested the other strategy: '... in those societies where domestic and public spheres are firmly differentiated, women may win power and value by stressing their differences from men.'[25]

This strategy tends to be applied in societies with severe sex segregation such as Japan. Women's resources and values, developed by women themselves, independent from the male world, may end up in hostility and in conflict with the male world, instead of constituting a harmonious complementarity, as Illich hoped for.[26]

No matter how paradoxical it may sound, Japanese feminists have some cultural resources that they can take advantage of for women's liberation. With the long tradition of the women's world, 'sisterhood' – such as Western sisters are desperately seeking – already exists.[27] Traditionally speaking, women in sex-segregated society have spent more hours and shared more experiences with those of the same sex than with their husbands. So long as they stayed in the women's world, their husbands did not care and would not disturb them. Meanwhile they developed women's networks in community activities, beyond nuclear families and these networks now provide them with confidence and identity.[28] This proves that women's networks can be a liberating resource for women, if working for an income like men is not necessarily a solution. Of course these network activities are only possible for those who are provided with the resource of time that their husbands can afford them. But ironically husbands often come to envy their wives when they compare their own meagre occupation-based networks with those which women establish among themselves; the former are less reliable than the latter.[29]

The other liberating resource could be the traditional family system itself no matter how paradoxical it may sound. Japanese wives, marrying into patrilineal households, are destined to take the responsibility of caring for their parents-in-law when they get old. Women usually face this stage after their child-rearing stage when they have started working outside the house. Today many have to leave their jobs for this reason.

It goes without saying that this family system has been maintained through the sacrifice of women. It is very clear that the government's family policy intends to exploit women's labour in order to maintain cheap welfare. Social welfare services being inadequate, family care of the elderly is too heavy a burden to carry for women only, but this cultural oppression of women could lead to a breakthrough in the modern sex-role assignment.

Care of the elderly differs from child-rearing in many ways. First of all, it is not an easy task in terms of the physical labour involved compared to child-rearing. For instance, bathing an old patient is hard work for a woman. Besides, considering the fact that the women are themselves relatively aged as their parents reach the length of the average lifespan, they need muscular help to carry out this burden. Secondly, whereas mothering is naturally accepted as a woman's role, in a patrilineal family setting a woman is called upon to care for her parents-in-law, not her own parents. There is a natural tendency for anyone to put a priority on his or her own parents rather than parents-in-law when they are in need, and the shrinking size of the family increases this responsibility of women. The decrease of the number of children threatens to change the family system of patrilineality to one of matrilineality:[30] a wife can tell her husband, 'It is your parent, not mine'.

Last of all, according to a cultural principle of seniority, men's participation in care of the elderly is more likely to be esteemed in the Japanese cultural context. It is easier for a Japanese man to take leave from work for parent-sitting rather than baby-sitting. Since the elderly people's care has become a serious family issue, some male-dominated labour unions have demanded paid leave in order to tend aged parents, something inconceivable ten years ago. Men will learn what their old age will be like by taking care of their own parents. Hopefully they will as a result realize how miserable is the retired life of a workaholic.

BEYOND THE 'FALSE QUESTION'

Thus Japanese women are trying to take advantage of their cultural disadvantage. The Japanese feminist strategy is such a paradox that women question the 'women's problem' itself. Forced to choose 'either work or home', they reject the question itself as false. By not accepting male standards of 'equality' which entail the 'masculinization' of women, women propose their alternative, that is, the 'feminization' of men.[31] From this perspective, the Japanese feminist strategy for 'both equality and protection' is no longer a paradox. Women require both 'equality and protection' to be extended to male workers so that they too will be free to become more involved in family life.

Women's liberation may appear to claim that women want to be free from the home and to be like men, but feminists aim at the deeper liberation of both sexes. They question the way of life of men, adjusted as it is to industrial society.

Industrial society was founded when the industrial system of production subjugated the family system of reproduction, and when both production and reproduction were put into the process of means and ends. Human labour came to be valued as productive only when it produced exchange value, and therefore women's domestic labour, mainly reproductive, was categorized as non-productive 'shadow work'. But the family is of consumatory rather than instrumental value; it requires time as well as personal involvement to maintain a family, because it is a human relationship itself that money cannot buy. If men want to come back home when they retire, they must find other ways of participating in family activities other than just bringing in money as breadwinners. They have to realize that their status, though valued by the industrial system, does not assure them any place in the home. Money will no longer give them any excuse for avoiding this responsibility in the family.

There is a serious misunderstanding on the part of Japanese men. When asked what is to be done to recover the father's place in the home, the Japanese businessman-husband is likely to answer that it is necessary to earn more money and reach a higher status, while on the other hand, his wife and children are likely to answer to the same question that he should return home to share more with them.

This is a challenge to industrial values which are mainly materialistic.

The women's movement is rooted in the counter culture which arose in opposition to middle-class values. With the support of the modern sex-role division, these middle class values gave rise to the 'miracle' of rapid economic growth in the 1960s. They also enabled Japanese house-wives to enjoy their share of prosperity. It is as if a 'pax feminina' developed under the oppressive 'pax Japana'. To a considerable extent, housewives are also guilty of appreciating the fruits of Japanese economic oppression of the Third World. They are not entitled to blame Japanese working women for participating in capitalist exploitation. Rather, Japanese women, the oppressed part of the oppressor, Japan, can join in solidarity with women of the Third World in the international women's movement.

The women's movement is, needless to say, a product of industrial society. It started as soon as industrialization was accomplished. However, in the process of struggling for equality between the sexes, women have come to realize that their goal can not be attained within the society that made them conceive the idea of equality between the sexes. The Japanese women's movement has raised many issues that question oppressive industrial values, and women are now opening the door towards alternative values in post-industrial society.

NOTES

1 Sōrifu, Tōkei-kyoku, (Statistics Department, Prime Minister's Office), *Fujin no genjō to shisaku: Kokuren kyōdō kikaku dai sankai hōkokusho* (Present situation and future policy for women in Japan: the third report for the UN decade for women), Tokyo, Gyōsei, 1983.
2 Sōrifu, ibid.
3 Betty Friedan, *The Feminine Mystique*, New York, W.W. Norton, 1963.
4 Fujieda Mioko, 'Uman libu' (Women's lib), *Asahi Journal*, 22nd February 1985.
5 Natalie Sokoloff, *Between Money and Love: The Dialectics of Women's Home and Market Work*. New York, Praegar, 1980.
6 Ivan Illich, *Shadow Work*, London, Martin Boyers, 1981.
7 Ueno Chizuko, *Shihonsei to kajirōdō: Marukusu shugi feminizumu no mondai kōsei* (Capitalism and domestic labour: a Marxist feminist problematic), Tokyo, Kaimeisha, 1985.
8 Inoue Teruko first translated 'women's studies' as 'joseigaku', with a definition that 'joseigaku' is a study of women by women for women. She intended to establish an autonomous discipline beyond mere interdisciplinary studies of women. Teruko Inoue, *Joseigaku to sono shūhen* (Women's studies and such), Tokyo, Keisō shobō, 1983.
9 Ito Masako (ed.), *Shufu to onna*, (Housewives and women), Tokyo, Miraisha, 1979.
10 Iwao Sumiko and Hara Hiroko (eds.) *Joseigaku kotohajime* (Introduction to women's studies). Tokyo, Kōdansha, 1979.
11 Umesao Tadao, 'Onna to bunmei (Women and civilization)', *Fujin Kōron*, May 1957.
12 Sōrifu, *op. cit.*
13 The terms *sengyō-shufu* and *kengyō-shufu* come from the terms applied to farmers. Japanese farmers have shifted from *sengyō-nōka* or full-time peasants, to *kengyō-nōka* or part-time peasants as they almost all engage in extra businesses other than agriculture. *Kengyō-nōka* is further divided into two categories: primary *kengyō* and secondary *kengyō*. A primary *kengyō-nōka* is a peasant family with farming as the major source of

income and another business as a supplement, whereas a secondary *kengyō-nōka* is a family whose main income source is wage labour with farming as a side business. Today more than 90 percent of Japanese peasant families are already secondary *kengyō-nōka*.

14 Of the three choices, they did not take the last because they were careful to avoid the risk of introducing immigrant workers to Japan's crowded, homogeneous society. Japanese capitalists were very aware of the fact that the cost of inviting foreign workers would be greater than the profit. Japan is notorious for accepting few boat people, political refugees, and foreign workers; and however sharply the government is condemned by international political opinion for its selfishness I don't believe it will relax this practice. Instead, Japan's multi-national corporations venture outside of Japan, which enables them to utilize cheap labour in developing countries and to avoid paying the cost of environmental protection in the homeland.

15 Sōrifu, *op. cit.*

16 Sokoloff, *op. cit.*

17 Higuchi Keiko, '"Shufu" to iu na no zaken' (Rights for housewives), *Sekai*, no. 478, August 1985.

18 Matsuda Michio, *Onna to jiyū to ai* (Women, freedom and love), Tokyo, Iwanami shoten, 1979.

19 Fukuda Tsuneari, 'Ayamareru josei kaihōron (The misleading women's liberation)', in *Fujin Kōron*, July 1955.

20 With this law, employers can excuse themselves by reorganizing the work system by new classifications of workers independent of any gender variable. For example, they can replace discrimination according to sex with a distinction between A-class and B-class employees: the former are those who are willing to devote themselves to the job for twenty-four hours, and the latter are those who work for limited hours and cannot be transferred. Since both men and women can freely choose either class, this employment system can no longer be seen as discriminating against women.

21 The difference between the two is that the former focusses the equality only of 'opportunity', whereas the latter intended a substantial equality in employment.

22 According to international statistics, American workers work for 1,851 hours per annum, and English ones for 1,887 hours in 1983.

23 Murase Haruki, *Kaiketsu hausu hazubando* (Viva! househusbands), Tokyo, Shōbunsha, 1984. Ochiai Emiko, 1985, 'Feminizumu no shochōryu (Current trends in feminism)', *Jurist*, no. 39, special issue on *Women: present and future*, Tokyo: Yuhikaku, 1985.

24 Michelle Z. Rosaldo, 'Women, culture and society: a theoretical overview', in Rosaldo and Lamphere (eds.), *Women, Culture and Society*, Stanford, Stanford University Press, 1974, pp. 17–42. see p. 36.

25 *Ibid.*, p. 37.

26 Illich, *op. cit.*

27 Ueno Chizuko, 'The individualist versus the communalist version of feminism', paper presented at NWSA 1984, held at Rutgers University (New Jersey).

28 Women's grass-roots activities such as consciousness-raising groups, consumer groups, communal child-rearing groups and so on, served as the base of these widely diffused women's networks. This is something different from traditional community activities, because such networks are based not just on neighbourhoods but on tastes, values, orientation and ideology.

29 Women's networks are helpful when women are in need during life crises such as divorce or their husband's death, whereas men's occupation-based networks usually do not help them with personal crises nor last after they leave the job.

30 With the decrease in the number of children, when only daughters are born to a family, the parents cannot put priority on sons as successors. As a consequence, there is a new tendency among Japanese parents to prefer the daughter's family to the son's family to live with in their old age. This is called a new matrilineality.

31 Ueno Chizuko, *Onna to iu kairaku* (A joy of womanhood), Tokyo, Keoso shobō, 1986.

10 Male homosexuality as treated by Japanese women writers

TOMOKO AOYAMA

Japanese literature has a long history of works on male homosexuality.[1] They range from the tales of *chigo* (page-boys and children attached to various temples or shrines) in mediaeval times, through works on *shudō* (the way of male homosexuality) by Ihara Saikaku and others in the Edo Period, to modern novels like *Forbidden Colours* and *Confessions of a Mask* by Mishima Yukio. All were written by male authors.

It is only recently that female writers began to deal with the topic in literature and in *shōjo manga* (Japanese girls' comics). This coincided with the rise of the feminist movement, although it was not female but male homosexuality that these writers dealt with.

This paper surveys such works by female writers of both creative fields since the early 1960s. Though conventionally *shōjo manga* is excluded from the category of literature, there are notable similarities and mutual influence, if indirect, between the two fields in their dealing with male homosexuality. The paper discusses why the topic was chosen and how the stream changed its course from pure fantasy toward reality and from pro-paternal to anti-paternal.

In the early 1970s there was a notable change of trend in *shōjo manga*. With writers such as Hagio Moto, Ōshima Yumiko, and Takemiya Keiko contributing their talents, and with the backing of the somewhat more understanding and 'liberal' editors, the genre achieved such a great improvement in its themes and techniques that claims were made that *shōjo manga* had reached its own 'Renaissance'. It appealed to a wider variety of readers not only among young girls but also among university students and even older age groups of both sexes. In 1981 a monthly periodical *Eureka – Poetry and Critique* published a special issue on *shōjo manga*.[2] Critiques of the 'Renaissance' invariably indicate that male homosexuality or love between boys has played a great role in the development of the genre.

Aaron Browning in Mizuno Hideko's *Fire*, (1968–71) is claimed to be the first male protagonist in *shōjo manga*.[3] Until then, the main stream of the genre dealt with girls exclusively; there were lots of Cinderellas and lots of female twins being tried either by their beastly step-mothers or jealous girls at a ballet school. Male characters were never anything more than the object of the heroines' love. *Shōjo manga* before *Fire* were certainly 'of the girl, for the girl' but not necessarily 'by the girl', for there were comparatively more male writers then than in the seventies, and all the editors were male.

Having a male protagonist was a revolutionary idea. A further turning point was marked by the writers' challenge to taboos in editorial codes. In order to have her own creation 'Aaron' accepted by the magazine *Seventeen*, Mizuno had to persuade the editor that the story dealt with the 'group sounds' (rock music in Japanese style) which was extremely popular among young people in the late sixties. In fact, *Fire* was created under the strong impact of the hippie movement rather than the commercialized 'group sounds'. The American setting of the story was necessarily more realistic than the vaguely Western settings of the innumerable stories of the past – including those by Mizuno.

Of the many male protagonists that followed Aaron, those created by Hagio and Takemiya gained the most popularity at the beginning of the 1970s. They were mostly boys much younger than Aaron, being in their early to mid-teens. This is partly due to the fact that the readers of the magazines that these two writers were writing for (mainly *Bessatsu shōjo komikku*) were younger than those of *Seventeen*. However, there seems to have been a positive intention on the part of the writers in choosing these younger heroes. Edgar Portsnell in Hagio's *Pō no ichizoku* (The Poe Clan, 1972–6) stopped growing in the mid-eighteenth century, when he became a vampire and member of the Poe clan at the age of fourteen. He lives an eternal life as a sort of *enfant terrible* throughout the centuries in many European cities. He is often disdainful of ignorant mortal adults, while also holding a grudge against the adult members of his own clan who deprived him of growing any older. All of these themes – the desire for eternity, the distrust and mockery of adults, and the admiration for childhood – were often shared by other boy protagonists, and without a doubt appealed to younger readers.

Another important point is that the boys were invariably pretty or beautiful. They were not the 'handsome boy next door' type who might some day fall in love with an ordinary girl, nor were they the flower children of the sixties. Instead, they took the form of a vampire (Edgar and Allan in *Pō no ichizoku*), an ingenious swindler (Tague Parisien in Takemiya's *Sora ga suki!* (I like the sky!)) and so on. Beauty and fantasy were emphasized over reality, though this was not a new concept in the world of *shōjo manga*. Hagio and Takemiya sought new modes of romanticism through science fiction, historical sagas and homosexuality.

Takemiya started writing short stories about boys in love with each other as early as 1970. The stories, 'Sanrūmu nite' (In the sun-room), 1970, 'Hohoemu shōnen' (The smiling boy), 1972, and 'Nijū no hiru to yoru' (Twenty days and nights), 1973, treat the physical beauty of the boys as something amorally fascinating. The rebellion against the conventional morality of heterosexual adults is at the same time still heavily dependant on rather hackneyed traditions of *shōjo manga*. The plot often depends on the twin/orphan situations and/or a fatal disease. The characters are drawn literally with starry eyes, often pearled with tears and surrounded by the vaguely Western settings typical of the genre. It is also notable that the girls in these stories are unable to participate in amoral fascination. They grow up too quickly ('Sanrūmu nite') or care too much about the boy's health ('Hohoemu shōnen') to stay in the aesthetic sphere.

Towards the mid-seventies, the longer stories dealing with love between boys appeared. *Tōma no shinzō* (*Thoma's heart*) by Hagio Moto appeared in serial form from May 1974 onwards, and was later reprinted in a three volume book form. Compared with its prototype, 'Jūichigatsu no gimunajiumu' (The gymnasium in November), 1971, by the same author, *Tōma no shinzō* is not only much longer, but also much more elaborate as far as plot, characterization, and graphic techniques are concerned.

The biggest change concerned the theme of 'love'. In the earlier story, Thoma's love for Erik, a newly arrived student, reaches its climax in a momentary hug. After Thoma's sudden death from pneumonia, it is revealed to Erik that Thoma was his long-lost twin brother. In the later version, however, Thoma is already dead when Erik first arrives at the gymnasium at the beginning of the story. Though the striking physical resemblance between the two boys is always expressed, the theme concerning the newly found and immediately lost twin brother is suppressed. Instead, the love of the three boys, Thoma, Erik, and Oskar is directed towards Julismor, a boy who has lost faith in himself and shut his mind to love. At the end of some 450 pages, love is the overall winner and Julismor opens his mind again to love before leaving the gymnasium for a theological school.

It cannot be denied that there are some weaknesses in this novel-like *shōjo manga*, particularly in the way it deals with the themes of death, religion, and diabolism. Nevertheless, it is a very well composed fiction. The setting of a German gymnasium works, because it is fictitious. In her interview with Yoshimoto Takaaki for *Eureka*, Hagio says: '... before writing this ('Jūichigatsu no gimunajimu'), I made two plans – one was about a boys' school and the other about a girls' school. Then I found the plan about the girls to be gloomy and disgusting ... Take a kissing scene, for instance ... as sticky as fermented soybeans'.[4]

The choice of young boys rather than girls, and a German gymnasium rather than a Japanese school would appear to have been made for the sake of free artistic expression to avoid the 'sticky' restrictions of realism.

Criticism has been made of the general tendency of *shōjo manga* towards superficiality and escapism. Kate Brady criticizes: 'Although many comic book writers no doubt research the historical periods in which their stories are set, historical stereotypes are the rule. Fantasy is the *shōjo manga*'s stock in trade: the more fantastic the better'.[5] She also points out that the exclusive use of male homosexuality – and no lesbian romances – 'is merely one more way of keeping reality at bay'.[6]

The poet, Suzuki Shiroyasu concludes his critique, '*Shōjo manga* – kibun no yōritsu' (Girls' comics – the crowning of sentiment):

Shōjo manga are aimed solely at creating sentiment and are allowed to cut all connections with reality for want of protecting that same sentiment. The structure of fiction and the graphic lines that support the development of the story are retained as the only real thing to the consciousness (...) As long as they are neatly composed and well drawn, *shōjo manga* are allowed to choose for their setting any period and any nation of mankind. No other mode of expression is allowed such liberty. And it is this very liberty that is unbearable to someone whose mind is bound to the requirements of reality.[7]

Hagio is fully aware of such criticism herself. In the interview with Yoshimoto she often calls herself an escapist. Yet, she says, her hand simply refuses to draw a face of someone revolting or something ugly. Yoshimoto, on the other hand, encouragingly disapproves of her self-criticism:

YOSHIMOTO: ... If you were not writing what you wanted to write because you were bound by such rules as: '*gekiga* (dramatic comics) should be like this' or '*shōjo manga* should be like that in order to appeal to the readers', then it could be called escapism. Such restrictions should be abandoned. You should deal with the muddy world or whatever in your work. If it so happens that you personally cannot draw what you dislike or that you naturally do not want to see something ugly, then I don't think it's escapism at all. I think it is your nature, your essential qualities as a writer, or should I say as a *gekiga* artist.[8]

Yoshimoto also claims that fantasy is shared by all art and literature, and that a fantasy is not escapism just because it does not deal with social problems. Indeed, as far as their attitudes toward reality are concerned, a parallel can be seen between some *shōjo manga* writers and aesthetic or anti-naturalistic writers such as Tanizaki Jun'-ichirō and Akutagawa Ryūnosuke. A prime example is Akutagawa's short essay 'Mukashi' (The past), in which the author discusses historical periods and foreign settings:

suppose I have grasped a theme and am going to write it into a novel. In order to express the theme artistically and powerfully, the introduction of some extraordinary incident might prove necessary. In such a case, the more extraordinary the incident is, the harder it becomes to write about it as something that has happened here and now in Japan. If I forced myself to write it in such a way, the readers would most probably feel it unnatural, and consequently the worthy theme would be killed in vain. In order to remove the difficulty, as is clear from the words 'hard to write as something that has happened here and now in Japan', there is no other way but to set it as something that happened either in the past (future would be a rare case) and/or in some other country.[9]

Akutagawa's words could apply as well to the extraordinary incidents and situations in *Tōma no shinzō*, which are designed for the sake of a powerful and artistic expression of the theme.

While 'love between boys' and its eroticism are highly spiritual in Hagio's works, Takemiya presents physical intercourse more explicitly. This does not mean, however, that Takemiya's work comes any closer to the reality of homosexual love.

Hensōkyoku (Variations) 1974–81, is a collection of Takemiya's stories in three volumes. Each story focusses on some of several main characters, who, with the sole exception of Annette, are male. The major portion of the book is told from Holbert Metschek's viewpoint. He is a music and art critic born in 1941. Through the eyes of this homosexual aesthete, the following three central figures are presented:

1   Wolfgang Richter (1950–71), a young musical genius.
2   Eduard Soltie (1950–), another muscial prodigy with perfect physical beauty.
3   Nino Alexis Soltie (1973–), a son born of Eduard and Annette, Wolfgang's sister.

Holbert's relationship with Wolfgang is based on an intellectual friendship, while with Eduard it starts out as a relationship between an ideally beautiful boy and his admirer. In order to approach Eduard, the twenty-three year old critic goes so far as to blackmail, though half jokingly, the fourteen year old boy, who is involved in a political movement. After their first affair, the black-mailer becomes the boy's protector, and a physical and intellectual friendship is established. In Nino's case, it is Nino himself who wants to begin a physical relationship when they meet in Copenhagen around 1990. Holbert, now nearly fifty years old, finds himself more and more attached to the seventeen year old boy: 'I, misanthropic Holbert Metschek, was attracted to Wolfgang Richter and Eduard Soltie because of their genius. But now I am beginning to be attracted to Nino Alexis for his humanity'.[10]

What he calls 'humanity' here is Nino's loneliness and his complex feelings about his father, Maestro Eduard. Eduard pours all his love and care onto a son of the late Wolfgang Richter rather than onto his own child. Holbert, who used to educate and protect the young Eduard, now gives his love and protection to Nino.

Fatherly figures in a mildly homosexual context have often appeared in other *shōjo manga*, particularly in some works by Ōshima Yumiko.[11] Yet the viewpoint was always that of their protégés and in many cases it was not too difficult to find an Electra complex, which had long been claimed to be one of the representative characteristics of *shōjo shumi* (girlish taste). Holbert Metschek, however, seems to appeal to the readers not as their ideal father, but as an ideal appreciator and protector of beauty and love – someone with whom the reader can identify. It is a metamorphosis of a Cinderella candidate into an aesthete like Lord Henry Wotton in Oscar Wild's *The Picture of Dorian Gray*. In the late 1970s, the 'Lord Henry syndrome' prevailed in *shōjo manga*. Even Ōshima Yumiko, the prime representative of *shōjo shumi*, has one of her female characters wish to be a male homosexual in order to love a beautiful boy.[12]

The subject of male homosexuality was never popular among female writers of novels and short stories as it was among *shōjo manga* writers. There are, however, a few writers who dealt with it. Mori Mari (1903–) was the earliest. In the period of less than a year between August 1961 and June 1962, three novelle by this author appeared in the literary magazines, *Shinchō* and *Gunzō*. Though they are independent stories, all invariably treat love between a wealthy intellectual in his late thirties and a beautiful young man as the main theme. There is no evidence that these novelle had any direct influence upon the *shōjo manga* writers discussed above. Yet there are remarkable similarities especially in their way of creating an alternative reality. It is as if Mori had predicted the popularity of homosexuality as a subject in *shōjo manga* in the 1970s.

Though it appeared between the two other stories, 'Nichiyōbi niwa boku wa ikanai' (I'm not coming on Sunday), *Gunzō*, December 1961, deals with the first stage of love. It tells how Sugimura Tatsukichi, a French literary critic turned writer, dissuades his literary disciple, Itō Hans, from marrying eighteen-year-old Yatsuka Yoshiko. She and her family, especially her mother, represent 'respectable society', with its promises of motherly love and care within the unimaginative context of 'commonsense'. Tatsukichi, on the other hand, offers the young man financial and intellectual protection against the horror of 'commonsense'. This is a careful, fatherly protection that arises from infatuation. In vain Yashiko's mother protests against Tatsukichi's disdain for her common-sense world. The broken engagement, followed by the accidental death of Yoshiko, indicates the victory of the paternal world over the maternal.

The wealthy intellectual in 'Koibito-tachi no mori' (Lovers' forest), *Sinchō*, August 1961, is named Guydeau de Guiche. Born of a wealthy Frenchman and a daughter of a Japanese diplomat, Guydeau has 'the honour of France and the voluptuousness of France within',[13] and a dauntless beauty without. Though this story, like the other two, is set in Japan, European flavour is as apparent as in *shōjo manga*. Just as *shōjo manga* writers create visual effects with their sophisticated graphic techniques, so does Mori create similar effects through words, in her exquisitely detailed description of luxurious props and characters. In fact, characters are treated almost as part of the extremely gorgeous and occidental props. Guydeau calls his young lover 'Paulo' instead of his Japanese name. When they are first introduced, all the Western names in the three novelle are written in Chinese characters with their pronunci-ations in *katakana* alongside, in the same way as some foreign words such as 'bed', 'divan', 'ham' etc. are chosen and written.[14]

The decorative forest of lovers has three other figures, each one a threat to the perfect couple of Guydeau and Paulo. Nashie, a young girlfriend of Paulo's, is the least powerful threat.. She is nothing but a faint remi-niscence of the motherly, commonsense world, which is now handled by the two lovers with calculated respect. A real physical threat comes from Mrs. Ueda, who has been having an affair with Guydeau for two years. The forty-eight year old woman desperately tries to cling to her own beauty and his love, both of which are fading quickly. Unlike the innocent Nashie, Mrs. Ueda does not belong to the motherly world, but is described as a typically 'feminine' character with her jealousy and burning hatred. She is childless and for this very reason has managed to retain until recently her slender figure, which used to look rather like Paulo's. In spite of Guydeau's careful camouflage she senses and finds out the truth about his relationship with the boy. Her physical threat is realized when she kills Guydeau. Yet the story does not end there. The third threatening figure, who makes his entrance at the beginning of the novella as a mysterious 'dark man', emerges after Guydeau's death as the next potential protector for Paulo. The story ends, with a beautiful boy, though still shocked by his lover's death, finding his natural self again. 'Then Paulo returned to Paulo. His heart, which had become sentimental and hysterically elevated in his love with Guydeau, returned to its own nature.'[15] He whistles a song which Guydeau once taught him and 'looks around him and up at the sky like a survivor who has just returned from somewhere far away'.[16] The ending suggests Paulo's strangely detached vitality. Only he escapes from the restraint and sentimentality of love which had carried away Mrs. Ueda's reason and Guydeau's life.

In 'Kareha no nedoko' (The bed of fallen leaves), *Shinchō*, June 1962, love's violent and destructive aspect is emphasized with greater depth. The dark, menacing figure, Tōta Olivio, plays a much more important role. This Italian is a heroin addict. He finds a victim for his sadism in Leo, a young lover of Guylan de Rochefoucaud's. Guylan, 'a born aesthete', found Leo when the latter was only fourteen years old. The Franco-Japanese writer saved the boy from a hooligan circle and has kept him under his love and protection. Guylan is, however, aware that the boy could be a danger to him.

Leo, since his sixteenth summer when he had tasted his forbidden fruit with Guylan in a hut on Mt Hodaka, became precocious, while remaining a child inside, and was transformed into an evil angel who seduces Guylan, sprinkling poisonous white powder. Knowing this, Guylan, who has been earnestly in love with Leo, makes him precocious in the extreme and gradually invites his own ruin.[17]

When he finds out that his young lover has had an affair, though half involuntarily, with the dark man, Guylan's sadistic infatuation with Leo deepens together with his fury and agony. He reflects: 'The shadow of Leo's masochism had already been there in our first kiss. Narcissism and masochism must be the two propensities of a beautiful boy. I, too, am a successor to Sade in my thought and sexual inclination'.[18] He knows that Leo is unconsciously attracted to Olivio and would sooner or later be captured by the dark man again. He wishes he could use a whip on Leo just as Olivio did, 'but he feels the soundness, which would never let him go into such insanity as Olivio's, still alive within him'.[19] His dilemma finally drives him to kill the boy with his hunting gun. Then he tries to complete his novel about a boy like Leo, but the dead boy lying in the bed of fallen leaves calls out to him. Rather than finish his last work, Guylan chooses to kill himself so he can be close to his boy, Leo. 'What is worthwhile is to be with Leo, to have him and never let him go, – to die for something valuable.'[20] Thus the image of an idealistic and intellectual figure collapses in pursuit of love.

While the idealized paternal image descends from the victorious confidence of 'I'm not coming on Sunday', down to the self destruction of 'The bed of fallen leaves', the beautiful young protégé grows sensually more and more tyrannical and, at the same time, masochistic. It is interesting that this transition coincides with a change in the spoken language on the latter's part. Itō Hans hardly utters anything at all – apart from the occasional 'yes' and some very brief answers. When he speaks, his language is not particularly feminine in spite of the fact that his physical and mental characteristics are very often described as being 'just like a woman's'. Paulo speaks a lot more, using young male language in front of Nashie, and female language in front of Guydeau. Leo's speech is almost totally feminine or infantile.

This transition to feminine/infantile is handed down to the elfish girl, Moira, in Mori's next novel, *Amai mitsu no heya* (The sweet honey room), 1975. The idealized fatherly figure is also treated in the novel in a much more straightforward way than in the three novelle, in the character of Moira's real father who is infatuated with his daughter. It seems as if the three novelle treating homosexuality were by-products for Mori in finding the 'sweet honey room' – where the Electra complex is almighty – for the father and his little daughter.

As we saw earlier, *shōjo manga* had its first male protagonist at the end of the 1960s. It was a sign of the urge for new subject matter in the genre on the part of both writers and readers. They were no longer satisfied with the persistent variations on the Cinderella theme. The new interest in boys was, therefore, due to the desire to explore masculinity or androgyny as opposed to the worn-out image of feminity. This revolt against conventions, however, is still heavily based on a paternalistic double standard: a pretty girl waiting for a handsome prince is boring, a pretty boy meeting another pretty boy is interesting. Female characters are still the same types; innocent, pretty sisters (Marybell in *Pō no ichizoku*, Annette in *Hensōkyoku*) or slightly jealous ones (Angel in 'Sanrūmu nite', Swena in 'Hohoemu shōnen') or beautiful yet helpless mothers (Erik's mother in *Tōma no shinzō*).

The same paternalism is also found in Mori Mari. Whilst her male protagonists live and die in their exclusive, aesthetic world of love, the female characters are excluded from joining them because of their innocence (Yoshiko in 'Nichiyōbi ...', Nashie in 'Koibito ...'), common-sense (Yoshiko's mother), and mad jealousy (Mrs. Ueda in 'Koibito ...'). The intellectual epicurean had to be a male in the early 1960s, for it would have been too hard to have a female 'paternal' figure with the same financial and social power. So had the young beautiful protégé to be a boy to maintain the anti-realistic decorum: it was too late for another Naomi to make a prey of an intellectual Jōji,[21] and a little too early for Moira to reign in her 'sweet honey room' with her father.

In spite of the double standard, it should be admitted that for the first time in the history of Japanese literature and comics, women had become the creator/admirer of physical beauty in the opposite sex in a context totally detached from marriage and family life leading to maternity. However naive or hackneyed the praises of the beautiful boys may have been in their conceptions, they were new in that they came from women, and the expression was fresh and had its charm.

In order to understand the mixture of rebelliousness and conventionalism found in these writers, Gilbert and Gubar's 'Metaphor of Literary Paternity'[22] is worthy of note. Although the history of Japanese women's literature does not parallel the Western one, as far as the treatment of homosexuality is concerned, the women authors started from longing for 'authorship/paternity/authority'.

Apart from the double standard found in the earlier homosexual stories by female writers, it is notable that there is no apparent influence from, and absolutely no reference to, Japanese classical works. Instead, Western classical or mythological figures are often mentioned, as well as more recent – especially nineteenth-century – literary works and figures. Obviously this reflects the general inclination towards Western culture since the Meiji era. The tradition of *nanshoku* (male homosexuality) was denied by Christian morality and obliterated like other 'old abuses' from the Edo period. On the other hand, as modernization proceeded, 'Western' homosexuality was often associated by the intellectuals with the glories of the Classical Age or the flower of decadence.

Paulo in 'Koibito-tachi no mori' is derogatorily called 'o-chigo-san'[23] among the intellectual guests who gathered at a party,[24] but these people, even those against Guydeau, 'cannot help admitting the reminiscence of a beautiful homosexual nobleman of the ancient time, and of a Narcissus'[25] in the couple of Guydeau and Paulo.

Although homosexuality was still considered antisocial and immoral in reality, it could be connived at or even admired when it was associated with the great tradition of Western paternalism. Also, now that the tradition of *chigo* and *shudō* is far away, it would be difficult to compare a homosexual couple in a modern story to representative figures from old tales or from Saikaku. In the first instance, they may simply not be recognized by the reader. Secondly, since *chigo* was in the Buddhist context and *shudō* was associated with *Bushidō* or *kabuki*, they would be too specific to compare with a modern homosexual couple. Finally, since most homosexual stories in *shōjo manga* are deliberately set in the West, it would only spoil the effect to refer to Japanese names. Thus Eduardo and Holbert in *Hensōkyoku* had to be 'just like Rimbaud and Verlaine' and could not be 'like Umewaka and Keikai'.[26]

Because these female pioneers of homosexual stories had a strong aesthetic tendency to keep a good distance from reality, their works hardly shared anything with those written by Mishima or by modern American writers such as Tennessee Williams, Truman Capote and James Baldwin. Homosexuality was such a perfect aesthetic sphere for these women writers that they would never allow anything ugly or grotesque to creep into it. None of the residents of the idealized world feels guilty about his being a homosexual, or has to seek his identity as did the residents of James Baldwin's *Another Country*. Even the sadomasochistic scenes found in the works by Mori, Hagio, and Takemiya are presented according to the highly formalized and conventional aesthetic codes. The choice of the three characters in Mori's[27] and Takemiya's[28] stories – an intellectual, a young beautiful man (boy) and an innocent girl – reminds us of the three characters in Mishima's *Kinjiki* (*Forbidden Colours* 1951): Shunsuke, Yūichi and Yasuko. However, the former groups of three are absolutely free from the latter's doubts and ugliness[29].

As the homosexual stories became a boom in *shōjo manga*, a few changes took place. Graphic techniques improved and the plots became increasingly more involved, if not always convincing. Some of the stories went on and on in more than ten volumes of about 200 pages each.[30] Scenes of sexual intercourse were presented more explicitly, sometimes in a violent context. In some cases it almost seemed like a kind of revenge on the part of women who had now become a spectator rather than a prey. New magazines such as *June* were published around 1980, dealing exclusively with homosexual love stories and praises of beautiful boys.

The commercial success, however, also brought the danger of smugness and stylization. The writers had to exploit new materials. Thus the settings and the backgrounds of the stories extended from the cherished nineteenth-century Europe to contemporary America (Yoshida Akimi, *Kariforunia monogatari* (California story)), the late sixth-century Japan (Yamagishi Ryōko, *Hi izuru tokoro no tenshi* (The prince of the country where the sun rises)), and the old Japanese high school at the beginning of the twentieth century (Kihara Toshie, *Mari to Shingo* (Mari and Shingo)). The May 1985 issue of *June* includes five short *manga* stories set in contemporary Japan, one *manga* and one serial novel set in the Edo period, and another *manga* about the old China. There are also articles about famous *kabuki* actors in the Edo period and boys as treated in modern *tanka* poetry.

Some of the writers purposely tried to break the perfection of the ideal homosexual world by introducing social problems, scepticism and frustration (*Kariforunia monogatari* and *Hi izuru tokoro no tenshi*), or by emphasizing comic factors (Aoike Yasuko, *Eroika yori ai o komete* (From

Eroica with love)). They intended in this way either to parody or to make more realistic the world of 'sweet homosexuality'.

Yoshida Akimi began her *Kariforunia monogatari* in 1978, when homosexuality was losing its shock value in the genre. The story, in eight volumes, is about a young man named Heath, originally from California and now living in New York City. It deals with the process of growing up and finding oneself, as the protagonist confronts and overcomes a variety of social and personal problems.

Though Heath himself is heterosexual, almost half of his male friends are either homosexual or bisexual. Here homosexuality is by no means a mere aesthetic symbol: it is treated as part of life's reality together with other social issues such as divorce, drug abuse, male prostitution, abortion, violence, robbery, police and naval corruption and the like.

The realistic approach is also apparent in Yoshida's graphic techniques. The size of eyes is much smaller than that in traditional *shōjo manga*. So is the body drawn closer to human anatomy – with muscles which were almost non-existent in most *shōjo manga* and are often too exaggerated in boys' comics.

The double standard found in the earlier works is disappearing. All the female main characters in *Kariforunia monogatari* are, as Hashimoto Osamu pointed out,[31] independent and intelligent. There is an extreme case of a pornographic film actress with a Ph.D. On the other hand, some of the male characters are jobless, uneducated and even a little retarded. The paternal figure with intellectual and social authority is represented in the character of Heath's father, a renowned lawyer. Both his wife and his son leave him in protest against this authority. All the potential fathers fail to enjoy father-son relationships in real family situations or in homosexuality. Heath's brother is killed in an accident before his son is born. One of Heath's bisexual friends has to ask his pregnant girlfriend to have an abortion for financial reasons. When Heath's girlfriend gets pregnant, she condemns him for being too selfish and therefore unfit to be a father. She miscarries their baby – in fact, a son – while Heath is in jail. Neither can Yves, a young homosexual friend of Heath's, find a fatherly protector in his real father or brother-in-law, or in Heath, whom Yves loves.

A similar tendency can be found in other works. Dorina in *Mari to Shingo* chooses to fight for her country's independence rather than marry Shingo and leave her country in Europe. Even in the historical setting of the late sixth century in *Hi izuru tokoro no tenshi*, princesses can be brave, intelligent and strong. Tojiko, married to Umayado no Ōji (Prince Shōtoku), thinks when facing the dead body of her sister:[32] 'Such a thing cannot be allowed to happen. This cannot be the right thing – women always preyed on by men!' On the other hand, princes, even the homosexual protagonist with supernatural power, Umayado no Ōji, may 'effeminately' suffer from love-sickness and jealousy.

Some of these changes in *shōjo manga* in dealing with homosexuality overlap with those in other forms of literature – from Mori's novelle to more recent works.

In 1979, *Mayonaka no tenshi* (*The midnight angel*) was published by Bungei shunjū. This novel in two thick volumes was written by a young critic-novelist, Kurimoto Kaoru (alias Nakajima Azusa, 1953–). It deals with male homosexuality in the world of show business. Imanishi Ryō, nicknamed 'Johnnie', is described as a teenage idol with the beauty of Narcissus and the cruelty of Salome. His manager, Taki, who discovered the boy, is prepared to do anything to make the boy a superstar. He rapes the boy, sells him to men and women of power, and even murders a princely song writer, Yūki Shūji, who falls in love with Johnnie. What drives Taki to all this, however, is not just his ambition in show business, he also feels that there is a strong bond between the boy and himself: 'It is a relationship that secretly has in itself all the conflicts between father and son, two brothers, believer and priestess, merchant and merchandise, husband and wife and two lovers.'[33] It is also emphasized that Taki finds in the boy the perfect raw material for his creation: 'I suppose a man encounters once in his life the material that he wants to try to complete as a work of art at the risk of his own life.'[34] Later in the novel, Yūki reproaches the relentless manager: 'The devil is not Ryō, no, not him. It's you. You are the Mephistopheles who made Ryō like that.'[35] Replying to this attack, Taki calls himself a Pygmalion and a Frankenstein. After many violent incidents and intrigues, the long novel concludes: 'The position of the work and its creator has now slowly yet completely inverted. From now on Ryō is the King – he is the God, who governs and manipulates Taki.'[36] Taki is certainly a crude and much drawn-out late twentieth-century version of Seikichi in Tanizaki's 'Shisei' (The Tatooer, 1910).

In spite of, or perhaps because of, the general mobilization of mythological and literary images used to describe the young star – Narcissus, Salome, Astarte, Tutankhamen, Alfred Douglas, Cleopatra, St. Sebastian and so on – there is hardly anything fresh and attractive about Johnnie. Yūki, a celebrity with beauty, intelligence, musical talent, wealth, good family background, and even sincerity, courage and all the 'manly' virtues, has no more substantiality than a Prince Charming. Just as Tanizaki's *Chijin no ai* (*A fool's love*, 1924) is the story of Kawai Jōji rather than of Naomi, so is the wretched and crazed Taki the true protagonist of *Mayonaka no tenshi*.

It is clear that the author, Kurimoto, ambitiously tries to combine the sub-culture of modern Japanese society with the aestheticism presented by Oscar Wilde, Tanizaki and others. Kurimoto could certainly be called a 'crossover' writer – from 'pure' literature to 'popular' literature and *shōjo manga*.[37] Whilst maintaining the idealized homosexual world in the relationship of Johnnie and Yūki, she sets the crooked, maniac love of Taki for Johnnie as a central theme of the novel. Through her 'author-ship', which is still dependent on aesthetic 'authority', the violence and weakness of 'paternity' is revealed.

While Mori Mari and Kurimoto Kaoru more or less followed traditional, and often rather stereotyped views on homosexual love, there were a few other women writers who dealt with homosexuality somewhat more peripherally, and from quite different angles.

Tomioka Taeko (1935–), who happens to be a friend of Mori Mari's,[38] takes a totally different attitude from the latter's when she writes a story of young, good-looking men. 'Wandārando' (Wonderland), 1976, is one of Tomioka's short stories collected under the title of *Tōsei bonjin-den* (the lives of contemporary ordinary people). As the book title indicates, all the characters are very ordinary people although sometimes they may look eccentric.

The young protagonist, Rui, used to share a room with Mino, who has now become a film star. Rui was cautious enough not to tell this to any stranger for fear of being mistaken for a homosexual. In fact he and Mino had never been lovers, yet they were always with each other 'just like two cherries bound together'[39] according to Sano-san, a single middle-aged successful business woman and former patroness of Mino's. They always enjoyed being together, going out, making jokes, and gossiping. The young film star, though just a mediocrity compared with the sensational superstar in Kurimoto's novel, rings up his former flatmate and makes a mock confession about his affair with an 'ideal woman' – 'ideal' as in his official answer in an interview for a women's weekly magazine: 'It's such a hustle, really, all this ... But then I'm a star, yes, aren't I? A twinkling little star! Oh, I do want to see you, Rui, but it's not allowed to a little star, you see. The "ideal" lady wouldn't let me, you know.'[40] In spite of the twofold 'gay' tone – happy, frivolous, young female language – in the conver-sation, Rui senses something gloomy and becomes doubtful about Mino's future as a star:

Mino should really make a sale out of that gloominess of his. He shouldn't show the inner part of the gloominess in his joke. He shouldn't ring his former flatmate and sneeringly call himself 'a twinkling star', either – as if he were an intellectual. The twinkling little star would be ferreted out by the masses before it knew. That's no good. That's his weakness, Rui thinks.[41]

Yet Rui is by no means a ruthless creator/manager of a superstar, and would never tell Mino what he thinks of him, for the former flatmate of the 'twinkling little star' is another fake 'gay' himself. He is attracted to Mimiko-san, a 'happily married' middle-class woman and girlfriend of his step-mother's, because he somehow feels sorry for her. He goes to a first-class hotel with her but fails to make love to her. When his old friends, 'Mine' and 'Fū-chan' come to his place, the real gay couple make fun of Rui's alleged love affair.

Watching Mine and Fū-chan, Rui begins to feel sad, feeling as if he were watching the Mino and himself of a few years ago. Surely Mine and Fū-chan may love each other, but we were only sharing a flat for the sake of economy. And yet, what a similar way of making fun! Always poking fun at somebody, always laughing and giggling about trivial things. It is the very laughter that is as if purposely made just to utter a sound, incessantly, desperately, and as if to kill time. It is not that they feel it's funny, but that they won't find anything funny unless they laugh.[42]

Half bored as he is, Rui still keeps company with them and appreciates their 'gay' jokes because they are much better than serious talks. On a later occasion, he listens to Mino grumble half jokingly about the danger of his being deflowered by homosexual actors. When his talk gets too serious, however, Rui silently criticizes his friend and suggests inviting the gay couple over. While waiting for them to arrive, Rui tells a false story about his love affair 'successful in a missionary position', just to cheer both of them up.

The frivolity of these 'gay' people – whether they are real gays or mock ones – is a criticism against straight-faced intellectuals,[43] and it is based on the morality of 'ordinary people': work hard in silence and talk nonsense with your friends.

At the same time it is a parody of the highbrow homosexuality as presented by Mori, Takemiya and others. The inhabitants of the 'Wonderland' are far from those living in the idealized 'Lovers' Forest'. They are 'ordinary people' with ordinary feelings of sadness and uneasiness which are consciously disguised beneath frivolity and gaiety. They are too modest to boast of their youth, beauty or intelligence. Instead of longing for 'authority/paternity', Tomioka's critical authorship mocks at the authoritative and paternal which is ironically represented by the business woman, Sano-san.

The same attitude of 'authorship' as opposed to 'authority/paternity' may be found in *Hana to mushi no kioku* (Memories of flowers and insects)[44] by Ōba Minako (1930–). The novel was first published serially in *Fujin kōron* from January 1978 to February 1979.

The narrator/protagonist of the novel is a young woman named Maki.

At the beginning of the novel she is introduced to Shōzaburō, an eligible young man, by her family friend Makiko.[45] As the story develops with various people surrounding the heroine, she is told by Makiko that her son Yū has left home and is now working at a gay bar somewhere in Tokyo. The perplexed mother asks the young confidante for help to find out where he is. Maki feels a 'mother's egoism' in Makiko's strange calmness: 'She (Makiko) is only concerned about her son's not losing himself in such a world. Perhaps she may be even thinking that his having a male lover is less unbearable than a female lover.'[46] She says to the mother: 'I suppose everyone has one's own way of living, others can't change it.'

Later, through the help of a neighbour who works at a 'host club', the heroine finds out where Makiko's son Yū is. She also discovers that Shōzaburō has been having an affair with the boy as well as with the mother, Makiko. The son boasts:

'I was the one that pushed him into her bed. I told him he had better have her ask her husband to make him a professor because her husband always believed in what she said. On the other hand, he had such confidence in himself. He believed she was obeying him while in fact he was doing whatever she told him to do.'[47]

His disdain and distrust of his parents overlap with Maki's feelings toward her poet father and housewife mother.[48] So does Shōzaburō's double affair with the mother and the son overlap with that of the heroine with Tadamasa, the father, and Fumimasa, the son. Maki listens to Yū's story, 'not knowing whether it was somebody else's story or her own, or a story of human beings which has been repeated again and again since time, long ago'.[49]

Two days later Maki and Shōzaburō meet again, this time as rivals who will soon be working together at the Los Angeles office of Tadamasa's company. She is not in the least interested in whom he has been sleeping with, but she thinks he is all the more interesting because of his relations with various people, for she believes 'nobody can exist alone'. Shōzaburō feels as if he had known her for years, though in fact they only met less than a week ago. Yet he would not like to admit they are alike: 'If you were the same as I am. I couldn't stand it. You must be different. You are a woman.'[50] To this Maki replies: 'We do share the same memories – as mankind. If you are a man and I, a woman, it doesn't make much difference. What men have felt must have surely been felt by women, and what women have felt, by men, too. We just pretend as if we didn't – as if we couldn't remember them. . . .'[51]

Inversion of conventional male/female attributes is often seen in the novel. Not only Makiko's son and Maki's neighbour, but also Shōzaburō is described 'as if he were a professional host'[52] in trying to please the ladies without much sincerity. He explains to Maki, a travel agent and Makiko, a medical doctor: 'Yes, certainly young men of today have come to believe that a woman who can earn a living is better than one that only sits in the house waiting for her husband to come home.'[53]

At the end of the novel, when Maki visits Makiko to say farewell before leaving Japan, the elder woman gives a very eloquent account of her family, her life, and men and women:

'When I was young, I was afraid, more than anything, of losing my self. I was a persecution maniac. First I raged because the man in reality was different from the man in my imagination. And then I decided never to lose myself for such a man – never to be an appendage. But in fact, because I was stingy in giving, I gained less, which always dissatisfied me – too little share for me ... All the men I met invariably disappointed me so much that I tried very hard to bring up my son to be different from such men.[54]

She holds herself responsible for her son's homosexuality. Then she advises the young heroine to live with others, to find herself in others instead of holding on to herself.

Male homosexuality is treated in the novel as something that reinforces the theme of human 'memories' equally shared by men and women. It is a picture of homosexuality viewed from a distance by two heterosexual women. It is also interesting that the novel has a scene suggesting, though very faintly, female homosexuality.[55]

The fictional world of male homosexuality, developed by women writers over the period of two and a half decades, was at first purposely detached from, and later merged, with this earthly world of men and women. At each stage the writers of both genres, *shōjo manga* and literature, sought the best way to deal with the topic in order to express their grasp of dream or reality. The 'authorship' was at first heavily dependent on aesthetic 'authority'. 'Paternity' was praised and used as a guard against 'maternity'. Later works tend to deny the attachment to 'authority/paternity', often in comic, sarcastic style, by introducing female and feminized characters as critics of patriarchy.

NOTES

1 Teruoka Yasutaka, 'Nanshoku kara shudō e' (From *nanshoku* to *shudō*), *Gendaigo-yaku saikaku zenshū* (Complete works of Saikaku in modern Japanese), vol. 3, 1976, pp. 279–301.
2 *Eureka*, Seidosha, vol. 13, no. 9, 1981.
3 *Gendai yōgo no kiso chishiki*, (Basic understanding of contemporary terminology), 1984, pp. 154–5.
4 *Eureka*, vol. 13, no. 9, 1981, p. 90.

5 'From fantasy to reality: magazines for women', *Feminist International*, no. 2, 1980, p. 6.
6 *Ibid.*, p. 6.
7 *Eureka*, vol. 13, no. 9, 1981, pp. 129–30.
8 *Ibid.*, p. 100.
9 *Akutagawa Ryūnosuke zenshū*, (Complete works of Akutagawa Ryūnosuke) vol. 2., Iwanami shoten, 1977, p. 124.
10 *Hensōkyoku*, (Variations) vol. 3, Sun Comics, Asahi sonorama, 1983, p. 151.
11 'Tsugumi no mori' (Forest of thrashes), 1973; 'Umi ni iru no wa' (The one who is at the sea is . . .), 1974; and 'Haine yonde' (Read Heine), 1976.
12 'Subete midori ni naru hi made' (Till all becomes green), *Bessatsu shōjo komikku*, Shōgakukan, February 1976.
13 *Koibito-tachi no mori* (Lovers' forest); Shinchō bunko, 1975, p. 87.
14 e.g. 半朱〔ハンス〕, 義童〔ギドウ〕, 巴羅〔バウロ〕, 殊里〔ジュリア〕, 礼門〔レイモン〕, 荔於〔レオ〕
15 *Koibito-tachi no mori*, p. 148.
16 *Ibid.*, p. 149.
17 *Ibid.*, p. 170.
18 *Ibid.*, pp. 208–9.
19 *Ibid.*, p. 216.
20 *Ibid.*, p. 237.
21 Tanizaki Jun'ichirō, *Chijin no ai* (A fool's love), 1924.
22 Sandra M. Gilbert and Susan Gubar, *The Madwoman in the Attic: The Woman Writer and the Nineteenth-century Literary Imagination*, Yale University Press, 1979, pp. 3–44: 'The queen's looking glass: female creativity, male images of women, and the metaphor of literary paternity'.
23 Originally a term of endearment for *chigo*, and often used derogatorily for favourite boys of homosexuals.
24 *Koibito-tachi no mori*, p. 137.
25 *Ibid.*, p. 142.
26 Umewaka and Keikai were characters in the famous *chigo* tale, *Aki no yo no naga-monogatari* (Long tales for Autumn nights), anon., early Muromachi period.
27 Tatsukichi, Hans, and Yoshiko from 'Nichiyobi . . .' and Guydeau, Paulo, and Nashie from 'Koibito' . . .
28 Holbert, Eduardo and Annette from *Hensōkyoku*.
29 Shunsuke is much older than Tatsukichi, Guydeau and Holbert, and his photograph on the cover of the prospectus is 'a picture of an ugly old man. That was the only way to put it.' (Mishima Yukio, *Forbidden Colours*, translated by Alfred H. Marks, London, Secker & Warburg, 1968, p. 5).
30 Takemiya's *Kaze to ki no uta*, (Songs of wind and trees), Flower Comics, Shōgakukan, is complete in 17 vols.
The following are still to be continued:
Yamagishi Ryōko, *Hi izuru tokoro no tenshi*, (The prince of the country where the sun rises), Hana to yume Comics, Hakusensha, 10 vols. Kihara Toshie, *Mari to Shingo* (Mari and Shingo), Hana to yume Comics, Hakusensha, 12 vols.
Aoike Yasuko, *Eroika yori ai o komete* (From Eroica with love), Princess Comics, Akita shoten, 11 vols.
31 Yoshida Akimi; 'Hiiro no aki' (Hero's autumn), *Eureka*, vol. 13, no. 9, 1981, p. 43.
32 *Hi izuru tokoro no tenshi*, vol. 10, Hana to yume Comics, Hakusensha, 1984, p. 75.
33 *Mayonaka no tenshi* (Midnight angel), vol. 1, Bungei shunjū, 1979, p. 250.
34 *Ibid.*, vol. 1, p. 12.
35 *Ibid.*, vol. 2, p. 189.
36 *Ibid.*, vol. 2, p. 292.
37 The cover illustration of *Mayonaka no tenshi* is designed by Takemiya Keiko, who has the same kind of commentary pages in the magazine *June* on readers' homosexual *manga* compositions as Kurimoto does on short stories.

38 Mori Mari, 'Masuo to Taeko to sono pāti (Masuo and Taeko and their party), 1966 and other essays. Also, Tomioka writes an introduction (*kaisetsu*) to Mori's *Koibito-tachi no mori* (lovers' forest).

39 Tomioka Taeko, *Tōsei bonjin-den* (The lives of contemporary ordinary people), Kōdansha bunko, 1980, p. 75.

40 *Ibid.*, p. 76.

41 *Ibid.*, p. 77.

42 *Ibid.*, p. 83.

43 'Warai Otoko' (Man who laughs) also in *Tōsei bonjin-den* is a severer attack on a talkative intellectual embodied in an able interpreter.

44 The title signifies: 1) the symbolic flower/insect imagery which is frequently and effectively used in the novel as in the memories of Maki and other characters. 2) the biological mechanism to survive as represented in the episode about *gekka-bijin* (a kind of cactus), (*Hana to mushi no kioku*, Chūkō bunko, 1982, pp. 273–5).

45 The similarity in the names Maki (万喜) and Makiko (槙子) suggests: 1) the similarity of their dispositions and way of thinking, and 2) pseudo-mother/daughter relationship as in the case of Tadamasa and Fumimasa (though these are real father and son).

46 *Hana to mushi no kioku*, Chūkō bunko, 1982, pp. 83–4.

47 *Ibid.*, pp. 249–50.

48 Similarity is also found in that one of the parents in each family (i.e. Maki's mother and Makiko's husband) is dead.

49 *Hana to mushi no kioku*, p. 251.

50 *Ibid.*, p. 270.

51 *Ibid.*, p. 271.

52 *Ibid.*, p. 19.

53 *Ibid.*, p. 23.

54 *Ibid.*, p. 292.

55 *Ibid.*, pp. 87–9 and p. 116.

11 Body politics: abortion law reform

SANDRA BUCKLEY

In 1985 a letter campaign was launched by sectors of the Japanese women's movement against the *Asahi* newspaper, following an article published by that paper on 22nd April. The article described the contents of an American film entitled 'The Silent Scream', which had already caused considerable controversy in the United States. It is a twenty-eight minute film narrated by Dr. Bernard Nathanson and sponsored by anti-abortion campaign funds. Nathanson, himself an ex-abortionist, subsequently declared himself 'pro-life'.

The controversy surrounding the film relates to one particular scene in which the film shows the abortion of a twelve-week foetus. Nathanson's narration describes the child moving serenely in the uterus ... The child senses aggression in its sanctuary ... We see the child's mouth wide open in a silent scream.[1] The *Asahi* article describes in considerable detail the content of this anti-abortion film. Despite its own admission that both medical and cinema experts have questioned the credibility of both the scene and Nathanson's reading of it,[2] the article concluded that: 'The practically indisputable fact that we learn from this film is that abortion is an extremely violent act of cruelly killing a small and helpless life.'[3]

The Japanese women's movement reacted angrily to the *Asahi*'s condoning of a film which had been publicly accused of using sophisticated editing techniques to create false evidence for the anti-abortion case. 'The Silent Scream' was one element of a concerted anti-abortion campaign mounted in 1985. Anti-abortionists in the United States have made maximum use of the media, more recently resorting to violence both to draw attention to their cause and to discourage potential abortion patients. There have been kidnappings of doctors and their families, bombings and assaults on clinic personnel.[4]

Japanese feminists are concerned that the wave of the anti-abortion activity in the U.S. might affect the state of abortion law in Japan. In both Japan and the U.S. abortion law reformists have thus far had little success in their political campaign for the reform of existing pro-abortion laws. However, in their determination to alter the present level of social tolerance towards abortion, they are now resorting to alternative channels of influence. Although the Japanese anti-abortionists have yet to resort to violence, the anticipated release of the Japanese version of 'The Silent Scream' is only the latest event in a long and intensive campaign aimed at turning around public opinion in Japan.

A major difficulty facing the women who are attempting to counter the force of the well-funded and politically well-connected reformists is the fact that they are denied access to the media outlets which have so readily served as the mouthpiece of the reformists. An American example of this was the recent CBC decision not to run the award-winning National Film Board documentary 'Abortion: Stories from North and South'. CBC claimed that it lacked journalistic balance.[5] No such criticism has been levelled at the overtly propagandistic 'The Silent Scream'. The latter, in fact, became something of a media event after Ronald Reagan, commenting on the distribution of copies of the film to all Congressmen, stated his view that: 'I feel sure that all the Congressmen who see this film will be converted to the anti-abortion cause.'[6] It is extremely difficult for the pro-abortionists to gain the same air-time as such a statement from the President of the United States.

Japan's opponents to abortion law reform are faced with the same wall of media silence. This paper itself was written in response to a request from members of the Japanese Anti-Abortion Law Reform Coalition to make available in English some description of the status of their campaign. Much of what follows has been drawn together from the research and reflections of Japanese women writing on abortion in their country.[7]

Two out of three women in Japan have had an abortion legal or otherwise.[8] A survey by the *Mainichi Shinbun* indicates that of those women who have had an abortion 69 percent have chosen this option more than once.[9] The high level of unwanted pregnancies is directly related to the lack of alternative methods of birth control. The pill remains unavailable except when prescribed for medical reasons other than contraception. Over 80 percent of Japanese women are dependent upon the man's use of a condom.[10] The next most frequent form of contraception is the rhythm method. Abortion cannot strictly be described as contraception, but it is the next most common form of birth control after the condom. One woman gynacologist put it simply when she explained that having an abortion in Japan is 'like having a tooth out'.[11]

Of the legal abortions performed in Japan 99.7 percent are approved under the economic reasons clause.[12] This clause was approved by the Occupation Administration for introduction into the Civil Code in

1948. Ironically this clause was approved by the U.S. Government in Japan years before many American states granted the same access to legal abortion. In the period immediately after the approval of the economic reasons clause the number of legal abortions performed in Japan soared, reaching 2 million a year in the early 1950s.[13] Today, at a time when women in other countries are working to achieve similarly liberal abortion laws, Japan's economic reasons clause is coming under attack.

The last decade has seen two major right-wing campaigns, in 1972–3 and in 1982–3, for the removal of the clause from the Civil Code. The most recent assault has seen an equally strong reaction on the part of Japanese women with a diverse range of interest groups uniting under the single banner of 'Umu umanai wa onna ga kimeru' (It's a Woman's Right to Decide). The recent cooperation across interest groups distinguishes the 1980s movement from previous political campaigns by Japanese women. The Anti-Abortion Law Reform Coalition has brought together environmentalists, consumer groups, anti-nuclear campaigners and many others. The single issue status usually attached to women's issues has been rejected; and women are questioning the potentially peripheralizing effect of the very category of 'women's issues'.

Women have turned to the evidence of history to argue that throughout Japan's modern period abortion law has not been concerned with the welfare of the mother or the right of women to self-determination, but with the political and economic objectives of state and capital. It is within this context that Japanese women have turned their attention away from the single issue of abortion to search out the underlying objectives that have motivated the recent campaign for amendments to the abortion law.

The first attempts to establish a government policy towards abortion date back to the late Tokugawa period. Various proclamations were issued condemning the practice of *mabiki* – thinning out – but there seems to have been no stringent application of these proclamations as law.[14] *Mabiki* refers not only to abortion but also to infanticide, both of which were widely practised. The slow level of population growth in Tokugawa Japan may be attributed in part to *mabiki*.[15] It is not surprising that during a period of increasing economic and political instability the late Tokugawa officials saw fit to ignore a major form of population control.

When *mabiki* was officially declared a crime in 1880, the law was adapted from the French criminal code – a legal system strongly influenced by the morality of Catholicism. Given that neither the Japanese as a whole nor their leaders underwent a sudden conversion to Catholicism in the 1880s it is necessary to look further for some explanation for the legislative proscription of abortion at this time. An article which appeared in a medical journal in 1907 may throw some light on the matter.[16] The article claims that the number of abortions performed in Japan at that time was three times the number in Europe or the United States. It suggests that if Japan could reduce both the number of induced abortions and the level of infant mortality one could predict a subsequent population expansion which could significantly enhance national strength. Other similar articles, written by doctors, politicians and other public figures, appeared throughout the Meiji period. The early 1900s saw a marked increase in the number of abortion cases brought to prosecution and penalties under the abortion law were tightened. At one level it might be possible to see this expression of anti-abortion sentiment as a gesturing towards Western Judeo-Christian morality. This would not be the only example of Meiji law reform aimed at cleaning up Japan's image in the eyes of the West. However, in the various 'her-stories' which address this period of women's 'history', a clear connection is constructed between the move to restrict women's access to abortion and an official desire for population growth as one dimension of the expansionist policies of the Meiji State.

The 1910s saw the emergence of a women's movement which was increasingly aware of the need, not to mention right, of women to control their own bodies. This took the form of growing demands for a more liberal abortion policy and the legalization of contraception. Margaret Sanger's controversial visit to Tokyo in 1922 brought the contraception debate to a head and culminated in the establishment of the first contraception clinic in Tokyo in 1930. The Tokyo clinic was followed closely by an Osaka branch and in turn smaller provincial clinics. As a natural corollary of these developments in the area of contraception, the Abortion Law Reform League was founded in 1932 to fight for the liberalization of the law.

The process of bringing such issues as abortion and contraception out of the private and into the public sphere was cut short in the mid-1930s. The objectives of the birth control movement ran counter to the expansionist policies of government. The nationalistic call 'Umeyo Fuyaseyo' – Bear Children and Strengthen the Nation – drowned out the women's demands. By 1938 both the Osaka and Tokyo clinics had been shut down.

In 1940 the government proclaimed its new National Eugenic Law. The new Japanese law was unashamedly based upon Germany's Sterilization Law of 1933.[17] There were two facets to it: firstly, the elimination of unauthorized abortions, and secondly, the active encouragement of sterilization and abortion in the name of eugenics. Those likely to produce handicapped children were to be sterilized while women carrying a high-risk foetus were to undergo abortions.

There were many parallels between the German and Japanese eugenic policies of the thirties and forties. The German call for 'One Wife Seven Children' became the 'Healthy and Abundant Households' policy in Japan – abundant here referring specifically to the number of children. The first 10,330 awards to 'Healthy and Abundant Households' were presented in 1940. The emphasis on population as a natural resource of a country at war precluded any tolerance of abortion other than for eugenic purposes. Although danger to the mother's health was cited as justifiable grounds for abortion, such operations came under close scrutiny. The Welfare Ministry actively campaigned to encourage women to marry younger and have more children. Throughout the war years the only group of women who had easy access to abortion were the army prostitutes.

After the Japanese defeat the abortion law underwent a complete turnabout. With the economy in tatters and the flood of expatriates returning from the surrendered territories, both the Japanese and the Occupation administration were quick to move for the passage of a liberal abortion law. In 1948 the Eugenic Protection Law was passed. In its initial form it still dealt mainly with eugenic issues but in 1949 the economic reasons clause was added to the legislation. A further amendment in 1952 streamlined the approval procedures such that the attending physician's consent became the only prerequisite.

The Eugenic Protection Law has stood without further amendment since 1952. According to the law there are five justifiable grounds for abortion:

(i) When the mother or her spouse are suffering from a mental disease, mental instability or exhibits signs of a mentally disturbed character, when the mother or her spouse have any hereditary sexual disease or sexual deformity.

(ii) When any blood relative of the mother or her spouse, not more than four generations removed, has suffered hereditary mental illness, sexual disease, sexual disturbance, or sexual deformity.

(iii) When the woman or her spouse has contracted leprosy.

(iv) When the continuation of the pregnancy or child-birth would be, for economic or physiological reasons, detrimental to the health of the mother.

(v) When the pregnancy is induced by menace or violence or when pregnancy occurs in a woman forced into adultery without opportunity to resist or reject.[18]

The first moves to eliminate the 'economic reasons clause' – the one feature of the law which distinguishes it from the wartime Eugenic Law – came in 1964, when the Minister for Welfare, Kobayashi Takeji, declared:

'I want to rid Japan of her reputation as the abortion paradise.'[19] Kobayashi was alluding here not only to the level of abortions among Japanese women, but also to the number of foreign women who were travelling to Japan to take advantage of the liberal abortion law and well-established abortion industry. From this same period onwards the religious sect of *Seichō no Ie* – House of Growth – began a political campaign to lobby for abortion law reform. *Seichō no Ie* has continued to play a central role in the anti-abortion campaign both politically and economically.

In 1968 the 'Eugenic Protection Law Diet Discussion Group' was established, followed closely in 1970 by the 'Eugenic Protection Law Reform League'. The latter was a coalition of Liberal Democratic Party politicians, members of the political arm of *Seichō no Ie*, and some conservative doctors. With the support and funding of this League an all-out campaign was launched which culminated in two attempts, in 1971 and 1973, to push through reforms to the abortion law. There were two facets to the proposed amendments – the elimination of the 'economic reasons clause' and the tightening of measures aimed at reducing handicapped births. Both attempts were defeated. The women's movement, together with the handicapped, fought against the amendments, but these particular campaigns of the early 1970s were single issue campaigns aimed solely at the defeat of the proposed reforms.

A second major reform campaign was mounted in the early 1980s and was again centered around the lobbying and funding of *Seichō no Ie*. This campaign saw a far higher financial investment by the reformists. *Seichō no Ie* not only funded national television advertisements under the slogan of 'Reverence for Life' but also published a book entitled *Is a Foetus not a Human Life?*[20] The official move came when a conservative member of the House of Councillors, Murakami Masakuni, called for the deletion of the economic reasons clause. He argued that given Japan's high level of economic development there could no longer be any need for Japanese women to undergo abortions for economic reasons. The Welfare Ministry then proceeded to draw up the proposed amendment.

Murakami, who was not only on the far right of the Liberal Democratic Party but also an influential member of *Seichō no Ie*, defended the need for reform saying: 'Why must we rush through the reforms? In order to save the Japanese race from extinction.'[21]

Despite the political influence of *Seichō no Ie* and the extensive media campaign launched by the anti-abortionists, the reform move was defeated. The 'Anti-Abortion Law Reform Coalition' rejected Murakami's basic economic argument, arguing instead that Japan's wealth, her 'economic miracle', was not shared by her women. Women remain economically dependent upon either father or husband most of their lives. When women do enter the workforce their options are severely limited, their long-term promotion prospects poor, and their wages are, on average, only 53 percent of male wages.[22] The women of Japan's amorphous 'middle-class' are themselves members of this socio-economic group through the salary and status of their father or husband. Women who through death, divorce or choice find themselves alone and outside the normal patriarchal structures of dependency, are thrown back upon an unsympathetic state system. Alimony orders are virtually unenforcable and social welfare benefits totally inadequate. With a monthly supporting mothers' benefit of $150, plus $25 for the first child and $10 for the second and third,[23] many women are forced back into an economically dependent role within the patriarchal nuclear unit simply in order to survive. The anti-abortion law reformists argue convincingly that the economic reasons clause remains a stark reality for all too many Japanese women.

The moral arguments against abortion put forward by *Seichō no Ie* are also rejected by the coalition. In the highly provocative work, *Is a Foetus Not a Human Life?*, one of the more outspoken anti-abortionists, Suzuki Kenji – an N.H.K. announcer and member of *Seichō no Ie* – made the following statement: 'If you are capable of conceiving a child in an act of hedonistic pursuit, killing that child though it has a life of its own, burying it and walking calmly away to continue with your life, with no regret, then you are the rarest kind of whore.'[24]

The language of Suzuki or Murakami is often indistinguishable from that of the Moral Majority or members of the 'Right To Life' movement in the United States. Much propaganda has flowed from the United States into the reformist campaign in Japan, and Suzuki and others are very aware of the emotional clout of their imported rhetoric.

The production of a Japanese version of the film 'The Silent Scream' is one example of the extent of the cooperation between the Japanese reform movement, in particular *Seichō no Ie*, and the American movement. Although the links are clear it would not be correct to think of the Japanese campaign as a mere extension or imitation of its American counterparts. In the Japanese case there is a far more obvious level of cohesion between proposed State reforms in the areas of defence and economic development and the proposed 'moral' reforms of *Seichō no Ie*. It is no coincidence that Murakami and other ultra-conservative reformists are members of both the Liberal Democratic Party and *Seichō no Ie*. Prime Minister Nakasone made no secret of his sympathies for the political platforms of *Seichō no Ie*. In recent years the religious group has gone from covert political activity through the funding of candidates to openly condoning and supporting the platforms of the ultra-conservatives. It has given its support to such policies as the return (from the Soviet Union) of the Northern Territories, the various administrative and constitutional reforms proposed by the Liberal Democratic Party (LDP), and the restitution of the Yasukuni Shrine to national prominence – an issue of particular concern due to the implications of this obvious challenge to the separation of state and religion.

The Coalition argues that the ultra-conservative assault on women's right to abortion is motivated not by moral or religious conviction but by political and economic objectives. It argues that the proposed abortion law reform has to be considered in conjunction with other major LDP objectives such as have been declared in the newly formulated 'Policy For The Strengthening Of The Family Foundation'. This policy clearly states that within a new Japanese style welfare state: '. . . the care of the elderly and the nurturing of children will be focussed upon the family'.[25]

In combination the proposed amendment to the abortion law and the new Japanese style welfare state amount to a life-long sentence to domestic and reproductive labour for Japanese women. Women are to bear children – unless they suffer from a mental or sexual disorder or leprosy – and to take on the welfare functions of the state in respect of children, the handicapped, the invalid and the elderly.

Some conservative voices within the women's movement have taken up a position which advocates the feminization of the male role as a step towards the return of the father to the nuclear unit where he would then share in the domestic/nurturing role.[26] The concept of the feminization of the male has achieved considerable currency as one dimension of a range of feminist positions. However, the objective of the male sharing of the domestic role cannot, in isolation from other far reaching economic and socio-political reforms, alter the existing dichotomy of productive/reproductive labour. Nor does it act to offer the female access to new or alternative activities – social or economic. The general reaction to this approach within the women's movement has been one of suspicion. The underlying pre-supposition of this theory of the domesticated male is that the 'typical' Japanese family is an economically stable, middle-class nuclear unit – soon to become a happy, self-supporting extended family and live happily ever after in the new Japanese style welfare state. Perhaps the strongest criticism levelled at this approach by the women's movement is that it comes dangerously close to complying with the objectives of state and capital e.g., the return of the welfare function to the family, the reduction of working hours in an increasingly automated workplace, the trend towards part-time, non-unionized labour. It is not surprising that this approach has received favourable media attention as a representative feminist position while alternative and more controversial viewpoints continue to find their only mouthpiece to be the *mini-komi* network.[27]

The Anti-Abortion Law Reform Coalition, together with many of the diverse interest groups which constitute the women's movement, has identified both the 'Policy For The Strengthening Of The Family Foundation' and the proposed amendments to the abortion law as integral expressions of a recent Japanese trend towards a comprehensive policy of ultra-conservative reform. The privatization of welfare and the reassertion of the reproductive, domestic/nurturing role of women are both interpreted as predictable facets of a coherent programme of conservative reform. In particular, women argue that the budgetary benefits of the new Japanese-style welfare state and Murakami's call for renewed population growth are key factors in an overall trend towards remilitarization.

The Coalition refuses to allow the abortion issue to be considered in isolation from the general political climate of the day, repeatedly bringing its arguments back to the objectives of capital and state:

Just as in the past the crime of abortion was founded upon the priorities of State and not the protection of the foetus, so it is today. Although these so-called reforms hide behind the call for reverence for life, the true objectives lurk distinctly in the background. In other words, faced with an aging population this is the guarantee of a new generation of young workers. Moreover, this reform movement aims at retooling the family (*ie*) system as an underpinning of a steady progress towards re-militarization. Women are to be house-bound, used as mere tools to bear and raise children. The new reforms are designed to recreate the old family system in order to clear the pathway to war.[28]

The Coalition of Women Against War makes a similar point in the introduction to the volume they published in support of the Anti-Abortion Law Reform Coalition. The pamphlet was entitled, *There Are Times When A Woman Can't Give Birth*. They write:

We are not satisfied when, apparently speaking in the name of love, they declare abortion an unforgiveable sin. Reverence for life goes without saying. That is why we are totally opposed to war which makes a mockery of life. We oppose all steps towards war and have formed this Coalition Of Women Against War. The signs of progression towards war are there for us to see with our own eyes – the Self-Defence Forces and remilitarization – but we believe this is not all. The progress also occurs in various other ways, not so easily discerned – the imposition of certain moral values, the strengthening of centralized controls, the textbook scandal with its arbitrary rewriting of history etc. – all of these serve to obstruct the freedom of the individual and the peace of society as a whole. They all mark the pathway to war.[29]

The Coalition points out that many of the ultra-conservatives who have backed the abortion law reforms are also active supporters of the proposal to increase defence spending beyond the existing ceiling of one percent of G.N.P.[30] This raises the obvious question of whether or not it is consistent for those who declare themselves 'pro-life' not to also be anti-rearmament and anti-war.

The renewed attempts to emphasize the eugenic function of the Eugenic Protection Law may also be seen as evidence of the link between abortion law reform and a general trend towards ultra-conservative policies. When the amendment to the economic reasons clause was first put forward in the 1970s, it went hand in hand with the eradication of handicapped births. Although both dimensions of the proposed amendments were defeated, Koga Akie[31] points out that policy developments in the area of mother-child welfare centres since the mid-seventies have achieved in practice what has yet to be legally condoned.

Family planning clinics have been receiving increased government funding since 1977. A stated objective of these clinics is to provide genetic counselling and screening services. In a detailed analysis of the range of services provided by Japan's family planning clinics Koga concludes that genetic screening is their primary function. The names of community action groups associated with the clinics would seem to confirm Koga's assertion: 'The Let's Not Bear Unhappy Children Group' or 'The Let's Bear Strong and Healthy Children Group' – 'Unhappy' here being a euphemism for handicapped.

The women's movement has joined with various organizations for the handicapped to fight the amendments. What both groups are demanding is a woman's right to choose for herself whether or not to carry a foetus to full term. Handicapped women or women carrying the child of a handicapped man come under intense pressure from the so-called family clinics to terminate the pregnancy and, in some cases, even to undergo sterilization. Once the pregnancy of a non-handicapped woman is registered with her local clinic she is scheduled into a comprehensive programme of on-going examinations. These examinations are structured around a continuous testing of the condition of the foetus. If any potential abnormality is identified then an abortion is strongly recommended, even into the third trimester.

Koga also observes that another function of these clinics is to educate women into the domestic-nurturing role. The courses and workshops run by the clinics are oriented towards encouraging women to devote themselves to these traditional functions. The clinics do not offer advice or support systems which would assist women in the pursuit of other options. Contraceptive counselling forms only a minor function of the clinics, genetic counselling, pregnancy monitoring, and motherhood education taking higher priority. Koga's reaction to the upgrading and expansion of these family planning clinics since the 1970s is simple and succinct: 'Wherever there is a system there will always be controls.'[32]

The traditional Japanese names for an abortionist are nothing short of nightmarish. *Oroshi baba* and *sashi baba* are two of the more explicit examples. *Oroshi* refers to the drawing down or dumping of a foetus while *sashi* comes from the verb *sasu*, to stab, and describes a primitive method of abortion whereby a sharp object is inserted into the uterus. Other common traditional methods of abortion practised in Japan included the application of extreme external pressure to the lower abdominal region, the stuffing of mulberry leaves into an opening in the uterus and the insertion into the uterus of a silk thread soaked in musk. In post-war Japan abortions are readily obtainable through private and public abortion clinics. Mulberry leaves and silk thread have been replaced by a highly efficient (and profitable) abortion industry. Government statistics indicate that the number of abortions performed annually is currently in the range of 600–700,000.[33] This figure is, however, a conservative estimate as it includes only officially recorded or legal abortions. It does not include all those abortions performed in private clinics under false procedural names, or illegal abortions.

Although abortion is legally available for economic reasons some women are still forced to resort to backyard abortionists – perhaps for reasons of privacy or cost or in cases where for legal or medical reasons a legal abortion has been refused. The problems of backyard abortions are similar world wide – inadequate sterilization methods, crude instrumentation, lack of emergency back-up (e.g., transfusions) and no medical follow-up. The risks to patients are high. The newsletters and magazines of the *mini-komi* often include heart-rending accounts of the hardships suffered by women who were forced to resort to illegal abortions. The opponents of the abortion law reforms argue that the demand for abortions will not decrease should the economic reasons clause be withdrawn. They argue that the action would only force all those women denied legal abortions to resort to backyard abortionists, with all the associated risks. Coalition representatives emphasize that the reforms are not accompanied by any move to improve sex education in schools or to make available more efficient methods of contraception.

Historically, both in Japan and elsewhere, the withdrawal of a woman's right to self-determination of her own sexuality and transfer of that power to the State has coincided with moves towards ultra-conservative and expansionist policies. The Coalition Against Abortion Law Reform has declared its intention to continue to police all attempts to legislate to restrict the freedom of women to control their own bodies. As an extension of the right to self-determination the Coalition is also campaig-

ning for the legalization of the pill. The family planning clinics are coming
under close Coalition scrutiny, and the government is facing increasing
pressure to redirect the activities of the clinics from the present preoccupation with genetic screening towards such areas as contraception
and family planning.

In Japan one still occasionally hears the expression *hara wa kari mono*,
meaning literally that the womb is something to be borrowed. Traditionally it was believed in Japan that the man's semen carried the new life
of the foetus which was merely implanted in the woman's womb. The
woman was thought to serve only as a vessel for the man's child. It was on
this basis that in the case of divorce the children were considered the
property of the father. Similarly, the children of a concubine could be
taken from her to be raised by the father's wife. The Japanese women who
are fighting today for the right to self-determination are adamant that
their bodies are their own; in the words of one feminist:

Our wombs are not for hire.

NOTES

1 *Asahi Shinbun*, 22 April, 1985, quoted in *Soshiren news*, no. 11, May 1985, p. 5.
2 During the controversy over this film in the United States, the CBS Network interviewed
   various experts on their opinion of the film. Their criticisms are summarized in *Soshiren
   news*, p. 5.
3 *Soshiren news*, no. 11, p. 5.
4 The extent and extremity of the violence brought the abortion issue into the public arena
   with extensive newspaper coverage, feature articles in major weeklies and even a segment
   on the CBS's '60 Minutes'.
5 Quoted in 'NFB pro-abortion film a Winner', *The Ottawa Citizen*, 10 November 1985.
6 *Soshiren news*, no. 11, p. 5.
7 There is a considerable literature now on the anti-abortion law reform movement. Three
   works I would recommend are: *Onna no sei to chūzetsu: yūsei hogohō no haikei*
   (Women's sexuality and abortion: Background to the eugenic protection law), Shakai
   hyōronsha henshūbu, Tokyo, 1983; *Onna ni wa umenai toki mo aru* (There are times
   when a woman cannot give birth), edited by Sensō e no michi o yurusanai onnatachi
   shūkai, Tokyo, 1982; *Shinchihei: Josei kaihō undō no genzai tokushū* (*Shinchihei* –
   special issue on 'Current state of the women's liberation movement'), August 1983. I am
   indebted to all the women whose ideas and opinions are expressed in these collections of
   articles.
8 'Women: A separate sphere,' *Time*, 1st August, 1983, p. 74.
9 James Trager, *Letters From Sachiko*, London, Abacus, 1982, p. 109.
10 *Ibid.*
11 'Women: A separate sphere', p. 74.
12 *Ibid.*
13 D. Robbins-Mowry, *The Hidden Sun*, Colorado, Westview, 1983, p. 125.
14 Nakashita Yūko, Chūzetsu sureba rōgoku yuki' (prison for abortion), in *Onna no sei to
   chūzetsu*, p. 46.
15 *Ibid.*
16 Quoted in Nakashita, pp. 47–8.
17 Samuel Coleman, *Family Planning in Japanese Society: Traditional Birth Control in a
   Modern Urban Culture*, Princeton University Press, 1983, p. 19.

18 Nakashita, p. 39.
19 *Ibid.*
20 *Taiji wa ningen dewa nai no ka* (Is a foetus not a human life?), compiled and published by Nihon kyōbunsha, Tokyo, 1982.
21 Yoshibu Kagako, 'Sensō e no michi o zukuri' (Forging a path to war), *Onna no sei to chūzetsu*, p. 33.
22 The figure varies considerably depending upon the source. A recent Department of Labour survey placed the figure at 52.8 percent. The range is between 52 percent and 55 percent. The overall trend appears to be a worsening of women's wages as a percentage of the male wage over the last five years.
23 Agora henshūbu, *Gekkan Agora*, No. 84, 1984, p. 31.
24 *Taija wa ningen dewa nai no ka*, p. 15.
25 'Katei kiban no jūjitsu ni kansuru seisaku yōkō' (Outline policy for strengthening the family foundation), *Shinchihei*, August 1983, p. 67.
26 See Ueno Chizuko, this volume.
27 *Mini-komi* is the communications network used by the women's movement as an alternative to *masu-komi* (mass communications).
28 Nakashita, pp. 54–5.
29 *Onna ni wa umenai toki mo aru*, p. 3.
30 'The Coalition of Women Against War' has lobbyed actively to bring the inter-relationship of these two policy areas to public attention.
31 Koga Akie, 'Kanri sareru sei: boshi hoken no jittai' (Controlled sex: The reality of mother and child welfare), *Shinchihei*, August 1983.
32 Koga, p. 90.
33 Awatani Mariko, 'Umu umanai no ketteiken wa onna ni' (It is a woman's right to decide whether or not to give birth), *Onna ni wa umenai toki mo aru*, p. 29.

12 Division of labour: multinational sex in Asia

VERA MACKIE

INTRODUCTION

In recent years, the problem of so-called 'sex tours' to South East Asia by Japanese tourists has been taken up by the Japanese media. Every year hundreds of thousands of Japanese tourists, predominantly male, travel to Korea, the Philippines, or other South East Asian countries. The main attraction is prostitution. Is this a new phenomenon, or is it merely part of a long history of the exploitation of female sexuality in the process of Japan's modernization?

Prostitution has long been an important part of Japanese culture. Although prostitution has existed from Heian times, the Tokugawa bakufu was the first to exercise control by instituting special licensed quarters. The licensed quarters continued to flourish until the Meiji period. As Japan's modernization gathered pace, the sale of daughters to silk mills, brothels, or as *Karayuki-san* (prostitutes for export throughout Asia) provided financial relief to struggling rural areas. The labour of these women contributed to the process of capital formation by the earliest entrepreneurs. As the influence of the Japanese state grew in both economic and military terms, not only Japanese women but Korean and Chinese women were conscripted as prostitutes. The phenomenon of large groups of Japanese men travelling to the Philippines and other countries in search of prostitutes is merely a logical development of this pattern.

Too often, however, prostitution has been discussed in purely aesthetic terms.[1] This is particularly so in the Japanese context where *geisha* have been elevated to the level of cultural artefact, along with Kabuki and Noh. Such romanticization ignores the economic and social factors which contributed to the institution of the licensed quarters. *Geisha* were rural women who had been *sold* to relieve the poverty of the families. They often met an early death through disease or maltreatment.

Here prostitution will be considered as a form of labour, the conditions of which can only be understood in terms of the relations of class and gender in a particular society. There is a logical line of development from the licensed quarters of Edo, to *Karayuki-san* and *Jūgun'ianfu* (military prostitutes), and the hospitality girls, bar hostesses and *kisaeng* of the present day.

In order to comprehend the issue of 'sex tours' it will be necessary to consider, first of all, the function of prostitution in modern societies, as a form of labour which shares many features with other types of labour commonly performed by women. In the South East Asian context, it is necessary to consider how relations of gender and class in a particular economy are shaped by international pressures. Finally, I will place this modern phenomenon in historical context by considering the place of women's labour in Japan's modernization.

WHAT IS PROSTITUTION?

Prostitution reflects the gender relations in a given society in their most extreme form. It is women's vulnerable position in the capitalist economy which makes prostitution an acceptable alternative to work in service industries or manufacturing. Analysis of both the workers and customers in this industry must be located in the context of class. Prostitution is by no means an aberration in capitalist society – it is simply a logical development of existing class and gender relations.

Prostitution is a form of labour whereby sexual services are exchanged for cash. This form of labour shares many features with other forms of labour commonly performed by women. Sexual service is one part of the 'nurturing' activities commonly assigned to women. Because prostitution is usually illegal, prostitutes are unlikely to be unionized. Prostitutes have no legal protection concerning working conditions, health and safety provisions, or remuneration. In addition, they must often face police harassment. The Prostitutes' Collectives which have sprung up in several countries attempt to force the community to think of prostitution in terms of labour relations.[2]

The institution of prostitution supports the family system and existing gender relations by providing a momentary sexual outlet free of emotional commitment, which poses no threat to family bonds. Prostitution reinforces a stereotypical view of male and female roles – male desire must be satisfied and it is the role of women to provide such satisfaction.[3] Women are also denied a complex role model which would integrate sexuality, nurturing activities and professional and intellectual competence. In considering prostitution, then, we must consider the ideologies which justify male and female roles.

Prostitutes themselves are often outside the nuclear family system,[4] and their existence is often put in opposition to that of 'good' wives and mothers. Social structures in capitalism are based on an implied split between public and private spheres, with 'productive' labour being carried out in the public sphere, and 'reproductive' labour being carried out in the private sphere.[5] Sexuality is generally thought to belong to the private sphere, but prostitution takes sexuality into the public sphere of commodity and exchange. This sexual component is present to a greater or lesser degree in much of the labour performed by women.[6] The ideological relegation of certain behaviour to the private sphere merely prevents discussion of such behaviour in terms of economics or labour relations, and permits the more efficient exploitation of women's labour. Prostitution is, above all, an *economic* relation, a *power* relation.

In the South East Asian context, relations of class and gender are shaped by international economic relations, as the local economy is integrated into an international division of labour. What does the international division of labour mean for a Philippine woman? What is the influence of the Japanese economy on the Philippines? A transaction between a Japanese businessman and the Philippine woman reflects relations of class, gender, and race in this region[7] and is linked to a long history of exploitation of female sexuality and labour in the course of Japan's modernization.

JAPAN AND SOUTH EAST ASIA

Until the 1970s the Japanese economy was largely based on manufacturing although service industries have steadily increased in importance. The economy was based on a dual labour market, whereby certain privileged workers enjoy security of tenure and superior working conditions, and casual or part-time workers provide a 'safety-valve' for capital in times of recession. Casual workers are often employed by sub-contractors who cannot offer the same security as larger companies.

Members of the privileged sector of the labour market – predominantly male university graduates – give loyalty to the company and develop technical expertise over the so-called 'life-time' period of employment. Routine clerical, service and assembly work is carried out by casual, part-time or temporary workers.

The 1970s saw a change in the composition of the labour market. For the first time most employed women were married. In 1978, for example, 65.7 percent of working women were married. There has been an increasing tendency for women to return to work in their late thirties and forties after child-care responsibilities have been eased. Thus, there are two peak working periods for women – between the ages of twenty and twenty-four before childcare responsibilities commence, and between the ages of forty and forty-four.[8]

The use of women in such part-time positions serves a dual purpose. Capital benefits from the existence of a deskilled, cheaper workforce which does not demand the fringe benefits or increments paid to permanent employees. At the same time increased fragmentation of the work process and deskilling of workers makes it difficult to organize workers in union activity.[9] The state benefits because women take the major burden of welfare. Women with only a temporary commitment to the workforce may look after children, the handicapped and the aged. Women's domestic labour also makes possible the extreme commitment of time and energy required of male members of the primary labour market.

Thus, the present structure of the Japanese economy can only be understood with reference to a sexual division of labour. Elite males are expected to commit so much time and energy to the workplace that it is impossible for them to engage in domestic labour. This division of labour is further supported by increments and allowances available only to married male employees. Early retirement for women, lower salaries for female employees, enforced retirement on marriage or pregnancy, and exclusion of women from promotional opportunities continue to persuade women that there is little alternative to the domestic role. The lack of adequate alimony and supporting mothers' benefits also keeps women in the nuclear family. Women are expected to give their primary commitment to the family, but engage in paid labour on a part-time basis where necessary.[10]

In many cases, however, the label of 'part-time' serves the ideological purpose of devaluing women's labour, rather than accurately describing their working conditions. The average 'part-timer' works thirty-three hours in a five-and-a-half day week. Two percent of part-timers surveyed in Osaka worked from forty to forty-eight hours each week.[11] As Paringaux points out, 'part-time' refers to the insecurity of these women's working conditions, rather than to their working hours. Occupations such as clerical, service, and assembly work are further devalued by being designated as 'feminine' occupations.[12]

This sexual division of labour is more complex, for women are also expected to provide sexual services. Women provide sexual services in marriage, primarily for reproductive purposes, but 'sexual labour' as it is referred to by Tanaka Yufuko,[13] has also been integrated into the capitalist economy. The services provided by bar hostesses and women in Turkish baths are often paid for at company expense. The ideology which designates women as 'mothers' who provide nurturing, and those others, who provide only sexual services, further divides women. It is interesting to note that Japanese women themselves are conscious of this division of labour. What this means is that there is no complex, integrated role for Japanese women – only men can move freely between the different spheres.

Since the late 1970s, however, the Japanese economy has undergone extensive restructuring, and can no longer be understood without reference to South East Asia. This has important implications for women in the Philippines and other South East Asian countries, as their labour provides an extra buffer for the dual economy of Japan.

Owing to increased labour costs in Japan, and increased concern about pollution, many manufacturing processes were moved 'offshore' in the late 1970s. This has been made possible because of improved transport and communications and increased fragmentation of the work process, which means that each process can be carried out by a relatively unskilled worker.[14]

It was particularly labour- and energy-intensive industries, and 'dirty' industries which moved to South East Asia. This is particularly true in the electronics industry. In one Kyūshū factory, for example, the total labour force was halved between 1970 and 1983, and the number of jobs available to women was cut by two-thirds.[15] This has happened because of mechanization and increased use of offshore production. Such Japanese companies which moved offshore were greeted with a cheap labour force which was often prevented from engaging in union activity, and local laws which provided favourable tax concessions and lax pollution and safety controls.

A possible result of this shift to manufacturing could have been competition between Japanese and South East Asian industry.[16] But Japanese industry has used this opportunity to restructure the domestic economy. Japan has moved from a manufacturing base to an information and technology base. This has important implications for workers in both Japan and South East Asia. Most new jobs created in Japan in this period have been in technical, clerical and sales positions – often the kind of part-time labour performed by women. At the same time, there is a trend away from 'life-time' employment, resulting in an intensification of labour market segmentation.[17] Japanese women continue to be exploited in technical, clerical and service occupations, while the 'dirtiest' jobs are performed by South East Asian women, as their labour is integrated into the international division of labour. The position of women in the Philippines, and the forces which make prostitution a viable choice for some women can only be understood by placing the Philippine economy in its international context.

Since the 1960s the Philippine economy has undergone great expansion, much of which is based on investment from Japan and the United States, and loans from the World Bank and other development agencies. Most of this development has been 'export-oriented' – directed at the export of primary products or manufactured goods, in the interests of overseas investors.

In agriculture, investment is often directed at increased centralization of ownership and commercialization of production. Reliance on machinery and fertilizers from overseas makes it difficult for small farmers to compete with foreign investors.[18] Land may also be appropriated for large-scale industrial projects funded by overseas money, or pollution from industry may make it impossible to carry on fishing or other local traditional industries.[19] All these factors combine to dislocate the rural population. Those who stay behind may carry out piece-work or other assembly work on a casual or outwork basis. Others may move to Manila or one of the Free Trade Zones in search of work.[20]

Industrial development has been encouraged in light industry – for example, the assembly of electrical goods or electronic circuits. There is private investment by Japanese and American companies, but World Bank loans also encourage this type of development[21] with the cooperation of local elites. Philippine workers in the electronics industry, for example, are integrated into an international division of labour whereby research and development, and capital- and technology-intensive processes are carried out in developed countries, while labour-intensive processes are carried out in third-world countries.

Such development does not necessarily benefit the economy as a whole, although certain members of the ruling elite may benefit. Rationalization of agriculture is not accompanied by programmes to find work for displaced rural workers, and 'export-oriented' development in the industrial sector does little to promote an integrated domestic economy.[22]

Because so much of industry is dependent on raw materials from overseas, and because these products are funded by overseas money, the country must accumulate foreign exchange. At the same time, demand is created for various consumer goods which must be paid for. One way of generating foreign exchange is the tourism industry. Tourism is the fifth-largest earner of foreign exchange in the Philippines, and tourism, in effect, means prostitution. An overwhelming majority of tourists travelling to the Philippines are male,[23] and tourist brochures advertise Manila 'night life'. The government promotes this industry by registering all 'hospitality girls' and arranging V.D. checks on a regular basis.

The tourist industry, however, is not free of foreign domination. Japanese tourists can travel to the Philippines by Japan Airlines, on a tour arranged by J.T.B., and stay in a Japanese-owned hotel. It has been estimated that between 40 percent to 75 percent of profit from the tourism industry goes back to the country which owns the hotel or travel firm.[24] An estimated 90 percent of bars in Mabini are foreign-owned, with a significant number owned by Japanese or Australian interests.[25] A 'hospitality girl' may receive $10 or $20 a night, but a staggering amount of money goes to tour operators and other middlemen.

Given the present economic situation in the Philippines, what are the implications for Philippine women? I have described the disruption of rural life. Given that women are considered the most 'expendable' members of rural society, they have several choices. They may stay in the rural community and seek casual work, or engage in piece-work at home, while continuing to make some contribution to agricultural labour. Or, they may migrate to the city in search of domestic or factory work.

It is not inevitable, however, that women must be the most 'expendable' members of a community. There must be some ideology which supports this view. In Philippine rural communities women's labour is devalued. Traditional ideas are overlaid with Catholicism and Latin machismo.[26] Men are expected to stay in rural communities to carry out the heaviest labour and carry on the family name. This is combined with a demand for cheap, 'docile' labour in the light industrial sector. The same ideology which devalues women's labour in rural society ensures that women will be underpaid in factory work.

Philippine workers are the cheapest in South East Asia. Compare the $60 per month received by the Philippine assembly worker with $70 in Malaysia, and $160 in Hong Kong.[27] Most electronics workers are female, between the ages of eighteen and twenty-five. Pineda-Ofreneo notes that 'migraine, headache and eyestrain are common among workers who have to look through a microscope all day long. Exposure to dangerous chemicals leads to nausea, lung trouble and various allergic reactions, sometimes leading to serious disability.'[28] Factory workers may also suffer from R.S.I. or cystitis. A woman who performs piecework producing goods for export or the tourist industry may receive A$4 for fourteen hours of work.[29] Domestic workers may also have to put up with severe restrictions on their freedom. All these workers are likely to be subject to sexual harassment, and the possibility of union activity is remote for out-workers or domestics. In this context, going to Australia or America as a 'mail-order bride' may seem like an attractive alternative.

The $200–$300 per month earned by the hospitality girl has to be compared with the $60 per month earned by the assembly worker in an electronics plant. Many women are sending money back to their families in rural areas, so the shift to prostitution is justified on the grounds that they must support their families. Many such women, however, will never be able to aspire to a stable family life, and may become outcasts in their own communities.

For prostitution to exist in any society, there must be a willingness to treat sex (or rather, women's bodies) as a commodity. There must also be an ideology which sees males as having a libido which must be satisfied at all costs, and a 'double-standard' which says that most women remain chaste while those women who provide sexual services are treated as 'deviant' women. Thus, in Philippine society, ideological pressures restrict women's choices, and economic forces reinforce this role division. The ideology which enforces the sexual division of labour on the Japanese mainland ensures that Japanese men are ready to become customers of the prostitution industry in the Philippines.

Gender ideologies in both Japan and the Philippines relegate women to the most vulnerable sector of the labour market, and expect women to provide domestic labour and sexual services. The sexual division of labour is thus articulated into the international division of labour. Not only the most menial and unpleasant factory work has been moved 'offshore' from Japan, but a large part of the prostitution industry has also moved 'offshore'.

There is, however, nothing new about the exploitation of women's sexuality for the purpose of accumulating foreign exchange, or capital. There are, of course, similarities between the 'hospitality girls' and the *Karayuki-san* – the rural women of the Meiji period, whose labour provided capital for some of Japan's first entrepreneurial ventures.

HISTORICAL PERSPECTIVE

Prostitution has existed in Japan since at least the Heian period, and the licensed quarters were established as early as 1600. It has been suggested that the licensed quarters served as a 'safety valve' during the Edo period, distracting townspeople from the repressive nature of their society, and providing a 'no-man's land' where *males* of different classes could mingle.[30]

There is some truth in this view, but it is also possible to link the licensed quarters with a general idea that women were somehow 'expendable' or 'exploitable'. Although women may indeed have had some power in traditional agrarian society, their contribution was steadily devalued by the Confucian and Buddhist traditions.[31] In rural communities, it was girl children who were 'thinned out' at birth. Even in warrior and aristocratic families, women were treated as commodities, who could be traded in order to gain political influence.

In such a society, it was not so strange that women could be sold to brothels in times of famine. (The same practice was common in rural China.)[32] Thus, the existence of the licensed quarters highlights two aspects of Tokugawa society – the devaluation of women, and the transition to an economy where even women's bodies could become a commodity.

Several writers have also referred to the increased 'samuraisation' of late Tokugawa society. Samurai wives were enjoined to be chaste, frugal and loyal, and merchant families followed this tradition. The limitation of women's roles in marriage necessitated the creation of another type of woman, who could provide all the qualities that wives lacked – sexuality, witty talk, and enjoyment free from Confucian ties of loyalty and obligation.[33] This division of women's roles is perpetuated in modern-day Japan in the dichotomy between wives and bar hostesses.[34] In the Meiji period, the practice of infanticide was prohibited, so that many more female children survived to adulthood.[35] At the same time, the new land taxes put greater pressure on rural areas, so that parents were still willing to indenture their daughters to textile mills or brothels in order to pay taxes or rent.

From 1868, Japan started to compete in the world economy, and the first major export product was silk. Young women from rural areas were indentured to silk mills. Payment was usually made in advance, so that the women had no choice but to work until the debt was paid off. The most important export industry was based on female labour, with women constituting 60 percent of the industrial labour force until well into this century.[36]

In the early Meiji period, the importance of women's labour was in working in spinning mills and textile factories, but in the late nineteenth century a new export industry came into being – the Karayuki-san.[37] Instead of working in local mills or factories, these women were sent to work in brothels in Singapore, Shanghai, or even as far away as Australia and Africa.[38] In this period then, prostitution took on a new meaning – women's labour provided capital for further overseas investment by entrepreneurs.

As Japanese expansion took on a military character, a new class of military prostitute, *jūgun'ianfu*, came into being.[39] Not only Japanese women, but also Korean and Chinese women were conscripted to serve the Japanese military. This parallels the situation in South East Asia today, where prostitution serves not only the tourist industry, but also the military bases in various parts of the Philippines and Korea.

The utilization of cheap women's labour in the process of industrialization is common to nineteenth-century Japan and present-day Asian countries. Some writers have explicitly compared the Meiji textile mills with the electronics factories in the Philippines and other Asian countries.[40] The 'choice' between factory work and prostitution is similar, as is the use of women's labour to gain foreign exchange. It is thus tempting to think that the Philippines is merely one step behind on the road towards moderniza-tion and will eventually be able to emulate Japan. However, Morris-Suzuki convincingly argues that the relationship between Japan and South East Asia is increasingly one of vertical integration. The influence of Japan and the United States on the Philippine economy does not encourage autono-mous development. Philippine manufacturing industries are dependent on research and development carried out in advanced countries. As most industrial development is oriented towards export, there is little integration of the economy, and little development of industries which would serve the domestic economy.[41] We must also remember that Japan was able to modernize with relatively little aid from overseas, and thus retained a degree of autonomy over the form economic development would take.

There is a further lesson for Philippine women in considering Japanese history. Despite Japan's increasing prosperity in terms of G.N.P. figures, women continue to be exploited in the workforce. In fact, it could be argued that Japan's recent prosperity has been based on the exploitation of women's labour. Even if the Philippine economy could emulate the Japan-ese model, a development model which values G.N.P. and export figures over the needs of significant sections of the population must be questioned. As discussed above, the Japanese economy is based on a sexual division of labour, supported by ideologies and structures which make it difficult for women to achieve an integrated role in society. Those who stay within the nuclear family are often denied full sexual expression, while those who provide sexual services can rarely aspire to full participation in family life.[42]

Thus, the issue of prostitution must be considered at several different levels. The existence of prostitution in any society reflects the class and gender relations of that society, supported by ideologies of women's role. 'Sex tours' must be considered in the context of the political economy of both the Philippines and Japan, and interaction between the two econo-mies. Structural inequalities in the two economies are articulated into an international division of labour. What then, are the implications for those seeking change in South East Asia?

POLITICAL IMPLICATIONS

In recent years, feminist theorists have analysed the ways in which women's labour is integrated into the capitalist economy. In addition to performing domestic labour and providing sexual services in marriage, women often perform waged labour.[43] These different types of labour have been discussed in terms of 'public' and 'private' spheres. Under such analysis, women's unpaid labour is not integrated directly into the capitalist economy, but supports this economy by allowing the more efficient exploitation of male workers. The ideology which assigns women's primary loyalty to the 'private' sphere of the family ensures that their labour will be devalued and underpaid when they enter the 'public' sphere of waged labour.

However, many of the activities designated as 'private' have been transferred to the public sphere of waged labour without challenging existing gender ideologies. Domestic servants perform housework, teachers are engaged in the socialization of children, and nurses look after the sick. The transfer of these activities to the sphere of waged labour is not threatening as long as patriarchal ideology is preserved – these occupations are still predominantly performed by women and thus devalued.[44] Strikes by teachers and calls for the professionalization of nursing are not opposed on purely economic grounds, but because such demands challenge ideas of appropriate feminine behaviour.

In this context, prostitution takes place when sexuality is transferred from the private sphere to the public sphere of commodity and exchange. The link between prostitution and marriage has been recognized since Engels[45] and Goldman,[46] but our understanding of this issue has in fact progressed very little since then. Prostitution is a service industry, a gender-typed occupation, lacking protection with respect to working conditions, safety, or remuneration. In this sense, prostitution is similar to many of the other occupations commonly performed by women. The association of prostitution with sexuality and 'deviant' behaviour prevents discussion of these issues. Thus, the relegation of prostitution and sexuality to the private sphere has an ideological dimension which allows the more efficient exploitation of women's labour.

In such a context, the most radical response to the issue of prostitution is in terms of labour relations.[47] Recognition of the similarity between prostitution and other forms of labour performed by women does not, however, imply acceptance. Libertarian demands for the freedom of women's right to choose prostitution as a viable alternative in capitalist society stop short of questioning the gender divisions which make this seem a viable alternative. Rather, prostitution must be seen as part of a larger pattern of inequality of class and gender.

Prostitution has been a concern of the Japanese women's movement since the Meiji period, when women questioned the institution of prostitution, and the existence of concubines.[48] This concern has continued to the present day, with several writers seeing links between Yoshiwara,

Karayuki-san, Jugun'ianfu, and 'sex tours'. This process has come full circle with the importation of South East Asian women into Japan to work as bar hostesses or prostitutes. The popular name for these women – '*Japayuki-san*' echoes the name '*Karayuki-san*'. Popular media representation of *Japayuki-san*, is however, often sensational, or even racist. The media tends to concentrate on the women as individuals, ignoring their clients or the operators who profit from their labour. Philippine women have even been the subject of 'scare' stories, linking them with the transmission of A.I.D.S. In 1986, there were also some stories about mail-order brides going to poor rural areas such as Yamagata.

The links between Japanese women and women in South East Asia are complex. Any improvement in the position of Japanese women which worsens the situation of South East Asian women must be rejected. The exploitation and repression of textile and electronics workers in South East Asia must be understood in this context. Although Japanese women enjoy a relatively privileged position among Asian women, there are economic and ideological structures common to all of these societies which exploit Japanese women in the same way. There are also historical links and similarities, as I have discussed above.

Recent activities by Japanese women's groups recognize these links. Campaigns against 'sex tours' are coordinated on an international basis. In 1973, women in both Seoul and Tokyo staged demonstrations against *kisaeng* tours. Philippine demonstrations against then Prime Minister Suzuki Zenkō in 1981 were linked with political activities in Japan. Discussion of 'sex tours' by feminist groups is always placed in the context of Japan's economic influence in South East Asia.[49] The Asian Women's Association and the Women's Christian Temperance Union have developed slide presentations to stimulate discussion of 'sex tours' and *Japayuki-san*. The W.C.T.U. now runs a women's shelter in Tokyo under the name of *Josei no Ie* (also known under the acronym H.E.L.P.). About half of the clients of the refuge are Philippine and Thai women escaping from prostitution.

In a world where capitalism has gone multinational, it is also necessary for feminists to forge 'multinational' links. Women in South East Asia must be conscious of how their situation is affected by Japan's economic influence, but they should also remember the historical lessons to be learned from the exploitation of Japanese women's labour and sexuality in the course of Japan's modernization and industrialization.

NOTES

1 See, for example, L. C. Dalby, *Geisha*, University of California Press, 1984. Ian Buruma, *A Japanese Mirror: Heroes and Villains in Popular Japanese Culture*, Penguin, 1985, also treats *geisha*, prostitutes and bar hostesses as primarily aesthetic phenomena, but is also sensitive to their political significance.
2 These collectives are discussed in C. Jaget, *Prostitutes: Our Lives*, Bristol, Falling Wall Press, 1980; R. Perkins, and G. Bennett, *Being a Prostitute*, Sydney, George Allen and Unwin, 1985; K. Daniels (ed.), *So Much Hard Work*, Sydney, Fontana, 1984.
3 Mary McIntosh discusses the ideological aspects of prostitution in 'Who needs prostitutes?', C. Smart, and B. Smart (eds.), *Women, Sexuality and Social Control*, London, Routledge and Kegan Paul, 1978.
4 Japanese feminist writers in particular often place women's oppression firmly in the family. A notable exception is Tanaka Yufuko, *Josei hishihaisha kaikyū no keizaigaku*, Tokyo, Gogatsusha, 1981. Tanaka is one of the few writers to address the issue of prostitution, although her argument could be developed in more detail.
5 For discussion of these concepts see Michele Barrett, *Women's Oppression Today*, 1980, pp. 19–29; Clare Burton, *Subordination*, Sydney, George Allen and Unwin, 1985; A. Game and R. Pringle, *Gender at Work*, Sydney, George Allen and Unwin, 1983.
6 Daniels, 1984, p. 13.
7 I have chosen to concentrate on the relationship between Japan and the Philippines, but much of this discussion is relevant to other South East Asian countries. See, for example, L. Caldwell 'The economics of sexual exploitation', *Feminist International*, no. 2, Tokyo, 1980; S. Hantrakul, 'Prostitution in Thailand', *Second Women in Asia Workshop Papers*, Monash University, 1980; W. Smith 'The impact of foreign investment and management style on female industrial workers in Malaysia', paper presented to the Conference on Women in the Urban and Industrial Workforce, Southeast and East Asia, Manila, 1982.
8 Much of the preceding discussion follows Kaji Etsuko, 'The invisible proletariat: working women in Japan', *Social Praxis*, 1973, pp. 375–87, and Ōhashi Terue, 'The reality of female labour in Japan', *Feminist International*, no. 2, 1980, Tokyo, pp. 17–22.
9 Phillipa Hall, 'Part-time labour and women', *Scarlet Woman*, no. 18, Autumn, 1984, p. 21.
10 These issues are discussed in more detail in S. Buckley and V. C. Mackie, 'Women in the new Japanese state', G. McCormack and Y. Sugimoto (eds.), *Democracy in Contemporary Japan*, Sydney, Hale and Ironmonger, 1986.
11 R. P. Paringaux, 'Japan's six-day week part-timers', *The Guardian Weekly*, vol. 21, April 1985.
12 Hall, 1984, p. 25.
13 Tanaka, 1981, p. 153.
14 R. Pineda-Ofreneo, 'Subcontracting in export-oriented industries: Impact on Filipina working women', *Second Women in Asia Workshop Papers*, Monash University, 1983, p. 18.
15 Committee for the Protection of Women in the Computer World, 'Computerization and women in Japan', *Ampo*, vol. 15, no. 2, 1983, p. 18.
16 T. Morris-Suzuki, 'Japan's role in the international division of labour – A reassessment', *Journal of Contemporary Asia*, vol. 14, no. 1, 1984, p. 66.
17 T. Morris-Suzuki, 'Sources of conflict in the information society: Some social consequences of technological change in Japan since 1973', paper presented to the 5th Conference of the Asian Studies Association of Australia, Adelaide, 1984.
18 R. Pineda-Ofreneo, 1983, p. 18.
19 Y. Kitazawa, 1978, 'Keizai shinryaku to josei', *Ajia to josei kaihō*, no. 3, p. 22.
20 Pineda-Ofreneo, 1983.
21 Bello, *et al. Development Debacle: The World Bank in the Philippines*, San Francisco,

Institute for Food and Development Policy, (1982) quoted in M. M. Turner, 'The political economy of the Philippines', *Pacific Affairs*, vol. 57, no. 3, Fall 1984.

22 'Philippines: Under attack and caught in a dilemma', *Far Eastern Economic Review*, 10th October 1985.

23 Asian Women's Association, *Asian Women's Liberation*, Tokyo, no. 4, 1981, p. 16; J. Florence and N. Pardy, 'The Australian connection' in *Scarlet Woman*, no. 19, Spring 1984, p. 30.

24 Ajia no onnatachi no kai, *Aija to josei kaihō*, Tokyo, no. 8, 1980, pp. 4–5.

25 David Hirst, 'Sun city empire crumbles for the ugly Australian', *The Weekend Australian*, 16th–17th February 1985.

26 For discussion of gender relations in Philippine society, see M. Horton, 'Filipino women: Toward an equal partnership', *Feminist International*, no. 1, Tokyo, 1978, V. N. Eviota, 'Women among men and other women in Philippine society', *Feminist International*, no. 2, Tokyo, 1980; J. Florence and M. Pardy, 'Makibaka: Free movement of the new Filipino woman', *Scarlet Woman*, no. 19, Spring 1984.

27 Ajia Taiheiyō shiryō sentaa, *Sekai kara*, vol. 17, Autumn 1983, p. 24.

28 Pineda-Ofreneo, p. 5.

29 Pineda-Ofreneo, p. 17.

30 Buruma, p. 79; Dalby, p. 271.

31 Joyce Ackroyd 1957, 'Women in feudal Japan', in the *Transactions of the Asiatic Society of Japan*, vol. 7, no. 3.

32 S. Gronewold, 'Beautiful merchandise: Prostitution in China', *Women and History*, vol. 1, Spring 1982.

33 Dalby, p. 169.

34 Laura Jackson, 'Bar hostesses', *Women in Changing Japan*, Westview Press, 1976.

35 M. Hane, *Peasants, Rebels, and Outcasts: the Underside of Modern Japan*, Pantheon, New York, 1982.

36 Kaji, pp. 379-80.

37 Morisaki Kazue, *Karayaki-san*, Tokyo, Asahi Shinbunsha, 1976: Yamazaki Tomoko, *Sandakan hachiban shōkan*, Tokyo, Bungei Shunjū, 1972; Yamazaki Tomoko, *Ai to senketsu: Nihon josei kōryū-shi*, Tokyo, Sanseidō, 1972.

38 D. C. S. Sissions, 'Karayuki-san: Japanese prostitutes in Australia', *Historical studies*, vol. 17, April 1976–October 1977, University of Melbourne; B. Mihalopolous, 'Karayuki-san: Japan's forgotten commodity', unpublished Honours Dissertation, University of Adelaide, 1983.

39 There are several works on military prostitutes. See, for example, Yamatani Tetsuo, *Okinawa no harumoni*, Tokyo, Banseisha, 1980; Senda Kako, *Jūgun'ian fu*, Tokyo, San'ichi Shobō, 1978; Ilmen Kim, *Nihon josei ai shi*, Tokyo, Tokuma shoten, 1980.

40 Kitazawa.

41 Morris-Suzuki, p. 80.

42 Jackson (see note 34 above), interviews several bar hostesses who are pessimistic about the possibility of marriage. Although Japanese society may be tolerant of the existence of turkish baths or hostess bars, the individual women themselves do not meet such tolerance.

43 There has been an extended debate on the issue of domestic labour. See, for example, M. Barrett.

44 Game and Pringle show how women's labour has been integrated into the capitalist economy without challenging gender divisions. They argue, on the contrary, that these gender divisions have been reinforced.

45 Engels, *The Origin of the Family, Private Property, and the State*, (1844) [1920].

46 E. Goldman, 'The traffic in women', *Anarchism and Other Essays*, Kennikat Press, (1910) [1969].

47 J. Aitken, 'The prostitute as worker', *Women and Labour Conference Papers*, May 1978; S. Jackson and D. Otto, 'From delicacy to dilemma: A feminist perspective' in Daniels (ed.), 1984.

48 Tanaka Kazuko, *A Short History of the Women's Movement in Modern Japan*, Femintern, Tokyo, 1980; S. L. Sievers, *Flowers in Salt: the Beginning of Feminist Consciousness in Japan*, Stanford, 1984, pp. 184–5.
49 These activities are discussed in Matsui, 'Asian women in struggle', *Second Women in Asia Workshop Papers*, Monash University, 1983.

V 'Modernization' and 'modernity': theoretical perspectives

13 Paths to modernity: the peculiarities of Japanese feudalism

JOHANN P. ARNASON

'MODERNITY': CONTEXTUALIZING THE JAPANESE CASE

Every analysis of specific patterns and results of modernization presupposes an implicit or explicit theory of modernity. With regard to the specific approach on which the following discussion is based, it should be noted that I prefer not to define modernity as a cultural project which has so far had only a limited and one-sided impact on social structures (Habermas) or as a socio-economic system that has gradually expanded around the globe and repeatedly shifted its centre of gravity, but functioned according to the same basic mechanisms from the beginning (Wallerstein).[1] In view of the complexity and variability of the underlying pattern, the notion of a configuration or a constellation would seem more appropriate than that of a project or a system. In more detailed terms, the phenomenon in question is a combination of heterogeneous elements and divergent lines of development that are only partly subsumed under a common denominator or coordinated within a coherent framework. The interconnected structures of a capitalist and industrial mode of production, a system of nation-states and an international market constitute dominant components of the modern configuration and a formative framework of modernizing processes. In the historical context they are inseparable from a cultural and political transformation that creates both preconditions for their development and premises for alternative perspectives. This second component of modernity involves – to mention only some of the most crucial aspects – a new attitude to social change that some theorists have described as 'reflexive awareness' and others as learning to regard it as normal,[2] the emergence and permanent organization of social movements, the institutionalization and radicalization of democracy, and, last but not least, the growing autonomy of different cultural spheres with regard to each other and to the social context. The most distinctive characteristic of the modern world would seem to be the emergence of new forms of the accumulation of wealth and power, situated within a broader field of both subordinate and countervailing forces.

To emphasize the double-edged – i.e. instrumental and contestatory – relationship of the political and cultural innovations to the mechanisms of accumulation is to oppose the recurrent attempts to reduce modernity to a functional pattern. The interpretation of modern society as the functional society par excellence has proved compatible with otherwise different paradigms and ideological orientations. From this point of view, the alternative theories that focus on the functional imperatives of capitalist development, the adaptation of social life to industrial production, or the growing ability of the social system to regulate its interaction with external and internal environments, are variations on a common theme and open to a common objection. They obscure the tension between the systemic patterns and the trans-functional context which is to a greater or lesser extent subsumed under them, but never entirely deprived of its capacity to transcend their functional logic and generate counter-projects. Some analyses of contemporary Japanese society might seem to suggest that it resembles the ideal type of a functional totality, centering on a symbiosis of state and capital, more closely than other advanced industrial societies. But in the light of the above perspective, this finding would constitute a question rather than an answer. A more critical approach would be concerned with the mechanisms of subsumption and integration which have made it possible to absorb or marginalize the trans-functional aspects to an exceptionally high degree.

Some recent theories have identified the trans-functional aspect with a specific factor and ascribed to it a normative content which sets it apart from other determinants of modernity. Such claims have been made on behalf of both cultural and political structures. Habermas' account of the modern differentiation between cognitive, moral and expressive spheres of culture is an example of the former; Agnes Heller's theory of the 'logic of democracy', distinct from the logics of capitalism and industrialization, is perhaps the most emphatic version of the latter.[3] In both cases it can be objected that the interrelations and tensions between functional and trans-functional tendencies recur within each specific sphere. In the present context, I shall confine myself to the issue of democracy. If it is analysed as a concrete complex of institutions, it is hard to deny that its 'logic' encompasses a wide range of divergent possibilities, and that their suppression or realization depends on the historical context. This has led some theorists to conclude that the notion of democracy is too elusive and value-laden to be included among the objective criteria of modernization.

If we want to retain it, it seems necessary to admit that it is susceptible to conflicting theoretical and practical interpretations which are based on different combinations of cultural premises, social forces and political institutions. In terms of their historical significance, some interpretations are circumscribed by the logic of capitalist development and state formation, whereas others are capable of challenging both. This approach seems more relevant to the peculiar history of democracy in Japan than either a normativistic definition or a sceptical dismissal would be. The gradual – but by no means uninterrupted – introduction and domestication of democratic institutions has clearly been an important part of the modernization of Japanese society. On the other hand, the ruling bloc has been singularly successful in adapting them to the imperatives of capitalist accumulation and bureaucratic administration.

The conception of modernity which I have briefly outlined allows for changing combinations of economic, political and cultural determinants; it thus leaves open the possibility of more or less distinctive versions within or alongside the dominant pattern. The transformations of Japanese society since it was drawn into the mainstream of the global modernizing process are a particularly striking case in point. The successive phases differ not only from Western models, but also – at first sight almost as radically – from each other. The developmental strategy of the Japanese ruling bloc after 1868 was characterized by an exceptionally close and effective coordination of state-building and interstate competition with the accumulation of capital; at the same time, the assimilation of Western forms of political organization was counterbalanced and its impact muted by a fundamentally authoritarian relationship between state and society. The interlude of limited but significant liberalization known as Taishō democracy (1912–26) led to an unprecedented development of contestatory currents, but the results were too ambiguous to justify the interpretation of this phase as the first 'take-off' towards full parliamentary democracy. During the next phase – the ultra-nationalist ascendancy and the Great Pacific War – the trend was drastically reversed: the nationalist imaginary[4] which had helped to integrate the ruling bloc and neutralize radical alternatives developed into visions of conquest and hegemony that stretched the ambitions of the Japanese state far beyond its resources and were bound to end in a total defeat. The post-war strategy of economic growth and economic conquest, supported by a de-militarized and developmentalist state, should be seen as a new variant rather than a new beginning; despite a far-reaching adaptation to the liberal-capitalist model, the relationship between state and economy in Japan still differs from the Western pattern in important respects.[5]

The contrasts between the successive phases are sufficiently sharp to show that the distinctive features of Japanese modernity cannot be satisfactorily explained in terms of a continuity of structural or cultural patterns. The most salient point is, rather, the ability of dominant social forces and structures to manage transformations, defuse challenges and adapt to a changing international context. Instead of emphasizing either continuity or discontinuity, the analysis should focus on the peculiar combination of both, which accounts for the very pronounced transformative capacity of post-1868 Japan.

On the other hand, historical research has now proved beyond doubt – even if the lesson has not been fully assimilated by sociologists – that the modernization of Japan did not begin in 1868 and was not simply a response to the challenge from the West. The domestic background of the Meiji Restoration cannot be reduced to the internal crisis of 'a feudalism which had been allowed to consume and rot the body politic';[6] the Tokugawa regime must be regarded as an important and formative stage of the Japanese road to modernity. While there is no denying its extremely repressive – in some ways even proto-totalitarian – character, it was certainly a much more complex, rationalized and dynamic social order than the idea of an artificially prolonged feudalism would indicate. It was based on a combination of feudal and post-feudal elements that was noteworthy not only for its durability, but also for some achievements which paved the way for later changes. This unique configuration was the outcome of a long-drawn-out historical process. At the very least, the conflicts and upheavals of the sixteenth century must be included in any account of the beginnings of Japanese modernity – and, as we shall see, some interpretations suggest an earlier starting point.

The Meiji Restoration and the profound changes which followed it were thus not the result of a 'fortuitous conjunction' of the 'two forces of internal decay and external pressure'.[7] Rather, a modernizing process that had been initiated in Japan long before and blocked in some respects but continued in others – for example with regard to urbanization and education – during the Tokugawa era was both accelerated and radicalized by the challenge from the West. There is no point in speculating about the chances and the likely outcome of a purely indigenous breakthrough; all we can do is try to establish what kind of interaction took place between Japan and the West in the mid-nineteenth century. The historical evidence strongly supports the idea of a convergence of two long-term processes – the increasingly global and radical modernization of the West on the one hand, the more isolated, limited and ambiguous transformations of Japanese society on the other – rather than that of a mutation forced upon a closed and decadent society by an open and dynamic one. This analysis is incompatible with the conception of modernity as a single 'great tradition', comparable with other cultural traditions (especially the world religions) in that it was invented by a particular society or group of societies and then adopted with more or less extensive modifications by others.[8] It seems more plausible to think of different roads to modernity as initially independent but at least partly comparable trajectories that later converge in a global context, dominated and shaped by one of them to a much greater extent than by the others. In the light of the overall results, the West and Japan stand out as the only two conclusive cases; the question of partial and abortive breakthroughs elsewhere (e.g. in China under the Sung dynasty) will not be discussed in this paper.

For further analysis of similarities and contrasts between Western and Japanese lines of development, it is essential to define the comparative frame of reference in precise terms. Japan has frequently been compared with other 'latecomers to modernization', within the West (Germany), on its fringe (Russia), or in its nearby environment (Turkey).[9] Each of the three perspectives is open to specific objections, but in the present context, their common shortcomings are more relevant. The distinction between pioneers and latecomers was most plausible when associated with a universal and normative model of modernization that was supposedly adopted more or less rapidly and successfully by various countries; with the recognition of different roads to and versions of modernity, the grounds for assuming structural similarities between 'latecomers' have become much less compelling. With regard to more specific issues, none of the three comparative strategies seems to have thrown much light on the unique trajectory of post–1868 Japan and the equally unique position it has come to occupy in the capitalist world system. Last but not least, the comparison of Japan with individual Western, semi-Western or Westernized societies obscures an essential point: as I will try to show, the Western – more precisely the European – state system as a whole is at least in some ways a more genuine counterpart.

The importance of the European state system that began to take shape in the fourteenth century and collapsed in the twentieth – first with the very short-lived rule of one European state over the whole continent and then with the more durable ascendancy of the two superpowers – as a matrix of modernization in general and capitalist development in particular, is widely recognized.[10] An East Asian state system with shifting power balances and fluctuating boundaries emerged earlier than the 'concert of Europe', and it may even have shown signs of developing into a world-system centering on South China during the Sung era. But two distinctive features of this system should be noted. On the one hand, the massive and constant preponderance of China has no parallel in other state systems;

on the other hand, this primacy was only partly and intermittently translated into direct imperial control over the weaker states, and, in contrast to Korea and Vietnam, Japan remained immune to all such attempts. Apart from the spectacularly unsuccessful Mongol invasion in the thirteenth century, no threat to Japanese independence was ever mounted from China. Conversely, pre-1868 Japan was never a serious threat to China. Hideyoshi's plan to conquer the Chinese empire was a wild flight of fancy, and although the foothold in Korea was important at the very beginning of recorded Japanese history, the attempt to re-establish it at the end of the sixteenth century failed; as far as the consequences for Japan were concerned, its only important result was to dissuade the Tokugawa regime from expansionist policies. Japan's links with China and the East Asian state system were thus primarily cultural. There is no comparable case of one country exerting on another a cultural influence as profound and lasting as that of China on Japan, without it leading to political subordination. As Marius B. Jansen puts it, China was for Japan 'a classical antiquity, a Renaissance Italy, and an eighteenth-century France all in one. It was a single cultural colossus, one that endured.'[11] But if the continuity of Chinese civilization and the distance between the two countries made it possible for the Chinese model to combine the force of a great tradition with that of a living example without being drawn into political and military conflicts of the kind that engulfed its European counterparts, the other side of the coin is no less important. As a result of its political autonomy and isolation, the development of Japanese society became in some ways as self-enclosed and self-sustaining as that of larger state systems elsewhere. On the analogy of Jansen's description of China, it might be suggested that Japan was its own Italy, Spain, Holland, France, England and Germany – in the sense that its modern transformations involved processes equivalent or comparable to those which in the European context were linked to the replacement of one dominant centre by another.[12]

To draw this parallel is not to answer in advance the empirical question of more concrete similarities or dissimilarities. Roughly speaking, it would seem that some phases of Japanese history were dominated by centrifugal tendencies to such an extent that we can legitimately compare them with the European pattern of interstate competition, whereas the strengthening of central authority was at other times carried far beyond what was possible in the West (i.e. on the scale of the European state system) and gave a very distinctive turn to otherwise parallel processes. For the first approach, the most obvious case in point is the Sengoku era (usually dated from 1490 to 1573); the Jesuits who came to Japan in the mid-sixteenth century and reported that the country was divided into sixty-six kingdoms were using the same label to characterize local rulers of very different strength and significance, but the description in terms of a plurality of states was not wide of the mark. The Tokugawa state represents the other extreme, and the contrast is all the more interesting because of the rapid transition. This historical turning-point highlights an important difference between Japan and the West: in Japan, the early modern phase of the process of state formation, initially stimulated by rivalries and conflicts between the warlords of the sixteenth century, culminated in a breakthrough to unification that proved impossible within the European state system. But it would be misleading to describe the Tokugawa ascendancy as a successful equivalent of the abortive imperial aspirations of some European states. Apart from all other differences, the ethnic and cultural homogeneity of Japan made the project of unification easier to realize and legitimize than the attempts to superimpose an imperial regime on the greater diversity of the West. Even if the latter had proved more successful, the resultant relationship between the centralized state and the socio-cultural context would – because of the underlying diversity – have been very different. On the other hand, the *bakuhan* system that was introduced after 1600 did not eliminate all similarities to the European state system; despite the effective and lasting pacification of the country and the concentration of power in the hands of the *bakufu*, the administrative autonomy of the *han* left some scope for differentiation and competition between them, and this was to prove highly relevant to the long-term perspectives of the Tokugawa regime.[13]

The following discussion will primarily be concerned with the pre-conditions and early stages of the Japanese road to modernity. But the above remarks may serve to indicate the broader perspective which is taken for granted and some related issues that would merit further exploration. In particular, the emphasis on the process of state formation and its socio-economic context should not be taken to imply a downgrading of cultural aspects. Given the limits of this paper, a selective approach is unavoidable, but it may help to pinpoint some problem areas where the relevance of other perspectives – e.g. a culture-centred one – is especially obvious.

THE MUROMACHI SHOGUNATE: JAPAN'S RENAISSANCE?

It has proved much easier to reach at least a working consensus on the chronological coordinates of modernity than on the relative weight of its determinants. Irrespective of the roles assigned to economic, political or cultural factors, there is a general tendency to date a decisive breakthrough to the sixteenth century. This is compatible with a special emphasis on such varying historical phenomena as the Reformation, the development of modern science and the transformation of the world-view initiated by the Copernican revolution, the beginnings of European expansion and the constitution of a world-system, the emergence of the absolutist state or the consolidation and diffusion of the capitalist mode of production. All these innovations must in turn be set against an historical background that includes the urban communities and social movements of the later Middle Ages, the Renaissance and the precursors of the Reformation.

If Japanese history is to be interpreted as an autonomous road to modernity, it must be possible to identify at least some landmarks analogous to those of European history. One of the most interesting recent attempts to do so is Kenneth A. Grossberg's analysis of the Muromachi shogunate (1336–1573). He sees the political system constructed by the first shoguns of the Ashikaga dynasty as comparable to the Renaissance monarchies in Europe:

In Japan during the Muromachi age, similar to occurrences in Europe during the quattrocento, a society arose which viewed the state as, if not exactly a work of art, at least a guided creation of human effort, 'the fruit of reflection and careful adaptation'. Strictly speaking, the Japanese Renaissance (like its European name-sake) was not a rebirth at all, but the result of a conscious art of political creation. As in the West, this Renaissance was also animated by the change from an overwhelmingly autarkic-agrarian economy to an increasingly capitalist-agrarian one, causing repercussions in governing the country ... While capitalist agri-culture is ultimately augmented by industrialization, most of the organizational technology – the ways of organizing human effort to carry on these economic activities and to govern the societies in which they are being carried on – is the creation not of the industrializing society but of its predecessor, the Renaissance state with its capitalist-agrarian economy. These are, after all, the governments which first experiment with professional bureaucracies and armies free of conflic-ting loyalties, fiscalism and the exploitation of commerce, absolutist ideology, and the idea of the state's primacy over its subunits.[14]

The innovations of the Ashikaga shoguns prefigured later developments: 'The union of samurai politics with the imperial capital carried the Japanese political experience into the early modern age and transformed the Bakufu from a decentralized feudal regime into the prototype of preindustrial autocracy embodied in the Tokugawa settlement of the seventeenth century.'[15] The analogy with the European Renaissance is further supported by the extraordinary cultural productivity of the Muromachi era, as well as by the disintegration of the traditional order: 'Like the European Renaissance, Muromachi Japan was rocked by internecine struggles and destabilizing social change but also expressed a spontaneity and a freedom in its various institutions that its successors seem to lack.'[16]

Grossberg's comparative perspective allows for parallels with both individual Western states and the European state system as a whole: the focus is on an institutional pattern which in the European case was from the outset embodied in a plurality of states, whereas its Japanese variant was first put into practice by a unified state and later reproduced – albeit on a limited scale and with major modifications – within the 'warring states' of the Sengoku era. 'The Ashikaga shoguns ... were feudal monarchs striving toward a postfeudal autocracy which is recognizably similar to that of the Valois in France or the York Lancaster in England',[17] but their political legacy was appropriated and divided by provincial warlords. The progressive fragmentation of the Muromachi state after 1441 raises some fundamental questions about Grossberg's thesis. But this objection would be pointless without a more detailed account of the argument against which it is directed. The reconstruction of the 'Muromachi synthesis', as Grossberg terms it, involves three main aspects: patterns of legitimation, administrative and military apparatuses, and the relationship to social forces.

According to Grossberg, pre-Muromachi Japan had known three different types of legitimate domination. The Taika reform in the seventh century had led to the development of a legal and bureaucratic order, originally patterned on China, but gradually adapted to the very different Japanese environment; from the ninth century onwards this model was overshadowed by the principles of aristocratic authority, centering on the court in Kyoto and embodied in a very elaborate rank system; finally, the rise of the military aristocracy in the provinces was associated with more distinctively feudal forms of authority that culminated in the establishment of the shogunate in 1192. All three patterns were invoked and used by the Ashikaga shoguns; more importantly, they were combined in a new way. The resultant synthesis was – in Grossberg's view – a more consistent model of monarchical authority than anything that had previously existed in Japan. The recognition of Ashikaga Yoshimitsu as 'King of Japan' by the Chinese emperor in 1402 was thus more than a diplomatic fiction. To consolidate their 'kinglike rule', the Ashikaga shoguns created a bureaucratic apparatus, more extensive and diversified than that of their Kamakura predecessors, and tried to strengthen both the hold of the central government on the country and their own personal authority within the government. At the same time they tried to build up a more centralized and reliable armed force that would make them less dependent on their vassals.

The socio-economic background of all these innovations was the development of a commercial agrarian economy in the Home Provinces around Kyoto. In addition to the shogunal domain and the land taxes, the government could thus draw on both the domestic market system and foreign trade to enlarge its financial basis. Similarly the links with new social forces – merchants and moneylenders, but also the Zen monasteries that were very actively involved in the economic life of the capital and its surroundings – broadened the social basis of the Muromachi rulers and enabled them to recruit officials outside the previously closed circles of the court nobility and the military aristocracy. In adapting to and taking advantage of this changing social context, the shogunate was gradually transformed from a feudal regime into an urban-based autocracy.

There is no reason to disagree with Grossberg's view of the Renaissance state as the first crystallization of European modernity. He has, moreover, shown that the Muromachi system is of quite exceptional interest for the comparative analysis of state formation. And there is no denying that some of the techniques of state-building used by the Ashikaga shoguns – in particular the attempt to create a bureaucratic and military apparatus that would constitute a counterweight to the feudal hierarchy – are comparable to those implemented by late medieval and early modern European monarchs, even if the results achieved in Japan fell much farther short of an effective monopoly of violence and taxation. Grossberg does not ignore this difference, but he suggests that we might regard the Muromachi *bakufu* as the least developed version of a model which was most adequately exemplified by the Valois monarchy in France during the late fifteenth century and adopted with less success by some other European states of the same epoch. The ultimate failure of the Ashikaga dynasty might even, as he sees it, have been due to conjunctural rather than structural causes: 'Yoshinori's death postponed the growth of absolutism in Japan for almost a century ...'[18] This claim is not convincing; as I will try to show – mainly on the basis of Grossberg's own presentation of the historical evidence – the different history of the allegedly similar projects is primarily to be explained in terms of structural differences, and the extreme fragility of the Muromachi state reflects its basic flaws and built-in tensions. The parallels underlined by Grossberg are real, but in the light of the broader context, they must be qualified much more strongly.

If Grossberg's concrete analyses cast some doubt on his more general interpretation of the Muromachi era, this applies no less to the perspective on Japanese history which he outlines at the end of his book. Given his emphasis on the similarities between the two Renaissances, Japanese and Western, it necessarily follows that major differences between the respective patterns of modernization must be interpreted as the result of later divergences. Grossberg's opinion is, in brief, that 'the components of the Muromachi synthesis did not all develop to the same degree or at the same rate, and some were even interrupted or terminated during the turbulence of the sixteenth century and the subsequent repression of early Tokugawa'.[19] On this view, the transition from the Renaissance to the absolutist state appears to have been a much more unequal development in Japan than in Europe; in particular, the unified and centralized state which grew out of Nobunaga's conquests and was perfected by the early Tokugawa shoguns blocked social mobility, suppressed social movements and imposed a much more closed and hierarchical social order than any European absolutism ever did. But if it can be shown that the analogy between the Muromachi era and the European Renaissance has been overdrawn, subsequent developments would seem both less anomalous and more distinctive. In other words: they would have less to do with a deviation from an initially common pattern than with enduring specific characteristics. This line of inquiry is inseparable from the much-discussed but still puzzling question of Japanese feudalism.

The discussion of the character of the Muromachi state should begin with at least a brief glance at its immediate background – a short but significant episode that has no parallel in European history. The Kenmu restoration (1334–6) was the last attempt of the court society, based on Kyoto and the Home Provinces, to reassert its primacy against the 'Eastern barbarians', i.e. the military society that had grown up in the eastern provinces and dominated the whole country since the founding of the Kamakura shogunate in 1192. It was a conflict between different forms of power and corresponding images of the social order, rather than a simple clash between two sections of a ruling class, and although the official aim was to restore a legitimate order and destroy an upstart social force, the assault on a power structure that had already created its own tradition was in some ways comparable to a revolutionary struggle. Conversely, the restoration of the shogunate by Ashikaga Takauji and his successors had – as Grossberg emphasizes – some revolutionary implications.

The Kenmu restoration was, of course, a very complex phenomenon; four main aspects should be distinguished. The initial success was due to an alliance between the imperial court and dissatisfied sections of the military aristocracy, but the superior strength of the latter very soon transformed the victory into a defeat. The short-lived attempt of the imperial loyalists to implement their own policies brought into the open the difference between the plans of Emperor Go-Daigo, aiming at the restoration of direct imperial rule, and the strategy of the court aristocracy, bent on recapturing the power and the privileges that had been surrendered to the provincial warriors in the twelfth century. Finally, the proto-nationalist ideas developed by the most articulate ideologist and most competent organizer among the loyalists, Kitabatake Chikafusa, who saw the continuity and supremacy of the imperial dynasty as essential to the collective identity of the Japanese polity and as a guarantee of its superiority over other civilizations, were a striking prefiguration of later developments.[20] It was to take many centuries before this idea could be revived with greater success in a thoroughly transformed social context. Chikafusa was, paradoxically, both far behind his time and far ahead of it. More generally speaking, there is no doubt that the restoration was bound to fail; the projects of the loyalists were hopelessly out of touch with the real distribution of power. But the fact that the attempt was made and the way in which it failed influenced the direction of later developments. On the one hand, the civil war which dragged on for more than fifty years after the decisive defeat of the loyalists at Minatogawa in 1336 stimulated the centrifugal and contestatory forces in Japanese society that were so much more conspicuous during the Muromachi era than either before or after. On the other hand, the Kenmu restoration had shown how explosive the combination of the imperial mystique with aristocratic resentment and feudal dissensions could be, and the Ashikaga shoguns drew some obvious lessons from this experience. The imperial centre was partly absorbed into the military society (by transferring the capital of the shogun from Kamakura to Kyoto), partly neutralized much more effectively than before. In view of the background and the utter powerlessness of the court at the end of the fourteenth century, the abolition of the imperial institution might seem to have been the easiest way out, but although Yoshimitsu may have toyed with this idea, it was never put into practice.[21]

The survival of the imperial institution raises some questions about Grossberg's reconstruction of the 'Muromachi synthesis'. He distinguishes three forms of legitimate power – bureaucratic, aristocratic and feudal-military – and claims that the stronger Ashikaga shoguns managed to combine them. But the legitimacy of all three types was ultimately derived from a fourth, more fundamental and less directly effective than others: the authority of the imperial institution. Since the Muromachi system never fully absorbed this ultimate source of legitimacy, the synthesis was at best incomplete. Grossberg uses the comparison with Europe to dispose of this objection; he likens the relationship between the shogun and the emperor to that between the rising European monarchies and the papacy, and the loss of authority suffered by the popes during their exile in Avignon to the degradation of the imperial court after 1336.

But the 'Babylonian captivity' of the popes lasted only seven decades, and it did not prevent some of the pontiffs involved from increasing the wealth and power of the church, whereas the imperial court in Kyoto was effectively excluded from political power for more than five centuries after the defeat of the Kenmu restoration. This discrepancy suggests that the analogy might be misconceived; it seems clear that it was both more necessary and more feasible to neutralize the imperial institution than the papacy. And this is easily explainable in terms of a fundamentally different relationship between sacred and secular power. The specific claims and credentials of the papacy were based on a distinction in principle between spiritual and worldly authority. The *tennō*, by contrast, embodied a fusion of the sacred and the secular which became the ultimate source of political power. The popes could, depending on circumstances, use their spiritual authority to guarantee a residual share in political power, participate on a more or less equal footing in the affairs of states and rulers, or even lay claim to the position of an ultimate arbiter. The sacred component of the Japanese imperial institution became an obstacle to direct imperial rule and a convenient justification for the delegation of power, but its political implications remained strong enough to constitute a potential threat to any specific form of the delegation of power, and to compel the successive regimes of non-imperial rule to devise more or less elaborate strategies of prevention. Although the strategy of the Ashikaga was in some ways very radical, it left the imperial institution intact, and a clear and stable definition of the relationship between shogun and emperor was never worked out. It seems likely that Yoshimitsu's decision to accept the title of 'King of Japan' from the Chinese emperor was an attempt to compensate for this intrinsic lack of legitimacy. Needless to say, the nominal allegiance to a foreign ruler was a double-edged solution that could easily play into the hands of adversaries of the regime. Yoshimitsu's successors did not follow his example but the connection with China remained important in another respect: the Muromachi *bakufu* preferred to import copper coins from China rather than mint its own currency. As Grossberg points out, this was clearly not because of a lack of adequate technology; it must have been the problematic legitimacy of the government that made a well-established foreign medium of exchange seem more reliable. This is a striking illustration of the shortcomings of the Muromachi power structure.

If the imperial institution and the imperial dynasty remained the ultimate and unquestioned – albeit temporarily eclipsed – point of reference for the political order, it is hardly justified to speak of 'a conscious act of political creation'. On this level the differences between the Muromachi shogunate and the Renaissance monarchies in the West are much more pronounced than the similarities. The self-thematization and self-theorization of the early modern European states, centering on the idea of sovereignty, was a constitutive part of their development.[22] The 'Muromachi synthesis' lacked this component, and it could not have developed without an explicit questioning of the imperial tradition which the Ashikaga shoguns preferred to disregard in practice rather than contest in principle. One of the most intriguing developments in Japanese history – the gradual transformation of the imperial institution, which the reforms of the seventh century had established as a strong political centre and as a link between indigenous traditions and imported models, into a cultural matrix that could survive the loss of political power and determine the premises of political thought – was clearly well advanced in the fourteenth century.

As developments during the Tokugawa era were to show, this did not necessarily exclude the emergence of ideological constructions which manipulated the imperial symbolism and neutralized some of its implications.[23] But the Tokugawa state could draw on a much richer historical experience and more developed cultural resources than its Muromachi predecessor. The failure of the latter to develop an ideological frame of reference is reflected in its political strategies. Grossberg's description of the Muromachi administration shows not only how easy it was for the most powerful feudal lords to appropriate the bureaucratic apparatus and transform various parts of it into hereditary possessions, but also how unclear the very principles of organization and demarcation were. One aspect of this confusion was the failure to develop a coherent system of taxation. Even more importantly, the regime had no consistent strategies with regard to the social forces with which it had to reckon. It could neither suppress nor coopt the urban and rural movements that flourished during the Muromachi era and found expression both in rebellions and in more permanent organizations of self-government and self-defence (*ikki*).[24] The process which Grossberg describes as the transformation of the shogun from a feudal lord into an urban prince seems in reality to have been blocked at a stage where the shogun was left with an insecure urban basis and an irreparably weakened position *vis-à-vis* the military aristocracy. The alliance with the Buddhist monasteries benefited the latter; by the end of the fifteenth century, the Buddhist establishment, although divided, had become a major social force and the shogun a mere figurehead. And the attempt to ally with the lower ranks of the feudal hierarchy against the provincial magnates (the *shogu daimyō*) was in the long run more conducive to further fragmentation than to a strengthening of shogunal power.

In short, the Muromachi system seems to have been a permanent – and increasingly desperate − balancing act, rather than a synthesis. It was surely one of the more innovative, but ultimately also one of the most incoherent exercises in state-building. Grossberg's allusion to 'the schizoid nature of Ashikaga power'[25] is in line with this interpretation, but more difficult to reconcile with some of his other statements.[26]

The most conclusive argument against equating the place and role of the Ashikaga shogunate in Japanese history with that of the Renaissance monarchy in the West is the unstoppable process of fragmentation after 1441. Grossberg compares the Ōnin war and its sequels to the breakdown of the French monarchy during the Hundred Years' War. But this would upset his chronological frame of reference. According to the main line of his argument, the policies of Yoshimitsu (shōgun 1368−94, de facto ruler 1379−1408) were in principle − if not in their degree of success − similar to those of Louis XI of France (1461−83), whereas the reference to the Hundred Years' War implies a comparison of Yoshimitsu and Philip the Fair (1285−1314) and thus a wholesale change of historical context. Apart from that, the fragmentation of France during the Hundred Years' War never came anywhere near the situation in Japan during the Sengoku era; and most importantly, the French monarchy survived as the indispensable focus of state formation, whereas the Ashikaga shogunate declined and disappeared, and the process of state formation had to begin again on a new basis. The long-term results of the Muromachi experiment were thus very different from those of the Renaissance monarchies.

But it is nevertheless misleading to suggest − as one of Grossberg's critics does − that 'Japan's experience is ... nearly the reverse of Western Europe's ...'[27] The triumph of centrifugal forces in Japan after 1441 did not lead to the same kind of fragmentation as the feudal dispersal which had preceded the reconsolidation of monarchical power in Europe. The point is, rather, that in the two cases analogous long-term trends were combined in different ways and in a different chronological order. In Japan, the culmination of the distinctively feudal trend towards a parcellization of power coincided with the transition to a more commercialized economy, a higher level of social mobility and a more developed administrative technology, whereas in Europe the centrifugal dynamic of feudalism had reached its highest point before the other developments began to make their mark. The combination worked against the Muromachi state. As to the positive results, a comparison with transformative processes in Europe reveals both contrasts and similarities. The upheavals of the Muromachi era did not result in a radical social or political revolution. Although there were some interesting experiments with self-government (e.g. the Yamashiro revolt, 1485−93) and the much more durable rule of the *Ikkō-ikki* in Kaga Province), there is no reason to disagree with Sansom when he writes that to interpret the overall change as a social revolution or a movement towards democracy 'is to debase the currency of political terms'.[28] In fact, the traditional Japanese term *gekokujō* ('the lower superseding the higher') seems in some ways a more appropriate interpretation of the process as a whole than any concepts derived from European experiences. Social mobility and political decentralization influenced the course of events more markedly than social movements, but the ultimate result was the emergence of new nuclei of state formation (the domains of the Sengoku daimyo), the unification of Japan and the reconsolidation of the hierarchical order by new leaders who had risen from lower ranks.

The domains of the Sengoku warlords were thus simultaneously the results of feudal fragmentation, the framework for a new combination of feudal principles with other structural components and the starting point for an astonishingly rapid process of reunification. This constellation throws some light on the general problematic of Japanese feudalism; I shall therefore conclude with a few remarks on that issue.

THE PECULIARITIES AND PARADOXES OF JAPANESE FEUDALISM

At the end of his classic work on European feudalism, Marc Bloch singled out Japan as the most obvious – and perhaps the only – comparable case outside Europe.[29] According to Bloch, the feudal epoch in Japanese history began in the eleventh century and its background was very similar to the European one: on the one hand a temporarily successful but in the long run abortive attempt to create a large and centralized state with effective mechanisms of control over its subjects (in Europe the Carolingian empire, in Japan the Taihō state), on the other hand the remnants of a society based on the principles of kinship, too weak to resist the process of feudalization but strong enough to influence various aspects of social life within the feudal framework. But whereas Bloch dates the beginning of the de-feudalization of European society from the thirteenth century, he leaves open the question of a comparable watershed in Japanese history.

The stress on parallels between Japan and Europe did not prevent Bloch from reorganizing several specific characteristics of Japanese feudalism. The network of suzerainty and dependence within the feudal military elite was, as he saw it, more sharply distinguished from the forms of subordination which applied to the peasants. The class character of the dominant group was thus more pronounced than in Europe. The organization of the Japanese manor, which rarely involved work on the lord's demesne, further increased the distance between the two poles of the feudal order. On the other hand, inequality within the ruling class was also more marked; the lord-vassal relationship was less contractual and involved a more one-sided emphasis on loyalty than the European version. The very size of the feudal elite – the samurai in Japan were more numerous than the nobility ever was in any European country – had similar consequences. The lion's share of power and property was reserved for the most privileged sections and centres (Bloch refers to barons and temples), and the bulk of the warrior class was much more modestly endowed. Later analysts of Japanese feudalism have discussed all these points and sometimes added major qualifications. It seems, for example, that the non-contractual character of the lord-vassal relationship, which has been invoked to explain the incapacity of Japanese feudalism to develop representative institutions comparable to those of Europe, was subject to more historical variation than Bloch thought; according to Peter Duus, the vassalage tie became more contractual during the Sengoku era than it had previously been, and he also argues that it was more vulnerable to betrayal and rebellion than in Europe.[30] Similarly, the question of the size of the feudal class and its impact on forms of organization and strategies of social closure must be related to the historical context. Joseph R. Strayer suggests that the large number of samurai 'required a degree of planning and organization which was never necessary in Europe. Japanese feudalism had to be more structured, more impersonal, more bureaucratic than European feudalism.'[31] But during the epoch to which he is primarily referring (the sixteenth century and the formative phase of the Tokugawa era), these structural constraints were enhanced by a more specific strategy: the ruling elite was striving to reaffirm the hierarchical distinction between the samurai, whose numbers had swollen during the Muromachi era, and the other social groups. The disproportionate size of the feudal class was to some extent a result of the very processes which its leaders were trying to block or reverse. The specific combination of bureaucratic rule with other mechanisms of control which eventually prevailed under the Tokugawa regime reflects this background and differs from other incipient or conceivable responses to it. No supposedly general laws of Japanese feudalism can explain its concrete features.

Apart from the brief 'cross-section of comparative history' at the end of the book, Bloch had in an earlier chapter pointed out an important contrast between the two types of feudalism: in Japan, feudalism co-existed with monarchy, whereas in Europe they penetrated each other. In his capacity as the ultimate sovereign, the Japanese emperor stood outside the network of feudal royalties; the hierarchy of lord-vassal relations did not extend beyond the shogun. Although Bloch did not develop this point further, the link with his comparison of the Carolingian state and the centralized system which grew out of the Taika reform is obvious: The remnants of the Carolingian state were integrated into the feudal system; in Japan, the feudal aristocracy created its own centre of authority and a corresponding administrative apparatus which gradually replaced the imperial institution without either abolishing or incorporating it. In the literature on Japanese feudalism, this peculiar pattern has obviously not gone unnoticed, but its implications have so far been explored less thoroughly than those of other distinctive features.[32] They are, as I shall try to show, particularly relevant to the comparative analysis of state formation. But the latter theme must first be located within the general problematic of feudalism.

In the present context, it is convenient to classify interpretations of feudalism on the basis of their answers to three questions. They differ, firstly, on the scope of the concept: should feudalism be defined in evolutionary or historical terms, i.e. as a universal phase of social development or as a societal type that is most clearly exemplified by medieval Europe but may have emerged with fundamentally similar features elsewhere? There is clearly a much stronger case for the latter alternative; the most serious attempt to defend the former was grounded in a unilinear version of Marxism, and recent Marxist criticism has undermined the general presuppositions as well as the specific arguments concerning feudalism.[33] The issue has, to all intents and purposes, been resolved in favour of the historical approach. Most of its advocates would agree that Japan was by far the most conspicuous and durable case of a non-European feudal society. While this does not rule out more partial and ephemeral parallels in other societies, the uniqueness of Japan in this context is no less pronounced than that of its modern transformation, and the search for connections between the two historical experiences is a prima facie promising strategy. But agreement on this point still leaves open a fundamental question: do the peculiarities of Japanese feudalism help to explain the difficulties and shortcomings of Japanese modernization, in contrast to the Western mainstream, or are they rather linked to a specific transformative capacity which found its most decisive expression in a distinctive road to modernity?

The second issue is more controversial. Is feudalism primarily a form of political organization or a complex of economic structures, or perhaps a combination of both? If medieval European society is regarded as the paradigmatic example, some of its most salient characteristics are to be found on the political level. Although this starting-point is acceptable to those Marxists who stress the political constitution of pre-modern class societies, the primacy of the political has been most strongly emphasized by non-Marxist historians and theorists of feudalism. Strayer and Coulborn proposed the following definition:

Feudalism is primarily a method of government, not an economic or a social system, though it obviously modifies and is modified by the social and economic environment. It is a method of government in which the essential relation is not that between ruler and subject, nor state and citizen, but between lord and vassal. This means that the performance of political functions depends on personal agreements between a limited number of individuals, and that political authority is treated as a private possession ... Since political power is personal rather than institutional, there is relatively little separation of functions; the military leader is usually an administrator and the administrator is usually a judge.[34]

The most obvious objection to a strictly political conception of feudalism is, of course, that the very idea of a strict demarcation between the political and the economic sphere is alien to this type of society. The political institutions do not simply 'modify the economic environment', but are defined and organized in such a way that the line which separates them from the latter is blurred. Strayer and Coulborn implicitly admit this when they refer to the conflation of public authority and private possession. Although they argue that vassalage, rather than the fief, is the essential element in feudalism, and that the former can exist without the latter, they add that the connection with landed property, established through the institution of the fief, 'gives stability to agreements which are otherwise fluctuating and impermanent and makes it possible to build a more solid institutional structure on the basic relationship of lord and man'.[35] In other words: if we are thinking of feudalism as a functioning system, rather than as an analytical model, its basic principle is obviously what the French historian Alain Guerreau terms 'a total assimilation of power over land and power over men'.[36]

If the feudal principles of organization are multi-dimensional and multi-functional, the interpretation of feudalism as a total social order might seem unobjectionable. But the arguments for and against this conception constitute the third major debate among theorists of feudalism. The most convincing version of the antithesis is a model of medieval European society which Georges Gurvitch and Fernand Braudel constructed on the basis of Bloch's work.[37] On this view, feudalism was a temporarily dominant component of a more complex totality rather than a closed and complete system; the development of feudal structures took place within the framework of a changing combination with four other social types. The relationship between lord and peasant, originally embodied in the manorial system, was incorporated into the feudal order and profoundly affected by it, but its basic aspects were, as Bloch shows, older and lasted longer than feudal society. The 'ecclesiastical society' of the church had its specific principles of integration and subordination and they differed from the feudal ones. The urban communities developed their most innovative forms of social life as an explicit challenge to feudal patterns; the most symbolic expression of the communal ethos was, according to Bloch, the mutual oath contrasting with the feudal oaths of subordination. Finally, the territorial monarchy should be regarded as a distinctive principle of organization. It antedated the feudal order and although it was temporarily eclipsed by the latter, it remained strong enough to become the focus of the long-drawn-out process of the 'reconstitution of the state' (Bloch) which culminated in absolutism.

This analysis is well founded, but not exhaustive. In *Feudal Society*, Bloch examines two further aspects of medieval society. On the one hand, the structures of kinship can be seen as a distinctive pattern of social organization, more diffuse and fragmented than the others, but analytically separable from them. In the early medieval world they could not be an adequate substitute for the disintegrating state, but they remained an indispensable ingredient of the social order. On the other hand, the combination as a whole can legitimately be described as 'feudal society', in the sense that although the non-feudal components are not reducible to feudal principles of organization, they were – during a whole historical epoch – more or less effectively subsumed under them. This is most obvious in the case of the relationship between lord and peasant, but the ecclesiastical, urban and monarchical structures were also conditioned and modified by the feudal framework. Feudalism is, in other words, a totalizing rather than a total social order; it is not a self-sufficient societal type, but it can combine with other types and shape the result to a greater degree than any other specific pattern.

This model has, to the best of my knowledge, not been applied to the comparison of European and Japanese feudalism. Within the limits of this paper, I can do no more than sketch the outlines of a possible application. The comparison would, if this argument can be sustained, centre on three contrasts between Japan and the West: a different structure of the feudal principle of organization, different forms of the subsumption of non-feudal components, and – as a result – different trends in the transformation of the feudal elite.

The pervasive influence of the feudal principle of organization in medieval Europe and its lasting impact on subsequent developments must be explained in the light of its internal complexity and its ability to synthesize different or even opposite elements. We have already noted the fusion of power over land with power over men. On the level of human relations, the feudal nexus combines the emphasis on hierarchy with a limited but important concession to the principle of mutual recognition.

And the hierarchical aspect was further qualified by the widespread practice of multiple vassal engagements; the institutionalized possibility of conflicting loyalties strengthened centrifugal forces, but it could also serve to highlight the contrast between feudalism and a political system with a stronger centre and more exclusive obligations. The structure of Japanese feudalism gave a significantly different twist to all three aspects. The personal link between lord and vassal was much more important than the link between the vassal and the land; in terms of Guerreau's interpretation, quoted above, the assimilation of power over land to power over men was more limited than in Europe. The grounding of feudal organization in landed property was slower to develop, remained more limited and proved easier to replace by other devices when the Tokugawa regime separated the warrior class from the villages.[38] Within the institution of vassalage, the emphasis on hierarchy remained – notwithstanding the above-mentioned historical variations – much stronger than in Europe. Finally, the lord-vassal relationship was more exclusive; the 'man of several masters', to whom Bloch devoted a whole chapter in *Feudal Society*, did not exist in Japan. At first sight, the combination of these characteristics might seem likely to make the hierarchical structure of the ruling elite more rigid, less dependent on the socio-economic context, and more immune to social change. But as the historical record shows, they could also facilitate its transformation. To understand this, we must look at the broader context, i.e. at the interaction of the feudal principles with the non-feudal structures that were subsumed under them but never fully absorbed.

On both chronological and structural grounds, the most appropriate starting-point is the relationship between the feudal order and the 'tie of kinship' (Bloch). In this respect, the main contrast between Japan and the West is easy to define: the Japanese pattern was characterized by a much higher degree of continuity, but this was a flexible and manipulable connection, rather than a permanently constraining factor. In the core zones of European feudalism, the cumulative impact of the Roman conquest, the decomposition of the empire and the barbarian invasions had dissolved the tribal order, and the fragments which survived could only play a secondary role. In Japan, the infrastructure of the Taihō state had retained a much stronger link to the familial or clan system (Western historians disagree on the translation of *uji*), and this facilitated the return to aristocratic rule, both in Kyoto and in the provinces. It has even been suggested that the military aristocracy was 'not a new class, but one with roots of great antiquity. They were the social, if not biological heirs of the old clan leaders'.[39] This could not be said about the European aristocracy. On the other hand, the more recently developed theory of '*ie* society'[40] suggests that the emergence of feudalism in Japan was associated with a new form of familial authority, more artificial and more systematically adapted to the demands of a comprehensive pattern of social organization. Whatever the merits of this specific thesis, it is in any case clear that the forms and images of familial authority were much more essential to the feudal order in Japan than in the West.

If feudalism was more closely and continuously linked to the structures of kinship in Japan than in Europe, its relation to the social framework of agricultural production was rather different. According to one of the most authoritative analyses of this problem, it has been a constant feature of Japanese history that 'the level of control over land production exercised by the Japanese elite has been relatively removed from the land itself. In other words, their main concern has been man and his labours rather than the land as such'.[41] The mechanisms of control over 'man and his labours' have varied, but on the whole they have involved less direct intervention in agricultural production and less direct coercion of the producer than in European feudalism. Perry Anderson's claim that 'a close replica of glebe serfdom'[42] prevailed in Japan seems very hard to sustain. The notion of serfdom is, of course, a highly controversial one, but even if we prefer a definition that would not stick too closely to specifically European and strictly legal criteria, it seems that no specific institution linked the lord-peasant relationship to the feudal order in Japan in the same way as the institution of serfdom in Europe. It is also more difficult to discern long-term trends on this level: 'Throughout Japanese history the peasant has been alternately freed and oppressed, and perhaps the most oppressive phase of all was the phase that followed the emancipation of the fifteenth century.'[43] The new forms of oppression, initiated by Nobunaga and Hideyoshi and perfected by the Tokugawa regime, were associated with a far-reaching transformation and a partial supersession of the feudal order. The strengthening of control over the peasantry coincided with an unprecedentedly radical separation of the samurai from the land; the restored hierarchical order was enforced by a more centralized state than in the early phases of feudalism, and it incorporated a subaltern form of peasant self-government.

If the Japanese version of the lord-peasant relationship differs from the European one, this applies *a fortiori* to religious institutions. The only approximate equivalent of the 'ecclesiastical society' of medieval Europe was the Buddhist establishment, and its history took a very different course. After the Taika reform, it was integrated into the imperial state, but as a subordinate component; the notion of a 'Buddhist state' is inappropriate, because it obscures the specificity and primacy of the imperial institution and its symbolism (the transfer of the capital from

Nara to Kyoto may have been a defensive measure against the rapidly growing power of the Buddhist monasteries). With the rise of the military aristocracy and the eclipse of the imperial court, the Buddhist establishment became a more autonomous force. The major monasteries accumulated wealth and power; they even built up their own armed forces and put up a fierce resistance to Nobunaga's unification policy in the sixteenth century. The last phase in the history of the Buddhist establishment as an autonomous power structure contrasts most clearly with the European pattern: an ecclesiastical institution became one of the main obstacles to a state monopoly or at least an effective central control over the means of violence. The result of its defeat was a far-reaching 'redefinition of the relationship between Buddhism and the state',[44] more radical than any measures taken by either the Catholic or the Protestant variants of absolutism in Europe.

In contrast to the temporary ability of religious institutions to compete with the political centre and obstruct the process of state formation, urban society played a much more subordinate role. In comparison with Europe three essential aspects were missing in Japan: the international commercial network which constituted the background to the development of European cities during the high Middle Ages, the project of the autonomous urban community as opposed to both imperial and feudal images of the social order (there was no Japanese equivalent of the principle *civitas sibi princeps*), and the cultural legacy of the city-states of the ancient world. Despite the upheavals of the fifteenth and sixteenth century, the urban component of Japanese society was thus – on the whole and in the long run – much more effectively subsumed under feudal principles of organization than its European counterpart. The other side of the coin was, of course, the ability of a centralized state which had grown out of the feudal order – the Tokugawa shogunate – to initiate and control an exceptionally rapid process of urbanization.[45]

It remains to consider the relationship between feudalism and monarchy. Bloch had, as we saw, characterized it as coexistence without interpenetration. The description is apt, but incomplete. The ascendant feudal society could not simply coexist with the imperial centre; the coexistence had to be stabilized and institutionalized. The feudal-military elite solved this problem by creating an alternative state which imitated the imperial institution and overshadowed it on the political level, but without ever achieving complete autonomy with regard to cultural legitimation. Bureaucratic techniques of administration originally developed by the Taihō state were adopted by the military state and adapted to a different context. The imitative relationship to the imperial institution went beyond this: to the extent that the later military regimes tried to generate their own legitimating principles, they competed – unsuccessfully in the long run – with the imperial institution in the sacralization of power. Many historians have credited the successive shogunates with advancing the secularization of the state. But the two aspects were not mutually exclusive, and they could even be simultaneously maximized. The most iconoclastic usurper among the military leaders, Oda Nobunaga, brought about a radical change in the relationship between the state and religious institutions, but he also went furthest in the direct divinization of his own power. In comparison with what he seems to have had in mind, later attempts by the Tokugawa shoguns were more systematic but less radical.[46] And the secularizing tendencies of the military regimes never gave rise to a new cultural definition of the state which could have replaced the imperial institution.

On the one hand, state formation in Japan thus involved the development of a political centre that was more purely feudal than the European state ever was, in the sense that it was a more direct expression and conscious creation of the feudal elite. On the other hand, it was also post- or rather trans-feudal from the beginning, because it had to emulate a paradigm which remained outside the feudal framework. The counter-state of the military aristocracy – first created by Yoritomo and his Hōjō successors in Kamakura, transferred to Kyoto by the Ashikaga and destroyed by a combination of structural flaws and excessive ambitions, rebuilt by Nobunaga and Hideyoshi and brought to perfection by the Tokugawa – was from the outset both feudal and centralized and its subsequent development led to further elaboration of both aspects. John W. Hall suggests that the peculiarities of the Tokugawa regime might be seen as the result of a historical paradox: the feudal class consolidated its rule over Japan at a time when the mechanisms and institutions through which it exercised authority were becoming less feudal.[47] But the above analysis suggests that the paradox of centralized feudalism was structural rather than historical; more precisely, the historical paradox was a particular aspect of the structural one.

As a result of a unique concatenation of historical processes, the dynamics of a feudal power structure – a process of fragmentation which gradually gave rise to a new trend towards monopolization – unfolded in the context of other historical developments which pointed beyond the feudal organization of society. While this situation confronted the feudal class with new problems, it also placed new resources at its disposal. But if the challenge can be described in historical terms, the response was a variation of a structural pattern which had been characteristic of Japanese feudalism from the outset. It was, in a sense, both more closed and more open to the non-feudal sectors of the social world than its European counterpart: it imposed its interests and principles more directly on the whole society, but in order to do so, it had to adapt itself more thoroughly to techniques and models of non-feudal origin.

On the other hand, the imperial institution was – in contrast to the other non-feudal components – never subsumed under the feudal order, and its survival constituted a potential threat to the feudal military state. The possibility of a revolutionary restoration, i.e. the use of the imperial institution to legitimize the overthrow of the military state, was built into the constellation described above, and it became more concrete when the imperial institution re-emerged as the unifying symbol of contestatory currents during the later Tokugawa era. But the successive metamorphoses of the military state could also be described as combinations of revolution and restoration. The terminal crises of the Kamakura and Muromachi shogunates show the intrinsic fragility of a power structure that was unable to overcome a problematic relationship to the central symbol of collective identity; in both cases, the hegemony of the feudal class was restored, but the methods and consequences of the restoration had some undeniably revolutionary aspects. The Tokugawa regime proved much more effective in neutralizing threats from inside and outside the feudal order, but the ultimate result of its efforts was to pave the way for a more radical variant of revolutionary restoration.

This outcome can, of course, only be understood in the light of the radical transformation of the samurai during the Tokugawa era: a process of pacification, urbanization, bureaucratization and intellectualization that has no parallel in the history of any other feudal class. The trajectory of the feudal elite in Europe – from military aristocracy to court society – was very different.[48] The last chapter in the history of Japanese feudalism is perhaps the most massive manifestation of its built-in paradoxes; the most structured and durable form of centralized feudalism led to the most far-reaching change in the character of the feudal elite. And if the direction and the long-term consequences of this change differed from the European pattern, the same can be said of the state which presided over it. Many historians have emphasized parallels between the Tokugawa regime and the absolutist state in the West. The opposite view was perhaps most succinctly formulated by Perry Anderson; as he sees it, the Tokugawa regime:

was the negation of an absolutist state. The shogunate commanded no monopoly of coercion in Japan ... It enforced no uniform law ... It possessed no bureaucracy with competence throughout the area of its suzerainty ... It collected no national taxation ... It conducted no diplomacy ... Army, fiscality, bureaucracy, legality and diplomacy – all the key institutional complexes of absolutism in Europe were defective or missing.[49]

But although the weaknesses underlined by Anderson were real, the Tokugawa rulers counterbalanced them through other mechanisms of control and an overall supervision of the social order. As far as the pacification of society and the monopolization of decisive power resources was concerned, its record compares rather favourably with that of other early modern European monarchies. To mention only the most obvious case in point: there was an interval of 136 years between the defeat of the Fronde and the beginning of the French Revolution, as against 229 years between the Shimabara Rebellion and the revolt of Chōshū.

The Tokugawa regime was, in short, neither an exact replica nor a self-limiting and inferior version of European absolutism. It was a distinctive political formation, grounded in a peculiar combination of feudal and non-feudal elements, and although some of its problems were similar to those of the absolute monarchies in the West, they were often solved by different means.

If the ingredients of the Meiji restoration – the social forces that had grown up during the Tokugawa era, the background against which they elaborated their strategies, and the imperial institution that was reactivated to serve as the legitimizing framework for a new phase of state-building – were structurally and historically different from the Western pattern, their combination was bound to inaugurate a highly distinctive line of development. The Westernization of Japan after 1868 was genuine and significant, but it was only the most imitative – and temporarily most obvious – aspect of a more complex and autonomous process of modernization.

NOTES

1 I have discussed the theory of modernity and its relation to the Japanese experience more extensively in another paper: 'The Modern Constellation and the Japanese Enigma', to be published in *Thesis Eleven*, no. 17, 1987.
2 For the first interpretation, cf. Joseph Gusfield, 'The Modernity of Social Movements', in *Societal Growth* (ed.), Amos H. Hawley, New York, 1979, p. 295. For the second, cf. Immanuel Wallerstein, 'Should we Unthink the Nineteenth Century?' Conference paper, Fernand Braudel Center for the Study of Economies Historical Systems and Civilizations, 1986.
3 Cf. Agnes Heller, *A Theory of History*, London, 1982.
4 On the concept of the social imaginary, cf. Cornelius Castoriadis *Crossroads of the Labyrinth*, Brighton 1984, and *the Imaginary Institution of Society*, Cambridge, 1986.
5 Cf. particularly Chalmers Johnson, *MITI and the Japanese Miracle*, Stanford, 1972.
6 E. H. Norman, *Origins of the Modern Japanese State*, ed. by John W. Dower, New York, 1975, p. 111. The unfinished and posthumously published text *Feudal Background of Japanese Politics* refers to 'the extreme decadence of Japanese feudalism' (*ibid.*, p. 323); the proposal to 'illustrate ... how *modern* was the political organization which characterized Tokugawa Japan and conversely, how *feudal* is the 'thought' control, the absence of a charter of liberties, the power of the bureaucrat, particularly of the military branch of government, in contemporary Japan' (*ibid.* p. 322) suggests a

rather different approach, but it seems significant that Norman intended to delete this paragraph.

7 *Ibid.*, p. 118.
8 S. N. Eisenstadt has defended this view in his more recent writings.
9 The most representative example of the first approach is Reinhard Bendix, 'Preconditions of Development: A Comparison of Japan and Germany', in R. P. Dore (ed.), *Aspects of Social Change in Modern Japan*, Princeton, 1967, pp. 27–70. On Japan and Russia, cf. particularly Cyril E. Black *et al.*, *The Modernization of Japan and Russia*, New York, 1975. On Japan and Turkey, cf. Robert E. Ward and Dankwart A. Rustow (eds.), *Political Modernization in Japan and Turkey*, Princeton 1964. The last-mentioned volume contains papers from a conference held in 1962; its avowed purpose was to compare 'two of the non-Western countries which have gone farthest in the direction of modernization – Japan and Turkey ...' (p.V). A quarter of a century later, it seems almost unbelievable that anybody should ever have thought of associating Japan and Turkey in this way. For a more specific attempt to compare the Meiji revolution and the Kemalist revolution, cf. Ellen K. Trimberger, *Revolution from Above: Military Bureaucrats and Development in Japan, Turkey, Egypt and Peru*, New Brunswick, 1978.
10 Cf. particularly Anthony Giddens, *The Nation-State and Violence*, Oxford 1986.
11 Marius B. Jansen, *Japan and Its World – Two Centuries of Change*, Princeton, 1980, p. 11.
12 The question of more specific Japanese analogies to the successive centres of modern European history cannot be discussed here, but it should be noted that institutional similarities and functional parallels do not necessarily coincide. The relatively high degree of urban autonomy achieved by the city of Sakai during the fifteenth and sixteenth centuries invites comparison with the Italian city-states (the Jesuits who visited Sakai in the sixteenth century compared it to Venice). But apart from the fact that the autonomy of Sakai was much more restricted and short-lived, its history was not (as was that of the Italian city-states) an 'experimental' prelude to later developments on a larger scale. In Japan, this role was rather played by the feudal domains of the Sengoku era.
13 Jon Halliday's interpretation of Tokugawa Japan implicitly suggests a much stronger analogy with the European state system: 'The Tokugawa regime cannot be characterized as a centralized autocracy, at least in its initial stages. It was more like a nation-wide truce, with a relatively centralized regime ruling about 20 percent of the country, surrounded by feudal lords of varying degrees of friendliness or hostility to the shogun' (J. Halliday, *A Political History of Japanese Capitalism*, New York, 1975, p. 4). But I find this description very misleading. The daimyo could neither marry nor travel without shogunate approval; they were required to finance various projects of the *bakufu*; they could not erect fortifications without permission; the shogun could transfer them from one fief to another, and they had to spend every second year in the capital and leave their families there as hostages during their absence. This surely suggests a degree of central control far beyond a 'nation-wide truce'. For a good summary of the shogun-daimyo relationship, cf. the introductory pages in Marius B. Jansen, *Sakamoto Ryoma and the Meiji Restoration*, Stanford, 1971.
14 Kenneth Alan Grossberg, *Japan's Renaissance – The Politics of the Muromachi Bakufu*, Cambridge (Mass.), and London, 1981, pp. 13–14. (The quotation in the first sentence is from Jacob Burckhardt, *The Civilization of the Renaissance in Italy*, New York, 1958, vol. 1, p. 107).
15 *Ibid.*, p. 16.
16 *Ibid.*, p. 115.
17 *Ibid.*, p. 2.
18 *Ibid.*, p. 124. It might perhaps be argued that the circumstances of Yoshinori's death (in 1441) are a caution against taking the comparison with his Western contemporaries too far. No Renaissance monarch in Europe was killed in full view of the public during a theatre performance.
19 *Ibid.*, p. 115.
20 It is worth noting that Chikafusa also had some understanding of the economic

dimensions of the conflict. He was clearly aware of the role played by feudal forms of land tenure in the disintegration of the Heian order, and according to George Sansom, *History of Japan*, vol. 2, Stanford, 1961, p. 30, he explained the conflicts of the fourteenth century as the result of 'the claims of an unlimited number of persons on a limited amount of land'. Sansom comments (*ibid.*): 'In that statement, although he could not know it, Chikafusa was summarizing the whole of Japanese history from the beginning down to modern times.'

21 According to Sansom's account of the civil war, some supporters of the Ashikaga favoured a more radical policy; for example, Ko no Moronao is said to have declared: 'If for some reason a King is needed, let us have one made of wood or metal and let all the live Kings be banished.' (*ibid.*, p. 141.) However remarkable this statement may seem in the context of fourteenth-century Japan, it would probably be a mistake to read too much into it. Moronao was clearly one of the most disreputable characters of his age, and there is nothing to show that his visceral rejection of the imperial court and the Kyoto elite had any constructive connotations.

22 Cf. the discussion of this issue in Anthony Giddens, *The Nation-State and Violence*, ch. 3–4.

23 Cf. Herman Ooms, *Tokugawa Ideology*, Princeton, 1985.

24 On the *ikki*, cf. David L. Davis, '*Ikki* in late Medieval Japan', in John W. Hall and Jeffrey P. Mass (eds.), *Medieval Japan – Essays in Institutional History*, New Haven and London, 1974, pp. 221–47.

25 Grossberg, *Japan's Renaissance*, p. 52.

26 Grossberg's book has so far not been widely discussed, but a few words should be said about two critical reviews: by Suzanne Gay in *Journal of Asian Studies*, vol. LXII, no. 3, p. 664–5, and by Peter J. Arnesen in *Journal of Japanese Studies*, vol. 9, no. 2, pp. 385–91. While it would be absurd for a layman to try to arbitrate a debate between specialists, it seems to me that some of the issues involved could be formulated in more general terms. For a comparison of Europe and Japan, it is essential to distinguish between three aspects:
a) the actual achievement of the Ashikaga shoguns; b) the intrinsic logic of their project, which might have been comparable to the strategies of their European contemporaries, even if the real results were much less impressive; c) the significance of both the project and the results in the broader context of Japanese history. The objections raised by the two reviewers are mainly related to the first issue, but they do not distinguish it clearly from the two others. They disagree on some points (for example, Gay claims that the *bakufu* enjoyed 'complete authority in judicial matters', whereas Arnesen says that its 'judicial authority was more closely circumscribed than that of even *the feudal* monarchs of England and France' (p. 391), but in both cases, the general thrust of the argument is to show that the power of the Muromachi shogunate was severely restricted by other power centres and social forces. But Grossberg can admit this without giving up the substance of his thesis; as he sees it, France was far ahead of the other Renaissance monarchies in the West, and Japan was well behind the weakest of them, but the structural affinities are nevertheless undeniable. The crucial question is, as I have tried to show, the character and the coherence of the project, rather than the solidity of the results.

27 Arnesen, *op. cit.*, p. 387.

28 Sansom, *History of Japan II*, p. 235.

29 Cf. Marc Bloch, *Feudal Society*, vol. 2, London, 1962, pp. 446–7.

30 Cf. Peter Duus, *Feudalism in Japan*, New York, 1969, pp. 71–3.

31 J. R. Strayer, 'The Tokugawa Period and Japanese Feudalism' in John W. Hall and Marius B. Jansen (eds.), *Studies in the Institutional History of Early Modern Japan*, Princeton, 1968, p. 6.

32 Cf. Bloch, *Feudal Society*, p. 382. Other authors have drawn attention to this phenomenon, but without adding much to Bloch's analysis. F. Jouon des Longrais (*L'Est et l'Ouest – Institutions du Japan et de l'Occident comparées*, Paris, 1958) begins his discussion of Japanese feudalism by describing the 'extra-feudal situation of the emperor

and his court' as the most distinctive characteristic, but his discussion of it is very brief, and in the concluding summary, he makes no mention of it.

33 Cf. particularly Perry Anderson, *Passages from Antiquity to Feudalism* and *Lineages of the Absolutist State*, London, 1974.
34 Joseph R. Strayer – Rushton Coulborn, 'The Idea of Feudalism' in E. Coulborn (ed.), *Feudalism in History*, Princeton, 1956, pp. 4–5.
35 *Ibid.*, p. 6.
36 A. Guerreau, *Le féodalisme – Un horizon théorique*, Paris, 1980, p. 180.
37 For the most concise summary of this argument, cf. Fernand Braudel, *Civilisation matérielle et capitalisme*, vol. 2, p. 433, Paris, 1979. Braudel draws on Gurvitch's reading of Bloch.
38 Cf. Jouon des Longrais, *L'Est et l'Ouest*, especially pp. 119–22.
39 Duus, *Feudalism in Japan*, p. 37.
40 Cf. Murakami Yasusuke, '*Ie* Society as a Pattern of Civilization', *Journal of Japanese Studies*, vol. 10, no. 2 (1984), pp. 281–363.
41 John W. Hall, *Government and Local Power in Japan, 500 to 1700*, Princeton, 1966, p. 11.
42 Anderson, *Lineages of the Absolutist State*, p. 413.
43 Sansom, *History of Japan*, 2, p. 206.
44 Neil McMullin, *Buddhism and the State in Sixteenth-Century Japan*, Princeton 1984, p. 5. The above summary is mainly based on McMullin's analysis.
45 On urbanization during the early Tokugawa era cf. James L. McClain, *Kanazawa – A Seventeenth-Century Japanese Castle Town*, New Haven, 1982.
46 Cf. Ooms, *Tokugawa Ideology*; on Nobunaga, pp. 21–45.
47 Cf. John W. Hall, 'Feudalism in Japan – A Reassessment', in Hall-Jansen, *Studies in The Institutional History of Early Modern Japan*, p. 48.
48 Cf. Norbert Elias, *The Civilizing Process*, vol. 1, New York, 1978; vol. 2, Oxford, 1982; *Court Society*, Oxford, 1983.
49 Anderson, *Lineages of the Absolutist State*, pp. 416–17.

14 The concept of modernization re-examined from the Japanese experience

KAWAMURA NOZOMU

Classical sociology consists of social dynamics and social statistics. In social dynamics, the sociological dichotomy based upon the two-stage model of social development gained currency. The only notable exception was Auguste Comte who proposed a three-stage model of development where human knowledge develops from the theological through the metaphysical to the empirical stage and society correspondingly transforms from the militaristic through the legalistic to the industrial stage. Herbert Spencer advanced a two-stage developmental model from militaristic to industrial society. This type of dichotomous model is discernible in H. J. S. Maine's model of development from status to contract, in Ferdinand Tönnies' model from *Gemeinschaft* to *Gesellschaft*, and in Emile Durkheim's model from mechanical to organic solidarity.

Further, Max Weber who interpreted the change towards *Gessellschaft* as a process of rationalization suggested the transformation from traditionalism to rationalism as a major trend and put forth the notion of *disenchantment*. In this sense, he too follows the tradition of the sociological dichotomous model. Talcott Parsons' model of pattern variables is another case in point. To the extent that it was based on Tonnies' dichotomy of *Gemeinschaft* (community) and *Gesellschaft* (association), a developmental direction – (1) from affectivity to affective neutrality, (2) from particularism to universalism, (3) from functional diffuseness to functional specificity, (4) from ascriptive orientation to achievement orientation, and (5) from collectivity orientation to individual orientation – was implied in the Parsonian model. In the case of Karl Marx, a five-stage model – composed of primitive communism, slavery, feudalism, capitalism and socialism (communism) – is well known. In this case too, however, the dichotomy of precapitalism and capitalism was significant. For example, in his *German Ideology*, Marx observes as follows:

Here, therefore, emerges the difference between natural instruments of production and those created by civilization. The *field* (water, etc.) can be regarded as a natural instrument of production. In the first case, that of the natural instrument of production, individuals are subservient to nature; in the second, to a product of labour. In the first case, therefore, property (landed property) appears as direct natural domination, in the second, as domination of labour, particularly of accumulated labour, capital. The first case presupposes that the individuals are united by some bond: family, tribe, the land itself, etc.; the second, that they are independent of one another and are only held together by exchange. In the first case, what is involved is chiefly an exchange between men and nature in which the labour of the former is exchanged for the products of the latter; in the second, it is predominantly an exchange of men among themselves. In the first case, average human common sense is adequate – physical activity and mental activity are not yet separated; in the second, the division between physical and mental labour must already have been effected in practice. In the first case, the domination of the proprietor over the propertyless may be based on personal relations, on a kind of community; in the second, it must have taken on a material shape in a third party – money.[1]

Marx proposed a two-stage development theory in which communal society is transformed into civic society, a society of alienation. He further suggested a transition from civic society to communist society where alienation is overcome in favour of humanitarianism. Accordingly, it would be more accurate to view the Marxist theory as a three-stage model from communal through civic to humanitarian society, the last phase being the revival of communities. On this point, the first draft of Marx's letter to Vera Zasulich dated 18th March 1881 is most suggestive.

In this draft, Marx viewed the Russian commune as a significant basis for the construction of Russian socialist society in the future. Generally, he suggested, the social institutions in opposition to the commune were facing a crisis. According to him, this crisis indicated that these institutions were doomed to be abolished and that modern societies have no choice but to return eventually to the communal 'pre-archaic form'. Marx argued in favour of an American writer, Morgan, that the 'new system' towards which modern societies were proceeding would be the revival of the 'higher form of pre-archaic society'.

In this connection, it should be pointed out that, when F. Tönnies suggested a transition from *Gemeinschaft* to *Gesellschaft*, he had in mind the rebirth of *Gemeinschaft* of a higher form in the future. In the vocabulary of Parsonian pattern variables, affectivity, particularism, ascription, functional diffuseness and collectivity orientation, which were played down in modern societies, would be resurrected in the future.

Let us examine this point focussing upon the diffuseness-specificity axis of the pattern variables. Marx, for example, in *Capital* makes a distinction between industrialization in general and its capitalist form.[2] When one refers to industrialization, one should distinguish capitalist industrialization. In contrast, modernization is not regarded as going through two different processes. Modernization in a narrow sense may be seen as identical in content with the movement towards capitalism and materialism. Capitalist 'modernization' here is not distinguished from socialist 'modernization'.

In the preface to the first edition of *Capital*, Marx made a reference to the fact that this study was based primarily upon the example of England. He wrote: 'In this work I have to examine the capitalist mode of production ... Up to the present time, the classic case in England.'[3] He further notes: 'If ... the German reader shrugs his shoulders at the condition of the English industrial and agricultural labourers, or in optimist fashion, comforts himself with the thought that in Germany things are not nearly so bad; I must tell him "This concerns you too"!'[4]

In *Capital*, Marx addressed himself to the law of the capitalist mode of production itself. His focus was not upon the level of development of that mode. As he put it: 'The country that is more developed industrially only shows, to the less developed, the image of its own future.'[5] He further argued:

In all other spheres, we, like all the rest of Continental Western Europe, suffer not only from the development of capitalist production, but also from the incompleteness of that development. Alongside of modern evils, a whole series of inherited evils oppress us, arising from the passive survival of antiquated modes of production, with their inevitable train of social and political anachronisms. We suffer not only from the living, but also from the dead.[6]

In Japan prior to the outbreak of World War Two, Marxist social scientists engaged in a debate over the nature of Japanese capitalism. They disagreed over the historical characteristics of the emperor system and the landlord system established after the Meiji Restoration. With regard to the emperor system, these scholars debated whether it should be regarded as a kind of bourgeois monarchy or as a kind of absolute monarchy. Concerning the landlord system, one position argued that the high rate of tenancy in kind resulted from excessive competition between small cultivators and should be classified as a form of capitalist rent. Another position viewed it as product rent, a form of feudal rent. The central focus of this debate was upon the under-developed nature of Japanese capitalism.

The so-called 'culture unique to Japan' and 'the Japanese national polity unparalleled in any other nation in the world' were praised during World War Two in the context of the rise of ultra-nationalism. A few militant materialists opposed this trend firmly. Tosaka Jun, one of these militants, criticized those Japanese nationalists who presented 'Japanism' as an *a priori* principle and argued in his *Nihon ideorogī* (Japanese ideology) that 'Japanese spiritualism is . . . like Bauchredner (the ventriloquist) – voice without substance . . . for Japanism, what is called the Japanese spirit, or even "Japan" itself, is not an object to be explained but simply the method or the *principle* which is used rather arbitarily to account for some other phenomena'.[7] On this point, Tosaka goes on to detail his argument as follows:

The situation in which 'Japaneseness' is used as the *principle of explanation* is entirely opposite to the situation in which it is considered as the specific problematic that ought to be accounted for by various other principles. When 'Japaneseness' is discussed, it tends to be treated as a principle of explanation where the abstract independent entity 'Japaneseness' is assumed to precede what is international – despite the fact that 'Japaneseness' should be examined as a concrete link in the chain of the international context. The crucial test to find out whether the 'Japaneseness' argument is conservative/reactionary or specifically progressive is where it stands in relation to what is international.[8]

In this argument, the national is appraised in the context of internationalism. That is, Tosaka regarded 'that which is Japanese' as 'progressive' in so far as it was a specific manifestation of 'that which is international'. The Japanese Left today tends to oppose the national anthem 'Kimigayo' while singing *The Internationale* and object to the national flag the 'Rising Sun' while flying the 'Red Flag'. The position of the Left in relation to nationalism was not clarified in pre-war years. Tosaka takes the position of universalistic internationalism when he criticizes the anti-technological and anti-materialist position of 'Japanese spiritualism'. For example, Tosaka points out: 'Various categories which Japanists are fond of using – such as Japan, the Japanese people, Japanese spirit, Japanese agriculture, the way Japanese gods, the emperor and all other miscellaneous expedient categories – may on the surface appear to be directly connected with the daily lives of the Japanese masses but in fact are neither associated with nor related to them.'[9] Tosaka accuses Japanism of being the ideology of 'primitivization'. He claims that 'Japanism which is itself the ideology product of this highly advanced monopoly capitalism attempts to make the present situation "ideologically primitive".'[10]

In a similar vein, Kōzai, in discussing 'Japaneseness', argued that what are described as national characteristics tended to develop historically. He maintained that the concepts of the nation and national characteristics were nothing but historical categories. He further observed:

Nevertheless, analysts often fail to take such a historical perspective in considering our national characteristics. For example, some find the trait of 'Japanese spirit' or

'Japaneseness' in the elegant mentality of the Japanese who have a 'capacity for feeling the pathos of nature' [*mono no aware*]. Others observe such characteristics in the courageous 'soul of Japan' [*yamato damashii*]. But, are these really the eternal essence of the Japanese? The answer must be firmly negative. These features are the abstractions of particular feelings existent in a particular social class under a particular historical condition in the name of the entire nation.[11]

Kōzai took the position of proletarian internationalism and historical materialism in opposition to the notion of 'Japaneseness' and made the following statement of protest:

The worldview based upon historical materialism will be neither transformed nor overwhelmed by 'Japaneseness'. For historical materialism is the philosophical science which is based upon a truly international foundation disregarding national discrepancies. Instead of being deformed by 'Japaneseness', historical materialism accounts for it. Historical materialism will be neither castrated nor nullified by 'Japaneseness'. This is because historical materialism arose from the agony of social contradiction in Japan and bears the responsibility of opposing, rather than being emasculated by, 'Japaneseness'.[12]

These statements represent the critical viewpoints of pre-war Japanese materialists and Marxist philosophers against 'Japaneseness' and 'Japanese culture'. What requires attention is that this position is methodologically congruent with the several criticisms levelled at Ruth Benedict's *The Chrysanthemum and the Sword* immediately after the war, despite the differences between the scholars in their world view and ideological orientation.

Take, for example, Minami Hiroshi, who wrote an article entitled 'From the position of social psychology' in the special issue of the journal, *Minzokugaku kenkyū* (Studies of ethnography) devoted to 'the lessons of Ruth Benedict's *The Chrysanthemum and the Sword*' (May 1950). He argued that the fundamental weakness of this work was that it was 'too static and a-historical to capture the real dynamics of Japanese social psychology'. He further maintained, 'in order to grasp the complex dynamics of the Japanese living in contemporary Japan, it is not appropriate to think of general "Japanese" nor is it adequate to consider, as Benedict does, the personality giant, as it were, that is the individual embodying the aggregate of personalities and behaviour traits of persons who belong to a variety of social classes and strata'.[13] This view represents Minami's position in 1950 and does not conform to the way in which he portrayed the 'Japanese' in his *Nihonteki jiga* (The Japanese self) published in 1983.

R. P. Dore also comments upon Benedict's definition of *giri* as being 'particularly Japanese', saying that '*giri* behaviour' and '*giri* relations' operate in Western societies. Dore presents an example: 'We really ought to go and see Auntie Mabel when we are in London. She is a bore, but she will be upset if we don't.' Dore argues that this is 'a perfect example of a *giri*-act and a *giri* relation'.[14] He notes further: 'the word *giri* has had many uses at different historical periods and in different sub-cultures in Japan' and accuses Benedict of ignoring these differences and seeking 'to create a unified "category" of moral obligation'. Dore argues: 'The method is a logical extension of her basic assumption – surely a mistaken one – that there is such an entity as a homogenous "Japanese culture" or "Japanese culture pattern" which persists through time and pervades all regions and all social classes.'[15]

At the time when Dore wrote *City Life in Japan* (1958), he assumed that 'the course of development was to make Japanese society in more and more respects like English society' and that 'the goal in the direction of which Japanese society might be thought to be moving is ... "a society which differs from none of the Western industrial societies more than they differ from each other"'.[16] Dore has since changed his view and in his *British Factory – Japanese Factory* (1973) suggested the possibility that the British system is moving 'in a Japanese direction'.[17] Britain 'began grudgingly to accept a new collectivism', but Britain remains differentiated from 'a society like Japan which jumped from feudal forms of corporatism to a modern form of corporatism without ever experiencing either the sturdy indifference to one's neighbour or a thorough-going laissez-faire market economy'.[18] Yet, it is noteworthy that Dore has, in this study, pointed to the possibility of the 'Japanization' of British society.

On the other hand, in his *City Life in Japan* published in 1958, Dore warned against an 'exclusive preoccupation with' social change which 'would run the danger of ignoring the particular differentia of Japanese society, in particular those characteristics of outlook and way of life'.[19] He noted that this 'book does attempt, if not to give an analysis of Japanese "national character", at least to convey something of the flavour, the "essential Japaneseness" of life in Tokyo'.[20] Specifically, he tried to take up those characteristics 'more amenable to a "cultural" than to a "structural" approach and as easily portrayed in a novel as in the form of scientific treatise, which make a Frenchman different from an Englishman, and a Japanese different from both'.[21] However, as long as Dore takes this approach, he should not be entitled to criticize Benedict for assuming that there is a 'Japanese culture' or a 'Japanese culture pattern'.

In this respect, C. Douglas Lummis, in his *A New Look at the Chrysanthemum and the Sword* (1982), pointed out that the conclusion of

Benedict's book was 'that it was a great benefit for Japan to have lost the war'.[22] This, on the other hand:

places Japanese who believe it in a terrible dilemma, since it stipulates that those who seriously seek freedom and democracy must turn away from their national tradition and look to the West, and particularly to the U.S.; whereas those who seriously seek to preserve Japan's national tradition and identity must adopt the anti-democratic ideology of the militaristic period. This kind of Occupation mentality has great power even today.[23]

When Kōzai took up the issue of *wakon* (the Japanese spirit), he attempted to integrate the establishment of the national tradition and identity with the position of democracy and freedom. In his 'Notes on *wakon*', Kōzai quotes his own sentences published during the war and confirms that ' "Japaneseness" was nothing but the sole ideological core which glorified and sanctified that war of aggression'.[24] At the same time, he advocates, 'the genuine *wakon* exists only on the side of the Japanese who fight for the realization of democracy and peace'.[25] Kōzai also points out that 'state power based on the emperor system since the Meiji period suppressed people's freedom and rights under the anti-popular slogan of *yamato damashii* (the Japanese soul) as opposed to mass-based *wakon*'.[26]

Wakon originally meant 'spiritual power' (*tamashii*) as distinguished from 'Chinese learning' (*karazae*). In *Konjaku monogatari*, one can find the phrase: 'a person with much learning but without *yamato damashii*'. Based on these observations, Kōzai interprets *wakon* not as pedantic ability but as 'sensibility and prudence in life'. According to Kōzai, as the phrase, 'ability of reading Japanese' (*wadoku no sai*), indicates pedantic ability including not only the Chinese version (*kansai*) but the Japanese version (*wasai*) as well.[27]

This dichotomy between *wakon* (Japanese spirit) and *kansai* (Chinese learning) was transformed into that between *wakon* and *yōsai* (Western learning) after Sakuma Shozan, a philosopher at the end of the Tokugawa period, used the phrase 'morality in the East and art in the West'. Kōzai points out that it is often alleged that 'these philosophers were modern with respect to science and technology but pre-modern and feudalistic with regard to morality'.[28] Opposing this allegation, Kōzai argues that 'their efforts were devoted to the eventual maintenance of the independence of the Japanese nation in confronting the European powers'. He observes that the philosophical substance of the expression, 'morality in the East', consists of the 'spirit of the independence of the Japanese nation', namely, *wakon*.[29] According to Kōzai, '*wakon* as advocated towards the end of the Tokugawa period included within itself the core of national resistance against the colonialism of the European powers'.[30]

Kōzai suggests that *wakon* did not mean pre-modern spirit nor did *yōsai* imply science and technology particular only to the West. Learning is nothing but the 'realization of modern and universal truth'. Kōzai argues against praising *yōkon yōsai* (Western spirit and Western learning) in opposing *wakon yōsai* (Japanese spirit and Western learning).[31] He expressed the following view: 'Is it not the case that *yōkon* (Western spirit) in *this context* represents the spirit of powerful capitalist nations which were impetuously dashing towards the imperialist stage by implementing the 'divide and rule' programme.'[32] Aruga noted: 'To the extent that *yōkon* is defined *in this sense*, it must be said that Japanese capitalism rapidly transformed itself into the mode of *yōkon* through the Sino-Japanese War and the Russo-Japanese War.'[33]

The foregoing discussion revolves around the contrast between Japan and the West and that between technology (learning) and spirit (soul). With the combination of these two axes, four types of orientation can be conceptualized: as Figure 1 shows, they are *wakon wasai, yōkon yōsai, wakon yōsai* and *yōkon wasai*. Furthermore, unless technology and spirit are assumed to progress separately, the Japanese situation is that of *wakon wasai* and the Western counterpart *yōkon yōsai*. However, when a country imports civilization and culture from abroad (from the West in the case of Japan), it tends to be assumed that it is easier to transplant technology and civilization than spirit and culture. On the basis of this assumption, the slogan of *wakon wasai* was advocated in the case of Japan. In contrast, *yōkon wasai*[34] becomes a central concern for those, like Ezra F. Vogel, who see the Japanese success as the possible lesson for Americans.[35]

The reason why Vogel emphasizes *wasai* is that he realized that 'Japanese success has less to do with traditional character traits than with specific organizational structures, policy programmes, and conscious planning'.[36]

Yet, irrespective of whether the catch-phrase may be *wakon yōsai* or *yōkon wasai*, these remain slogans of a transitional period and will eventually be uplifted to a new stage of *wakon wasai* or *yōkon yōsai*. One cannot ignore the significance of the concepts as long as *wakon yōsai* has been an issue in the context of 'modernization' and *yōsai* represented modern universal civilization. If *wasai* is pre-modern and traditional technology and *yōsai* is modern and universal technology, *wakon* must be pre-modern and traditional spirit or culture which has produced *wasai* and should be distinguished from *yōkon* which is assumed to be the modern and rational spirit. At least, *wakon* is regarded as being qualitatively different from M. Weber's 'the spirit of capitalism'.[37] While Kōzai ignored this point completely, *wakon yōsai* in this context presented itself

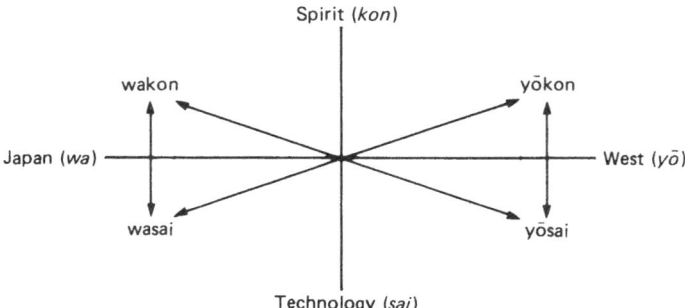

Figure 1 Four types of orientations concerning 'technology' and 'spirit'

either as a transitional state to progress into *yōkon yōsai* or as an opposition to it.

In January 1983, Robert N. Bellah delivered a public lecture entitled 'Cultural Identity and Modernization of Asia' at an international seminar commemorating the one hundredth anniversary of Kokugakuin University. In this lecture, Bellah supported the slogan of *wakon yōsai* as a position to provide a harmonious solution to the contradictory requirements of 'modernization' and 'tradition'.[38] According to him, *wakon* is the sole agent determining national purpose and 'modernization' is simply a means to achieve it. Bellah argued that 'modernization is what Weber called rationalization, namely, increasing effectiveness of means, but has nothing to do with the goal'.[39] Opposing the position which views 'modernization' as imperative, Bellah maintained: 'Probably, we are coming close to the age when tradition determines the goal and modernization provides means to achieve it. Where modernization threatens the goal, modernization itself must be regulated.'[40]

However, it is difficult to grasp 'modernization' only as means. We have already questioned Kōzai's position where *yōkon* is regarded as the spirit of imperialism to facilitate colonization programmes and *yōkon yōsai* is seen as technology which has universal attributes. As long as *yōkon* represents a modern universal and rational spirit, 'modernization' must mean a process in which *yōkon* is firmly established. In this sense 'modernization' which corresponds to what Weber calls 'rationalization' and 'disenchantment' comprises the universalistic content of civil autonomy. 'Tradition' in this context manifests itself in each locality or nation.

In this respect, one must positively evaluate Nakane Chie's attempt to make a distinction between such universal processes as 'modernization' and 'industrialization' on the one hand and the cultural characteristics particular to each nation on the other and to highlight the relative autonomy of the cultural features. Nakane accounts for differences in the theories of the social sciences between Japan and the West by attributing them metaphorically to differences in the criteria of scaling between the Western centimetre measure and Japanese *shaku* measure. She then opposes the mode of analysis in which Japanese realities are examined by the use of Western theories and methods.[41] Nakane points out that, when Japanese *kimono* cloths are measured with a centimetre cloth measure – that is, when Western theories are applied to Japanese society – some fractions, namely inexplicable portions, must be dealt with. Nakane says: 'Japanese social scientists have so far expediently handled these parts by claiming that they represent Japanese *feudal remnants* or Japanese *backwardness*.'[42]

Nakane maintains that the same level of 'industrialization' does not necessarily entail the same patterns of social structure. Despite the common large-scale transformation associated with 'industrialization', she argues, 'surprisingly significant differences are discernible between societies' in the area of substantive interpersonal relations and these differences are attributable to differences in the tradition of each society'. In summarizing these points, Nakane makes the following observations:

> In modernization theories advocated so far, the view that the so-called base determines superstructure gained currency. Accordingly, the predominant assumption was that, as the level of Japanese industrialization reaches that of the West, her social arrangements must become akin to that of the West as well. Thus, those social phenomena which are not observable in the West tended to be *grouped together* and to be labelled as Japanese backwardness or the remnants of feudalism.[43]

> Admittedly, interrelations between base and superstructure do exist. However, it would be too simplistic to assume that, when Japan industrializes, the Japanese ways of thinking and the Japanese mode of interpersonal relationship change into those of the West, or even come closer to them.[44]

According to Nakane, 'modernization' should be equated with 'westernization'. If there is British-style 'modernization', there must be Japanese-style 'modernization' as well. However, the central point is that, while 'modernization' is an abstract and universal process, its concrete manifestation takes different forms depending upon the country. The crux of the matter is not how to make a distinction, if any, – as Nakane attempts – between the part which has undergone changes as a consequence of 'modernization (or industrialization)' and the part which has remained intact despite such processes. Nakane, for example, points out that 'the Japanese working in all types of modern organization are significantly different from their counterparts in the "West" and have not essentially changed since the Meiji period'.[45] In so claiming, she appears to believe that she has refuted the Marxist theory, that the base conditions the superstructure. It is rather odd that, while Nakane argues in favour of 'essential Japaneseness', she qualifies it with the proviso of 'at least since the Meiji period'. When one talks about a difference between the Japanese language and the English language or between the Japanese people and the English people, one would assume that an essentially identical distinction can be made independently of time – be it pre-modern or modern. No matter how 'modernization' may be defined, it does not change the colour of Japanese eyes into that of English nor does it make English the mother tongue of the Japanese.

Nakane draws attention to 'the *relationships* between individuals, between an individual and a group and between groups composed of individuals', and argues that 'these *relationships* are the most unchangeable of various elements constituting a society (or culture)'.[46] Unfortunately, Nakane provides neither theoretical nor empirical proof of the unchangeability of these 'relationships' and 'social structure' in their totality. Further, when reference is made to such notions as 'elements' or 'social structure', what they designate differs depending upon whether they are grasped as analytical or substantive concepts. Nakane tends to describe such notions as 'elements' and 'relationships' in substantive terms while they are the products of her analytical scheme which she uses as a point of departure. Incidentally, the Marxist concepts of base and superstructure should never be used as the substantive ones she makes them out to be.

Nakane defines 'basic principles' embodied in a particular society as 'social structure' and regards social anthropology as the academic discipline to examine such an unchangeable entity. According to Nakane, the concept of 'social structure' as defined in social anthropology is different from that as used in sociology, economics and history; the former is 'far more abstract' and the latter refers simply to the 'institutional fabric' of society. Here, her position of assuming 'basic principles' *a priori* must first be questioned. Nakane argues that 'social anthropology does not use "Western societies" as the yardstick in cross-societal comparison in contrast with sociology where theories derived from "Western societies" are always used as the standard for comparison'.[47]

Nakane's definition of 'social structure' and her description of the academic characteristics of 'social anthropology' are both extremely inarticulate. However, her perspective should be positively assessed to the extent that it provided Japanese studies with an approach different from the position in which Japanese society was reduced to a universal and abstract capitalist society and the central focus of analysis centred upon the magnitude of gap in development stages between Japan and advanced capitalist societies. However, a pre-war Japanese sociologist made the same points in a more precise way and on the basis of empirical studies, although Nakane makes no reference to his work. The sociologist, Aruga Kizaemon, who published *Nihon kazoku seido to kosaku seido* (Japanese family system and tenancy system) in 1943, observed:

Too many scholars have hitherto tended to regard the establishment of a Japanese capitalist economy as being essentially Western. Even those who attempted to bring some Japanese ingredients into analysis were apt to regard them openly as late developing variants or as unorthodox modes.

Culture mediates between the universal and the particular ... With regard to the capitalist economy, an analytical distinction must be made between its economic functions and the characteristics of social relations which operate the economy. The economic functions represent the aspect in which differences between particular nations are increasingly ironed out. The characteristics of social relations point to the aspect in which such differences are retained.[48]

The 'social relations' mentioned here are concrete realities observable through specific life styles. Aruga regarded what has persisted despite expanding capitalism as the national characteristics of 'social relations'.

Aruga starts his analysis by studying the compulsory labour of *nago*, a category of tenants particular to Japan. He observes that 'the compulsory labour with which *nago* provided *nanushi* (landlords) did not originate as a form of ground rent but as a form of labour within the framework of the relationship between *oyakata* (literally, parent side) and *kokata* (child side) in the extended family business. In this form of labour, *kokata* supplied *oyakata* their labour service. Its most primitive form was, in substance, intra-family labour'.[49] According to him, the service labour

was not the remnant of the payment of feudal land rent by labour ... but the direct residue of family labour cultivators prior to the implementation of land rent in the form of labour. Since family labour originated from the *kokata*'s service to the *oyakata* in the *dōzoku* (family) group, the characteristics of this social relationship were conditioned by the national characteristics of Japan.[50]

For Aruga, the national characteristics of social relations in Japan lie in the fact that the land ownership relations between landlords and tenants take the form of the *dōzoku* genealogical relationships between *oyakata* and *kokata* or between the head family and the branch family. Aruga clarified that the landlord-tenant relationship originated from an arrangement within the extended family business where *kokata* provided labour for *oyakata*. In the process in which *kokata* became increasingly independent, a transfer of exclusive land usage rights took place in such a way that each registered *kokata* cultivator of land was supposed to pay rent. In this context, the complex extended family system of *oyakata* was dissolved, *kokata* acquired independence, and land rent came into being. Therefore, Aruga argues, feudal rent in Japan originated not from the form of labour rent but product rent. For example: 'The tenancy system was initially based on the transfer of exclusive land usage rights. But, the system gradually shifted to a lease relationship. This was because the *oyakata* extended family business, from which the tenancy system originated, collapsed.'[51]

Aruga characterizes the Tokugawa feudal system in terms of the independence of *kokata* tenants and the collapse of the extended family business managed by *oyakata*. In contrast to the Marxist historians and economists who regarded the *oyakata-kokata* relationship (i.e., Japanese landlord-tenant relationship) as a feudal relationship based upon labour rent, a form preceding product rent, Aruga regarded it as based on *dōzoku* kinship or genealogical relations (the foundation of which is the *oyakata* extended family business). Aruga would not have argued so forcefully over the national characteristics if he had not observed the persistence of the *dōzoku*-type social relations between *oyakata* and *kokata*, which was supposed to be a pre-feudal social relation according to the Marxist model. His empirical inquiry has shown that these relations did exist in Tokugawa Japan despite the establishment of the feudal system, the class conflict between feudal lords and owner farmers, and the spread of the local network of owner farmers across villages and *kumi* groups.

In this sense, for Aruga, the so-called 'Japaneseness' does not point to what Marxist scholars saw as the remnants of feudalism still present in Japanese capitalism. Rather, Aruga saw 'Japaneseness' in the remnants of pre-feudalism, that is, those of the so-called 'Asiatic' elements, which served as an obstacle to the establishment of full-fledged feudalistic social relations.

In a special issue on 'The lessons of Ruth Benedict's "The Chrysanthemum and the Sword"' of the journal *Minzokugaku kenkyū* (Studies of ethnography) published in May 1950, Aruga contributed a commentary entitled 'The problem of the stratification system in Japanese social structure'. In this article, he maintained that *on* derived from the characteristics of social relations in the Japanese clan system. In order to understand what Aruga means by 'national characteristics', the following passage is particularly relevant:

The superordinate-subordinate relations existed within each of several strata. In the lowest stratum, the landlord-tenant relations operated. In the medieval period, the landlord belonged to the clan of his lord of the manor who in turn subordinated himself to the clan of his superior lord. In this chain of superordinate-subordinate relations, the subordinate rendered service to the superordinate who, in return, provided them with compensations. Since tenancy implied the internal and tentative supply of landownership, it took on the character of the distribution of feudal rights.... *On* covered the entire range of compensations, both material and non-material, which the master gave to his servants and which were closely associated with the stratification system ... This practice did not originate from the warriors' society but from the clan society which was more ancient.[52]

Furthermore, Aruga accounted for the emperor being situated at the apex of the hierarchy from the viewpoint of Japanese national characteristics. According to him, the relationship between the emperor and the people after the Meiji Restoration derives from the 'characteristics of the Japanese clan system'. Aruga maintained that 'unless the characteristics of the clan system were maintained in the social structure, it would have been impossible for the belief to persist, the belief in which the god at the Ise Shrine was placed at the top of the hierarchy of various clan gods. This belief system was akin to that in which the private god of the head family of a clan was interpreted as the public god of the entire clan.'[53] Aruga found that the clan-based characteristics of Japanese social relations was most conspicuous in that the vertical hierarchical order was conditioned by the clan or *dōzoku* genealogical relations.

Aruga, of course, recognized not only vertical hierarchical social relations but also horizontal lateral social relations. He paid special attention to the *kumi*-type associations, as exemplified in the five-person *kumi* system (*gonin-gumi*), which were based on geographical communities. In Aruga's view, however, the egalitarian *kumi* relationship tended to be conditional and temporary. The central thesis of Aruga was that, once a vertical hierarchical order was established in the social relationship between families, the order was inevitably characterized by the *oyakata-kokata* relationship operative in *dōzoku* groups. According to Aruga, these national characteristics are observable within each capitalistic enterprise in Japan and in the hierarchical social relationship between those enterprises. In the last section, an attempt will be made to assess Aruga's argument in the context of modernization theories.

E. H. Norman pointed out in his *Japan's Emergence as a Modern State*: 'In Japan there has been a time-lag between the adoption of a new mode of life and the full maturity of its cultural and psychological expression.' At the same time, he observed that, with the lag being eliminated, 'it will scarcely leave any space for the patriarchal and often genial traditions of its medieval past'.[54] He maintained:

The ideal of feudal loyalty, the patriarchal system, the attitude toward women, the exaltation of the martial virtues, these have acquired in Japan all the garish luster of a tropical sunset. This metaphor is used to suggest that there is a waxing and a waning even in what often appears to be the inherent and inalienable spiritual or cultural tinting of a nation.

If the sunset had displayed the last garish luster in 1941, it should be the case as of 1985 that the sun has completely gone down, the glare has entirely disappeared, and it is pitch-dark. In this sense, Norman's prediction was not completely accurate. Still, this sun is certainly destined to sink sooner or later. Accordingly, if the process of 'modernization' from the pre-modern phase is the only process of any significance, there would be little to be added to the above quoted passage of Norman though he might not have been accurate in his prediction of the exact time point of the sunset.

The crucial point is that the so-called process of 'modernization' includes not only the process from the pre-modern to the modern phase but also that from the modern to the post-modern phase. The Japanese situation is complicated, as Dore's recent view suggests, by the fact that it does not take a two step process *from* the complete dissolution of communities *through* the creation of the autonomous independent individuals *to* the formation of associations of such free individuals. The Japanese pattern is characterized by the tendency for some elements of pre-industrial communities to be revitalized in the process of new group formation in the post-modern phase. In addition, the revitalization is facilitated by two competing positions, that of conservatives and the Right on the one hand and that of progressives and the Left on the other.

Benedict's *The Chrysanthemum and the Sword* shares Norman's sunset analogy. It is often alleged that the title of this book suggests that the 'sword' symbolizes Japanese militarism and the 'chrysanthemum' the culture of new born Japan. On the contrary, the 'chrysanthemum' is the symbol of 'simulated wildness', that of the 'culture of shame' which is about to disappear. The chrysanthemums as portrayed by Benedict are 'grown in pots and arranged ... with each perfect petal separately disposed by the grower's hand and often held in place by a tiny invisible wire rack inserted in the living flower'.[55] On the other hand, according to Benedict,[56] the second generation Japanese in the United States 'have already lost the knowledge and the practice of the Japanese code'.

On the basis of this observation, Benedict argues:

The Western world can neither suppose that the Japanese can take these (new assumptions and new virtues) on sight and make them truly their own, nor must it imagine that Japan cannot ultimately work out a freer, less rigorous ethics. . . . The Japanese in Japan can, in a new era, set up a way of life which does not demand the old requirements of individual restraint. Chrysanthemums can be beautiful without wire racks and such drastic prunings.[57]

The 'sword', on the other hand, symbolizes self-restraints in the transitional period; it is a 'symbol they (the Japanese) can keep in a freer and more peaceful world'. 'The Japanese have proposed "to lay aside the sword" in the Western sense. In their Japanese sense, they have an abiding strength in their concern with keeping an inner sword free from the rust which always threatens it.'[58] The 'sword' is not the symbol of militarism or aggression.

Viewed this way, the reason why Benedict was able to grasp 'Japanese culture' as a pattern is that it was, for her, the setting sun which was doomed to disappear. For Benedict, the American belief in liberty and equality pointed to 'modernity' while the Japanese 'confidence in order and hierarchy' was nothing but 'pre-modern'. However, the point which deserves attention is the fact that Benedict completely ignored the important role which blood relations and pseudo-blood relations played in *ie* social relations and denied the existence of the patriarchical extended family, *dōzoku* and clan groups in Japanese history, while she defined the vertical hierarchical order as the 'pattern of Japanese culture'.

For example, Benedict maintains that clan organization cannot develop without surnames or their equivalents. She argues that 'one of these equivalents in some tribes is keeping a genealogy. But in Japan only the upper classes kept genealogies' and 'only noble familes and warrior (*samurai*) families were allowed to use surnames'.[59] According to Benedict, even when they kept the genealogical record, they did so 'backward in time from the present living person, not downward in time to include every contemproary who stemmed from an original ancestor'.[60] Benedict also makes the point that 'Japan was a feudal country. Loyalty was due, not to a great group of relatives, but to a feudal lord.'[61]

Moreover, Benedict observes:

in Japan there is no cult of veneration of remote ancestors and at the shrines where 'common people' worship all villagers join together without having to prove their common ancestry. They are called the children of their Shrine-god, but they are 'children' because they live in the same territory.... Even in the cemetery ... the identity even of the third ancestral general sinks rapidly in oblivion. Family ties in Japan are whittled down almost to Occidental proportions and the French family is perhaps the nearest equivalent.[62]

This type of error can be found in the following passage where Benedict attempts to account for 'filial piety'.

Japan's veneration is of recent ancestors. A gravestone must be relettered annually to keep its identity and when living persons no longer remember an ancestor his grave is neglected. Nor are tablets for them kept in the family shrine. The Japanese do not value piety except to those remembered in the flesh and they concentrate on the here and now.[63]

It is surprising that, while Benedict took up the 'pattern of Japanese culture', she failed to understand the nature of *ie* and *ie* relations. In this process, Benedict moved in a direction different from Aruga's in addressing herself to the same issue of hierarchical order. In this respect, it is interesting to note that Aruga argued that the establishment of Japanese capitalism was 'induced by the development of entrepreneurs' family business under the guidence of the government',[64] a process different from the Western European pattern in which the Protestant ethic was a key element. Aruga portrayed the process in which 'large merchants took over the administration of government-run firms and managed them as their family business by incorporating the *dōzoku* formula into it'. Aruga stressed that 'the government instructed these large merchants to display their loyalty to the state by following such a process'.[65]

What Bell calls 'post-industrial society' is a society in which knowledge and information become more predominant than industry. It is also called 'information society' or 'high knowledge society'. Bell conceived of 'industrialization' and 'post-industrialization' as partially overlapping and a continuous process. In contrast, societies which A. Toffler thinks are based on the 'third wave' are regarded as being in opposition to societies based on the 'second wave'. Toffler argues that the forthcoming civilization will pave the way for establishing our new code of behaviour, overcoming the restrictions of the 'second wave' industrial societies such as standardization, synchronization and centralization, and subjugating the concentration of energy, wealth and power.[66] In other words, the 'third wave' civilization is described as a society opposite to the nuclear family, standardized mass education, large corporations, large labour unions, a centralized nation state, a pseudo-representative system of politics. It is clear, then, that the 'third wave' society and civilization must be based upon the recognition of the diversity of traditional civilization which has persisted against standardization.

In Japan, on the other hand, conservatives began to argue in favour of the 'age transcending modernity'. This argument is congruent with the theme of the 'return to Japanese tradition' and connected with the claim for the increasing national integration by the emperor, which is in tune with the wartime argument for the 'conquest of modernity'. On this point, the report entitled 'The Age of Culture', compiled by the policy study group organized by the late Prime Minister Ōhira, made the following suggestion:

Post-Meiji Japan has defined itself as being backward and underdeveloped in all respects and made every attempt to westernize, modernize, and industrialize by pattering itself after Western advanced industrialized countries.... However, new demands (i.e., 'demands for culture') are now being called for in pursuit of better conditions in the future.[67]

Mr Ōhira who addressed the group in their first meeting claimed that 'material civilization based on modern rationalism has arrived at the point of saturation. We are now shifting away from the modern age to the age which transcends it — away from the economy-centred age to the culture-centred age.'[68] In this argument, the 'age of Japanese culture' implies the end of 'modernization', 'industrialization' and 'westernization'. The 'age transcending modernity', that is the 'age of culture', is the age in which Japan 'has achieved modernization and taken its rank with Western advanced countries'.[69]

Japan is the only capitalist society which has attained a high level of development without sharing the cultural tradition of Western Europe. As a late-developer capitalist nation, Japan partially consists of under-developed aspects — namely, not only 'feudalistic elements' but also 'Asiatic elements', while being at the same time a modern capitalist country which comprises Asiatic cultural characteristics different from those of Western Europe. Accordingly, the Japanese experience in the process of modernization will be useful in considering an alternative type of modernization different from the West European type.

Moreover, Japanese modernization which is the late-developer type presents problems associated with the situation in which communal relations of various kinds have survived without being completely dissolved. In this regard, two aspects are significant. One is the aspect in which large corporations and monopoly capitalists make positive use of the remnants of communal relations. The other is the aspect in which these relations may serve as a momentum to facilitate the formation of a new communal society. To the extent that modernization in a broad sense comprises the process of post-modernization, Japanese modernization may present new elements to it. If modernization in a narrow sense means capitalistic industrialization, the process through which such industrialization brought about post-industrialization may find a corresponding process with respect to modernization. That is, modernization in a broad sense is pregnant with the process of post-modernization. The Japanese experience will be of use in examining this process as well.

NOTES

1 Karl Marx, *Pre-capitalist Economic Formations*, edited and with an Introduction by E. J. Hobsbawm, London, Lawrence and Wishart, 1964, p. 71.
2 Marx, *Capital*, London, William Glaisher, 1918, chapter 13.
3 *Ibid.*, p. xvii.
4 *Ibid.*

5 *Ibid.*
6 *Ibid.*, pp. xvii-xviii.
7 Tosaka Jun, *Nihon ideorogi-ron* (On Japanese ideology), 1935, in *tosaka Jun zenshū*, vol. 2, Tokyo, Keisō shobō, 1966, p. 292.
8 *Ibid.*, pp. 279–80.
9 *Ibid.*, p. 285.
10 *Ibid.*, pp. 325–6.
11 Kōzai Yukishige, *Gendai tetsugaku* (Contemporary philosophy), 1937, 1964 edition, Tokyo, Mikasa shobō, pp. 35–6.
12 *Ibid.*, pp. 101–2.
13 Minami Hiroshi, 'Shakai shinri-gaku no tachiba kara (From the viewpoint of social psychology)'. *Minzokugaku kenkyū*, vol. 14, no. 4, Tokyo, Seki shoin, 1950, pp. 11–12.
14 Donald Dore, *City Life in Japan*, Berkeley, University of California Press, 1958, p. 254.
15 *Ibid.*, p. 253.
16 *Ibid.*, p. 5.
17 Dore, *British Factory – Japanese Factory*, Berkeley, University of California Press, 1973, p. 419.
18 *Ibid.*, p. 420.
19 *Ibid.*, p. 7.
20 Dore, *City Life in Japan*, 1958, p. 7.
21 *Ibid.*
22 Douglas Lummis, *A New Look at the Chrysanthemum and the Sword*, Tokyo, Shōhakusha, 1982, p. 77.
23 *Ibid.*
24 Kōzai, *Wakon-ron nōto* (Notes on *wakon*), Tokyo, Iwanami shoten, 1984, p. 5.
25 *Ibid.*, p. 93.
26 *Ibid.*, p. 97.
27 *Ibid.*, p. 96.
28 *Ibid.*, p. 32.
29 *Ibid.*, p. 34.
30 *Ibid.*, p. 84.
31 *Ibid.*
32 *Ibid.*
33 *Ibid.*, pp. 83–4.
34 Ezra F. Vogel, ' "Wakon yōsai" no "jidai" (The "age" of "*wakon yōsai*")', *Chūō kōron*, September 1979.
35 Ezra F. Vogel, *Japan as Number One*, Cambridge, Harvard University Press.
36 *Ibid.*, p. ix.
37 Max Weber, *The Protestant Ethic and the Spirit of Capitalism*, London, Unwin University Books, 1968.
38 Robert N. Bellah, 'Bunka-teki aidentitī to ajia no kindaika' (Cultural identity and modernisation in Asia), in Kokugakuin Daigaku Nihon Bunka Kenkyū-sho (ed.), *Ajia no saihakken* (Rediscovery of Asia), Tokyo, Kōbundo, p. 15.
39 *Ibid.*, p. 14.
40 *Ibid.*, p. 24.
41 Nakane Chie, *Tate shakai no ningen kankei* (Human relations in vertically structured society), Tokyo, Kōdansha, p. 15.
42 *Ibid.*, pp. 15–16.
43 *Ibid.*, pp. 17–18.
44 *Ibid.*, p. 19.
45 *Ibid.*
46 *Ibid.*, p. 23.
47 *Ibid.*, p. 21.
48 Aruga Kizaemon, *Nihon kazoku seido to kosaku-seido* (Family system and tenancy system in Japan), Tokyo, Kawade shobō, 1943, postscript 2.
49 *Ibid.*, p. 612.

50 *Ibid.*, pp. 597–8.
51 *Ibid.*, p. 685.
52 Aruga Kizaemon, 'Nihon shakai kōzō ni okeru kaisō-sei no mondai (Problems of the stratification system in Japanese social structure)', *Minzokugaku kenkyū*, vol. 14, no. 4, Tokyo, Seki shoin, 1950, p. 22.
53 *Ibid.*, p. 18.
54 E. Herbert Norman, *Japan's Emergence as a Modern State*, New York, Institute of Pacific Relations, 1940, p. 9.
55 Ruth Benedict, *The Chrysanthemum and the Sword*, Boston, Houghton Mifflin, 1946, p. 295.
56 *Ibid.*
57 *Ibid.*, pp. 295–6.
58 *Ibid.*, p. 296.
59 *Ibid.*, p. 50.
60 *Ibid.*
61 *Ibid.*
62 *Ibid.*, pp. 51–2.
63 *Ibid.*, p. 122.
64 Aruga Kizaemon, *Bunmei, bunka, bungaku* (Civilisation, culture and literature), Tokyo, Ochanomizu shobō, p. 25.
65 *Ibid.*
66 Alvin Toffler, *The Third Wave*, New York, Morrow, 1980, pp. 20–1.
67 Bunka no Jidai Kenkyū Gurūpu, *Bunka no jidai: Ōhira Sōri no seisaku kenkyū-kai hōkoku 1*, (The age of culture: policy study group of Prime Minister Ōhira, Report no. 1), Tokyo, Ōkura-shō insatsu-kyoku, 1980, p. 5.
68 *Ibid.*, p. 21.
69 *Ibid.*, p. 3.

Glossary

abure jigoku あぶれ地獄 — unemployment hell
aidagara 間柄 — interpersonal relations
Aikokusha 愛国社 — Society of Patriots
akebono nippo あけぼの日報 — Dawn Daily
amakudari jinji 天下り人事 — appointment of retired officials to positions

anpan あんパン — sweet bean paste filled bread
Anpo (Ampo) 安保 — Japan–U.S. Security Treaty
ashahi shimbun 朝日新聞 — Asahi News
Ashikaga 足利 — late fourteenth to sixteenth century

asobi 遊び — play
bakufu 幕府 — feudal government
bakuhan 幕藩 — bakufu and domain
banzuke 番付 — ranking list (in sumo)
Beheiren ベ平連 — Peace in Vietnam Committee
Bessatsu shōjo komikku 別冊少女コミック — Comic for girls (separate volume)
bijaku-denpa 微弱電波 — very low power airwaves
Binbō monogatari 貧乏物語 — Tales of Poverty
boke ボケ — passive role in manzai dialogue
bōshin (site work control) 棒心 — site work control
Bukkyō kakushin renmei 仏教革新連盟 — Buddhist Renovation League
Bukō nenpyō 武江年表 — Chronicle of Events in Edo Period
bunka-ron 文化論 — theory of culture
bunmei-ron 文明論 — theory of civilization
buraku 部落 — village
bushidō 武士道 — martial arts

284

busshin ichinyo 物心一如	unity of mind and body
chigo 稚児	page boy
Chōsen rōsō 朝鮮労総	General alliance of Korean Labourers in Japan
Chōshū 長州	Name of Yamaguchi prefecture in feudal times
Chōya shimbun 朝野新聞	Choya News
Chūgoku densan 中国電産	Union of Electrical Workers of Chugoku Region
chūsei 中世	medieval
daimyō 大名	feudal lord
dashijako 出雑魚	small dried fish for fish stock
Datsu-a nyū-o 脱亜入欧	Dissociate from Asia; join Europe
Datsu-ou nyū-a 脱欧入亜	Dissociate from Europe; join Asia
Denryoku rōren 電力労連	Confederation of Japanese Electrical Power Workers
Densan (nippon denryoku sangyō kumiai) 電産（日本電力産業組合）	Confederation of Japanese Electrical Industrial Workers' Union
dezura 出面	a day's wage
dohyō-iri 土俵入	sumo wrestler's ceremonial entry into ring
doya-gai どや街	slum area where day labourers are concentrated
dōzoku 同族	extended family
Edo 江戸	Edo (Tokyo)
eejyanaika ええじゃないか	popular protest movement in pre-Meiji Japan
eta hinin 穢多非人	outcaste minorities
fu 府	urban prefecture
Fussesshō kai 不殺生戒	the 'do not kill' precept
gaijin 外人	foreigner
geisha 芸者	geisha
Gekidan mingei 劇団民芸	Folk Art Theatre Co.
gekiga 劇画	theatrical pictures
gekokujō 下克上	inferior overthrowing superior
genba 現場	site
Genpatsu bunkai 原発分会	Nuclear Power Workers' Union
gentō 現闘	site struggle committee
giri 義理	moral obligation
gohei 御幣	sacred paper

gonin-gumi 五人組	five family neighbourhood unit
Gunzō 群像	Gunzo (magazine)
hamo no kawa はもの皮	sea eel skins
han 藩	domain
hanba 飯場	(construction) site
Heian 平安	period late eighth to late twelfth century
Heiwa kyōgikai 平和協議会	Peace Council
hitei no ronri 否定の論理	theory of negation
Hiyatoi zenkyō 日雇全協	National Council of Day Labourers
Hōchi shimbun 報知新聞	Hochi News
Hokushinsha 北辰社	a late nineteenth-century political society
horumon yaki ホルモン焼	grilled offal dish
ie 家	home, family
Ikkō-ikki 一向一揆	a sixteenth-century rebellion
jimukyokuchō 事務局長	director
jinkaku 人格	character
jisha bugyō 寺社奉行	commissioner of shrines and temples
jishu kōza 自主講座	'self-management seminar'
Jōdo-kyō 浄土教	Pure Land teaching
Jōdoshū minshuka dōmei 浄土宗民主化同盟	Pure Land Sect Democratization League
jōhō-asobi 情報遊び	information play
jōhō-shi 情報誌	information magazine
jōi 攘夷	'expel the barbarian'
joseigaku 女性学	women's studies
jū-hachiban 十八番	hobby
jūgun ianfu 従軍慰安婦	army prostitute
jūjitsu 柔術	jujitsu
kabuki 歌舞伎	traditional drama form
kaikoku 開国	'open the country'
kaimei 開明	enlightenment
kaiseki ryōri 懐石料理	(formal) light meal
Kamakyō 釜共	Kamagasaki Joint Struggle Committee
kanjinzumō 勧進相撲	fund-raising sumo
karaoke カラオケ	singing to recorded orchestra backing

karayuki-san からゆきさん	Japanese prostitute sent to China or Southeast Asia
karazae 唐才	Chinese learning
katakana 片仮名	form of Japanese syllabary
Keichō 慶長	late sixteenth-century reign period
keimō 啓蒙	enlighten (ment)
ken 県	prefecture
kengyō-shufu 兼業主婦	part-time (working) housewife
Kenmu 建武	reign period from 1334 to 1336
Kenyūsha 硯友社	a late nineteenth-century political society
keshō mawashi 化粧廻し	decorative loin-cloth
kiebutsu 帰依仏	reliance on the Buddha
kiehō 帰依法	reliance on the dharma (law)
kiesō 帰依僧	reliance on the sangha (religious community)
kimono 着物	kimono
kindai no chōkoku 近代の超克	'transcending the modern'
kinkonkan 金魂巻	title of recent best-selling novel
Kisaeng 妓生	(Korean) geisha
kodai 古代	ancient
kōgai 公害	pollution
kokata 子方	'child part'
kokkai no kumitate 国会の組立	'Construction of a National Assembly'
koku 石	a measure of rice
kokugi 国技	national sport
Kokumin seikatsu jikan chōsa 国民生活時間調査	Survey of how people spend their time
kokutai 国体	national polity
komusubi 小結	a sumo ranking
kon 魂	spirit
Konminto 困民党	Party of the Poor
konnyaku こん蒻	Japanese yam
konomi 好み	preference
ku 区	ward
kuchikomi くちコミ	oral communication network
Kumamoto shimbun 熊本新聞	Kumamoto News
kumi 組	class, group, section

Kurosumi 黒住　　　　　　　　a nineteenth-century religious sect
kusa zumō 草相撲　　　　　　village sumo
kyōri 教理　　　　　　　　　　teaching
kyūdojin 旧土人　　　　　　　former natives = ainu
kyūdōsha 求道者　　　　　　　seeker after the truth
mabiki 間引　　　　　　　　　'thinning out' (infanticide)
maegashira 前頭　　　　　　　a sumo ranking
mago-uke 孫請　　　　　　　　sub-sub-contractor
Mainichi shimbun 毎日新聞　　Mainichi News
Man'yōshū 万葉集　　　　　　ancient collection of poems
manga 漫画　　　　　　　　　comic
manzai 漫才　　　　　　　　　vaudeville
marubi ㋓　　　　　　　　　　'poor'
marukin ㋜　　　　　　　　　'rich'
matsuri 祭　　　　　　　　　　festival
Meiji 明治　　　　　　　　　　period from 1868 to 1912
Meirokusha 明六社　　　　　　The Meiji Six Society
mikoshi みこし　　　　　　　　portable shrine
Minamata 水俣　　　　　　　　Minamata
mini-komi ミニコミ　　　　　　small-scale (subscriber) journal
minkan hōsō 民間放送　　　　　commercial radio
minpō 民放　　　　　　　　　　commercial broadcast
minpon shugi 民本主義　　　　　people-first-ism
minshū bunka 民衆文化　　　　popular culture
minshushugi 民主主義　　　　　democracy
minyō 民謡　　　　　　　　　　folk song
minzoku gaku 民俗学　　　　　ethnology
moto uke 元請　　　　　　　　original contractor
Muromachi 室町　　　　　　　period from 1392 to 1490
Muryōgikyō 無量義経　　　　　Sutra of Infinite Meaning
musō 無想　　　　　　　　　　meditation without any thought in
　　　　　　　　　　　　　　　mind

nago 名子　　　　　　　　　　serf
nanshoku 男色　　　　　　　　male homosexuality
nanushi 名主　　　　　　　　　village headman
NHK hōsō seron　　　　　　　Opinion Polls Institute of the
chōsajo NHK 放送世論調査所　　Japan Broadcasting Corporation
Nichi Nichi shimbun 日日新聞　Nichi Nichi newspaper
Nichirenshū 日蓮宗　　　　　　Buddhist sect based on Nichiren's
　　　　　　　　　　　　　　　teachings

Nihon ideorogii 日本イデオロギー	Japanese ideology
Nihon shoki 日本書紀	The Chronicle of Japan
Nihon rōdō kumiai hyōgikai 日本労働組合評議会	General Council of Trade Unions of Japan
Nihon heiwa suishin kaigi 日本平和推進会議	Japan Council for the Promotion of Peace
Nihon shūkyōsha heiwa kyōgikai 日本宗教者平和協議会	Japan Peace Council of Religionists
Nihonzan myōhōji 日本山妙法寺	Myohoji Head Temple
Nikkan yūkō kyōkai 日韓友好協会	Japan-Korea Friendship Society
ninpu dashi 人夫出	supplier of day labourers
ninsoku yoseba 人足寄場	navvy's camp
Nippo shinbun 日報新聞	Nippo News
Nippon sōgō kenkyūsho 日本総合研究所	Japan General Research Institute
Nissōkan 日想観	prayer facing the sunset
Nitchū yūkō kyōkai 日中友好協会	Japan-China Friendship Society
noh 能	Noh
ochi 落ち	point of a joke
Okinagusa 翁草	title of the eighteenth-century collection of essays
okusan 奥さん	housewife
Ōnin 応仁	civil war from 1467 to 1477
oroshi baba 堕し婆	old woman specializing in abortion
oyaji 親父	father
oyakata 親方	boss
ōzeki 大関	sumo champion
pachinko パチンコ	Japanese pinball
proresu プロレス	professional wrestling
rakugo 落語	traditional Japanese comic story
rijichō 理事長	president
ringi 稟議	Japanese bureaucratic consultative process
Risshisha 立志社	Aspiration Society
risshō ankoku 立正安国	the establishment of righteousness and the security of the country
Rōdō kumiai kiseikai 労働組合期成会	Society for the Formation of Labour Unions
Rōdō zasshi 労働雑誌	Labour Review (magazine)
Rōmu kyōkai 労務協会	Personnel Management Association
Rōmu hōkoku kai 労務報国会	Association for Service to the State through Labour

Rōnōtō 労農党	Labour-Farmer Party
sakaki 榊	evergreen tree in a camellia family
sakariba 盛り場	amusement quarters
Sangyō hōkoku kai 産業報国会	Association for Service to the State through Industry
Sanjirō 山自労	Sanya District Autonomous Labour Union
Sankirei 三帰礼	three-fold refuge
sashi baba 刺し婆	old woman specializing in abortion
sasu 刺す	to pierce, to stab
sechie zumō 節会相撲	Sumo wrestling at a court banquet
Seichō no ie 生長の家	a religious organization
Seigisha 正義社	Justice Association
Seikyōsha 政教社	Political Education Society
seitai 政体	polity
sekai no bunmei 世界の文明	civilizations of the world
sekiwake 関脇	sumo champion
Sengoku 戦国	civil wars
sengyō-shufu 専業主婦	full-time housewives
Shakaitō 車会党	Rickshawmen's Union
shaku 尺	traditional Japanese measurement unit
shikona 四股名	sumo wrestler's ring name
Shimabara no ran 島原の乱	Shimabara Rebellion
Shin Gan Hoe 新幹会	late nineteenth-century Korean political association
Shinchō 新潮	Shincho Magazine
Shintō 神道	Japanese indigenous religion
shita-uke 下請	sub-contractors
shizoku 士族	descendants of samurai families
shōjo manga 少女漫画	comics for girls
shōjo shumi 少女趣味	girlish taste
shokuan 職安	public employment security office
shōtai 小隊	platoon
Shōwa 昭和	period from 1925
shūdō 衆道	male homosexuality
Shufu 主婦	housewife
shufu no tomo 主婦の友	Housewives' Friend (magazine)
Shūkan Bunshun 週刊文春	Weekly Bunshun (magazine)
sōhyō 総評	General Council of Trade Unions of Japan

Sōhyō mindō 総評民同	Democratization League of Sohyo
sokkurisan そっくりさん	a person's double
Suiheisha 水平社	Levellers Association
sukiyaki すき焼	a Japanese stew
sumō bugyō 相撲奉行	sumo wrestling magistrate
sumō denshō 相撲伝承	sumo folklore
sushi 寿司	a sour rice dish
tachimochi 太刀持	sumo grand champion's sword
tai 鯛	sea bream
Taihō 大宝	period from 701 to 703
Taika 大化	period from 645 to 649
Taishō 大正	period from 1912 to 1926
taishū bunka 大衆文化	mass culture
tako no odori 蛸の踊り	octopus dance
tako-yaki 蛸焼	a savoury dish
takobeya 蛸部屋	shack for forced labourers
tamashii 魂	spiritual power
tanabata 七夕	Festival of the Weaver
Tandai shōshinroku 胆大小心録	essay published in 1808
tanka 短歌	Japanese verse of thirty-one syllables
tate 縦	verticality
Tayū Saizō den 太夫才蔵伝	Story of Master and Nitwit Servant
tehai shi 手配師	day labourers' boss
tempura 天婦羅	Japanese deep fried food
Tendaishū kakushin dōmei 天台宗革新同盟.	Reform League of the Tendai Sect of Buddhism
tenkō 転向	ideological conversion
tennō 天皇	emperor
tennō-sei 天皇制	emperor system
tobi 鳶	scaffolder
Tōdai 東大	Tokyo University
Tokugawa 徳川	feudal period from 1603 to 1867
tonichirō 東日労	Metropolitan Day Labourers' Union
toshiyori 年寄	sumo manager
tsuji zumō 辻相撲	street matches of sumo wrestling
tsukkomi 突っ込み	aggressive role in manzai
tsuyu harai 露払	clearing the way (in sumo)
uchikowashi 打毀	demolition of properties

Uchinānchu ウチナーンチュ	Okinawans
udon うどん	Japanese noodles
uji 氏	lineage
ukeoisha 請負者	sub-contractor
ukezara 受皿	receptacle
Utsubo monogatari 宇津保物語	20-volume work published in the tenth century
wadoku no sai 和読の才	ability at reading Japanese
wakon 和魂	Japanese spirit
wasai 和才	Japanese technology
yakuza やくざ	Japanese mafia
yamato damashii 大和魂	Japanese soul
Yasukuni 靖国	shrine dedicated to Japan's war dead
Yayoi 弥生	period from the second century B.C. to the third century A.D.
yōkon 洋魂	Western soul
yokozuna 横綱	sumo grand champion
Yomiuri shimbun 読売新聞	Yomiuri News
yanaoshi 世直し	fundamental social reform
yōsai 洋才	Western learning
yoseba 寄せ場	slum area occupied by day labourers and others
Yūaikai 友愛会	Friendship Society
Zen 禅	Zen
Zenkoku bukkyō kakushin renmei 全国仏教革新連盟	All Japan Reformist League of Buddhists
Zenkyō 全協	National Conference of Trade Unions of Japan
Zennichi jirō 全日自労	Sanya Independent Amalgamated Labour Union
Zentorō 全都労	All Metropolis United Self-employed Labour Union

Index